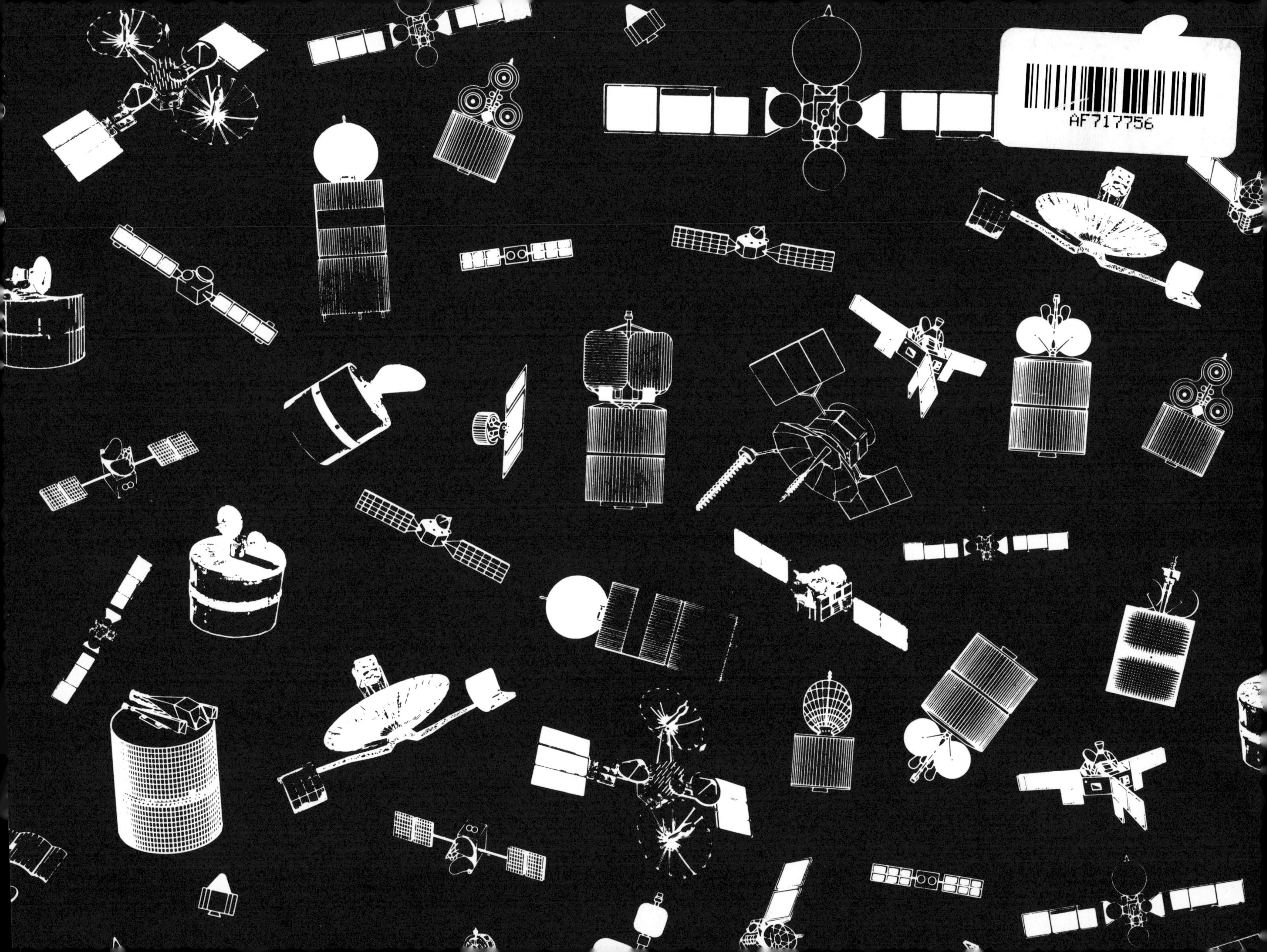

COMPENDIUM OF COMMUNICATION AND BROADCAST SATELLITES 1958 to 1980

OTHER IEEE PRESS BOOKS

Optical Fiber Technology, II
Edited by C. K. Kao
Protective Relaying for Power Systems
Edited by S. H. Horowitz
Analog MOS Integrated Circuits
Edited by P. R. Gray, D. A. Hodges, and R. W. Brodersen
Interference Analysis of Communication Systems
Edited by P. Stavroulakis
Integrated Injection Logic
Edited by J. E. Smith
Sensory Aids for the Hearing Impaired
Edited by H. Levitt, J. M. Pickett, and R. A. Houde
Data Conversion Integrated Circuits
Edited by Daniel J. Dooley
Semiconductor Injection Lasers
Edited by J. K. Butler
Satellite Communications
Edited by H. L. Van Trees
Frequency-Response Methods in Control Systems
Edited by A.G.J. MacFarlane
Programs for Digital Signal Processing
Edited by the Digital Signal Processing Committee
Automatic Speech & Speaker Recognition
Edited by N. R. Dixon and T. B. Martin
Speech Analysis
Edited by R. W. Schafer and J. D. Markel
The Engineer in Transition to Management
I. Gray
Multidimensional Systems: Theory & Applications
Edited by N. K. Bose
Analog Integrated Circuits
Edited by A. B. Grebene
Integrated-Circuit Operational Amplifiers
Edited by R. G. Meyer
Modern Spectrum Analysis
Edited by D. G. Childers
Digital Image Processing for Remote Sensing
Edited by R. Bernstein
Reflector Antennas
Edited by A. W. Love
Phase-Locked Loops & Their Application
Edited by W. C. Lindsey and M. K. Simon
Digital Signal Computers and Processors
Edited by A. C. Salazar
Systems Engineering: Methodology and Applications
Edited by A. P. Sage
Modern Crystal and Mechanical Filters
Edited by D. F. Sheahan and R. A. Johnson
Electrical Noise: Fundamentals and Sources
Edited by M. S. Gupta
Computer Methods in Image Analysis
Edited by J. K. Aggarwal, R. O. Duda, and A. Rosenfeld
Microprocessors: Fundamentals and Applications
Edited by W. C. Lin
Machine Recognition of Patterns
Edited by A. K. Agrawala
Turning Points in American Electrical History
Edited by J. E. Brittain
Charge-Coupled Devices: Technology and Applications
Edited by R. Melen and D. Buss
Spread Spectrum Techniques
Edited by R. C. Dixon
Electronic Switching: Central Office Systems of the World
Edited by A. E. Joel, Jr.
Electromagnetic Horn Antennas
Edited by A. W. Love
Waveform Quantization and Coding
Edited by N. S. Jayant
Communication Satellite Systems: An Overview of the Technology
Edited by R. G. Gould and Y. F. Lum
Literature Survey of Communication Satellite Systems and Technology
Edited by J. H. W. Unger
Solar Cells
Edited by C. E. Backus
Computer Networking
Edited by R. P. Blanc and I. W. Cotton
Communications Channels: Characterization and Behavior
Edited by B. Goldberg
Large-Scale Networks: Theory and Design
Edited by F. T. Boesch

COMPENDIUM OF COMMUNICATION AND BROADCAST SATELLITES 1958 to 1980

Edited by
Martin P. Brown, Jr.

Communication and Broadcast Satellite Systems Committee
of the
Satellite Systems Panel
IEEE Aerospace and Electronic Systems Society (AESS)

Published for the IEEE Aerospace and Electronic
Systems Society by the IEEE PRESS

IEEE PRESS

The Institute of Electrical and Electronics Engineers, Inc., New York

Sole Worldwide Distributor (Exclusive of IEEE):

JOHN WILEY & SONS, INC.
605 Third Avenue
New York, N.Y. 10016

IEEE Order Department
445 Hoes Lane, Piscataway, NJ 08854

International Standard Book Number: 0-87942-153-3
Library of Congress Catalog Card Number: 81-81858
IEEE Order Number: PC01461
WILEY Order Number: 0-471-86198-7

Printed in the United States of America

FOREWORD . . .

One of the goals of the Communication and Broadcast Satellite Systems Committee is the publication of material which is useful to its membership, as well as to interested parties outside the Society. This Compendium of Communication and Broadcast Satellites is another contribution* toward fulfilling this goal and is intended to assist individuals associated with this expanding field in obtaining a quick overview and comparison of most of the satellites falling into this category. As a means of presenting this overview the reader is provided with the following information in a uniform manner for each satellite:

- A full page picture of the satellite— containing a man (or woman) as a means of size comparison.
- A block diagram of the communications payload.
- A frequency plan showing the communication and broadcast bands utilized.
- A listing of all major transmission parameters, physical characteristics, and general information such as: number of satellites built, launch vehicle, prime contractor, etc.
- Date of introduction into service and designed lifetime.

It should be noted that this Compendium has been prepared through the voluntary efforts of many individuals who have researched a variety of sources. The information presented does not attempt to be all inclusive, nor indicate preferences; the Compendium is simply a handy reference source. We assume no responsibility for typographical errors, but would appreciate having them called to our attention and we solicit suggestions for additional material for inclusion in future editions.** The views expressed are not necessarily those of the organizations whose satellites are depicted (nor those of the author's organization).

Sincerely,

Martin P. Brown, Jr.
Chairman
IEEE Communication and Broadcast
Satellite Systems Committee

---

*A booklet is also available: "Abbreviations, Acronyms, and Symbols common to Communications and Broadcast Satellite Systems," 20 August 1977. Cost $1.00 members, $1.50 nonmembers. Checks payable to: IEEE Abbreviations. Send to: IEEE/AES Communication and Broadcast Satellite Systems Committee, c/o 3517 Alton Place, N.W., Washington, DC 20008, Attn: Roby S. Malandrucco.

**Correspondence should be sent to the above address.

ABOUT THE AUTHORS

The concept of compiling a compendium of satellites was presented to the AES Board of Governors almost four years ago. They accepted the idea at a time when it was felt that the compendium could be put into a small booklet and be completed in relatively short order. What none of us anticipated was that we were embarking on our project at a time when the rapidly expanding satellite business would begin to produce satellites faster than the meager resources of the Communications and Broadcast Satellite Committee could keep pace with. Since the efforts of the authors were to be of a voluntary nature, very few demands could be made to hurry things along. Progress was made on a slow but steady basis—in most cases "after hours." Many of the authors are members of the Communications and Broadcast Satellite Committee; others were selected due to their knowledge of a particular satellite. The number of authors was intentionally kept small in order to reduce the amount of coordination to manageable proportions.

AUTHORS

Pier L. Bargellini
COMSAT Laboratories
22300 Comsat Drive
Clarksburg, Maryland 20734

Raymond A. Berg
The Aerospace Corporation
P.O. Box 92957
Los Angeles, California 90009

Khodadad Betaharon
INTELSAT
490 L'Enfant Plaza, S.W.
Washington, D.C. 20024

Marcel R. E. Bichara
Institute of Physics
University of Geneva
32, Blvd. D'Yvoy
1211 Geneva 4, Switzerland

Martin P. Brown, Jr.
INTELSAT
490 L'Enfant Plaza, S.W.
Washington, D.C. 20024

William W. Curry, Jr.
Satellite Business Systems
P.O. Box 908
8003 Westpark Drive
McLean, Virginia 22101

N. George Davies
Communications Research Centre
Canadian Department of Communications
P.O. Box 11490, Station H
Ottawa
Ontario, Canada K2H 8S2

Hans Dodel
DFVLR-FBNE
8031 Oberpfaffenhofen
West Germany

Richard W. Duesing
INTELSAT
490 L'Enfant Plaza, S.W.
Washington, D.C. 20024

Gerald L. Dutcher
Hughes Aircraft Company
P.O. Box 92919, Airport Station
Building S21, Mail Station E339
Los Angeles, California 90009

European Space Agency*
Washington Office
955 L'Enfant Plaza, S.W.
Suite 1404
Washington, D.C. 20024

Barry G. Evans
Department of Electrical Engineering Science
University of Essex
Wivenhoe Park
Colchester CO4 3SQ, England

Richard G. Gould
Telecommunications Systems
Suite 1001
1725 K Street, N.W.
Washington, D.C. 20006

Laurence R. Gray
COMSAT
950 L'Enfant Plaza, S.W.
Washington, D.C. 20024

T. Michael Kelley
INTELSAT
490 L'Enfant Plaza, S.W.
Washington, D.C. 20024

Yoshitaka Kurihara
National Space Development Agency of Japan (NASDA)
2-4-1, Shiba-Hamamatsu-cho
Minato-Ku, Tokyo 105
Japan

Gustavo D. Moreno
INTELSAT
490 L'Enfant Plaza, S.W.
Washington, D.C. 20024

Walter L. Morgan
Communications Center of Clarksburg
52-B Green Valley Road
Clarksburg, Maryland 20734

Kunihiko Niwa
INTELSAT
490 L'Enfant Plaza, S.W.
Washington, D.C. 20024

Ramaswami Parthasarathy
INTELSAT
490 L'Enfant Plaza, S.W.
Washington, D.C. 20024

Jack F. Phiel, Jr.
INTELSAT
490 L'Enfant Plaza, S.W.
Washington, D.C. 20024

Michael J. Robusto
INTELSAT
490 L'Enfant Plaza, S.W.
Washington, D.C. 20024

Joji Tabata
National Space Development Agency of Japan (NASDA)
2-4-1, Shiba-Hamamatsu-cho
Minato-Ku, Tokyo 105
Japan

Jahangir A. Tehrani
INTELSAT
490 L'Enfant Plaza, S.W.
Washington, D.C. 20024

William W. Ward
MIT Lincoln Laboratory
Room C-270, Box 73
Lexington, Massachusetts 02173

*The MARECS satellite contribution was submitted by ESA without a specific author.

AUTHORS OF SATELLITE DESCRIPTIONS

	Satellite	Author
I.	**INTERNATIONAL**	
	1. INTELSAT I	Michael J. Robusto
	2. INTELSAT II	Michael J. Robusto
	3. INTELSAT III	Michael J. Robusto
	4. INTELSAT IV	Martin P. Brown, Jr.
	5. INTELSAT IV-A	Martin P. Brown, Jr.
	6. INTELSAT V	Martin P. Brown, Jr.
	7. INTELSAT V (MCS)	Martin P. Brown, Jr.
	8. MARECS	European Space Agency
	9. MARISAT	Hans Dodel
	10. MAROTS	Marcel R. E. Bichara
II.	**INTERNATIONAL/DOMESTIC**	
	1. AEROSAT	Hans Dodel
	2. LOUTCH	Walter L. Morgan
	3. MOLNIYA I	Martin P. Brown, Jr.
	4. MOLNIYA II	Martin P. Brown, Jr.
	5. MOLNIYA III	Martin P. Brown, Jr.
	6. STATSIONAR	Walter L. Morgan
	7. RĄDUGA	Walter L. Morgan
	8. STATSIONAR-T (EKRAN)	Walter L. Morgan
	9. GORIZONT	Walter L. Morgan
	10. VOLNA	Walter L. Morgan
III.	**REGIONAL**	
	ECS (European)	Marcel R. E. Bichara
IV.	**DOMESTIC**	
	1. COMSTAR	Jack F. Phiel, Jr.
	2. INSAT	Ramaswami Parthasarathy
	3. PALAPA A	Hans Dodel
	4. PALAPA B	Hans Dodel
	5. SATCOM	Michael J. Robusto
	6. SBS	Laurence F. Gray
	7. TELESAT (ANIK A)	Jahangir A. Tehrani
	8. TELESAT (ANIK B)	Jahangir A. Tehrani
	9. TELESAT (ANIK C)	Jahangir A. Tehrani
	10. WESTAR	Richard G. Gould
	11. WESTAR (ADVANCED)/TDRSS	Richard G. Gould
V.	**MILITARY**	
	1. DSCS II	William H. Curry, Jr.
	2. DSCS III	Richard W. Duesing
	3. FLTSATCOM	William H. Curry, Jr.
	4. GALS	Walter L. Morgan
	5. IDCSP	Khodadad Betaharon
	6. LEASAT	Gerald L. Dutcher
	7. LES 1 and 2	William W. Ward
	8. LES 3	William W. Ward
	9. LES 4	William W. Ward
	10. LES 5	William W. Ward
	11. LES 6	William W. Ward
	12. LES 7	William W. Ward
	13. LES 8 and 9	William W. Ward
	14. NATO II	Raymond Berg
	15. NATO III	Raymond Berg
	16. SKYNET I	Richard W. Duesing
	17. SKYNET II	Richard W. Duesing
	18. TACSAT	Martin P. Brown, Jr.
VI.	**EXPERIMENTAL**	
	1. ATS 1 thru 5	Richard W. Duesing
	2. ATS 6	Richard W. Duesing
	3. BSE (Yuri) or BS	Yoshitaka Kurihara
	4. COURIER	T. Michael Kelley
	5. CS (Sakura)	Yoshitaka Kurihara
	6. CTS (Hermes)	N. George Davies
	7. ECHO	T. Michael Kelley
	8. ECS (Ayame)	Joji Tabata
	9. ETS II (Kiku II)	Kunihiko Niwa
	10. OTS	Barry G. Evans
	11. RELAY	T. Michael Kelley
	12. SCORE	T. Michael Kelley
	13. SIRIO	Pier L. Bargellini
	14. SYMPHONIE	Marcel R. E. Bichara
	15. SYNCOM	T. Michael Kelley
	16. TELSTAR	T. Michael Kelley
	17. WEST FORD	William W. Ward
VII.	**AMATEUR**	
	1. OSCAR 1	Gustavo D. Moreno
	2. OSCAR 2	Gustavo D. Moreno
	3. OSCAR 3	Gustavo D. Moreno
	4. OSCAR 4	Gustavo D. Moreno
	5. OSCAR 5	Gustavo D. Moreno
	6. OSCAR 6	Gustavo D. Moreno
	7. OSCAR 7	Gustavo D. Moreno
	8. OSCAR 8	Gustavo D. Moreno
	9. OSCAR 9	Gustavo D. Moreno
	10. RS 1 and 2	Gustavo D. Moreno

ACKNOWLEDGEMENTS

The Communication and Broadcast Satellite Systems Committee is indebted to the following individuals who have contributed toward the completion of this Compendium:

To Richard G. Gould, who steadfastly furnished the guidance and leadership required for a voluntary undertaking of this magnitude. The Communications and Broadcast Satellite Systems Committee is but one of five Committees reporting to the Satellite Systems Panel of the IEEE Aerospace and Electronic Systems Society; Mr. Gould heads this very active Panel.

To several Technical Advisors who gave their guidance over the many years it took to assemble the Compendium:

Pier L. Bargellini
COMSAT Laboratories

Sidney Metzger
COMSAT

Jack L. Dicks
INTELSAT

Emeric I. Podraczky
INTELSAT

To several Editorial Advisors who freely gave their time toward reviewing and assisting in the preparation of its contents:

Martin P. Brown, Sr.
M. P. Brown Engineering Co.

Aurora Purgal
INTELSAT

Diane Franks
INTELSAT

Shizuo Itohara
INTELSAT

Laurene Hassard
INTELSAT

G. H. Wootton-Woolley
COMSAT

To the INTELSAT Librarian and Research Assistant who helped in numerous ways toward completion of the Compendium:

Rosa Liu	Ann Marshall
INTELSAT, Librarian	INTELSAT, Library Research Assistant

The Editor would like to particularly express his thanks to Professor Marcel R. E. Bichara for his faithful encouragement.

Please note that the contribution of each author does not necessarily represent the views of his organization, or the organization who funded the satellite described. Further, it does not represent official policies, either expressed or implied, of any Government. The information presented has been compiled by the authors using material in the open literature and is not intended to contain details which may be considered sensitive to the spacecraft sponsor or his contractor. Readers who desire more detailed information should contact the sponsoring organization directly.

TABLE OF CONTENTS

PART A: OVERVIEW
COMMUNICATION AND BROADCAST SATELLITES

		Page
I.	BEFORE WARC-79, COMMUNICATION AND BROADCAST SATELLITE FREQUENCY BAND ALLOCATIONS	3
II.	AFTER WARC-79, COMMUNICATION AND BROADCAST SATELLITE FREQUENCY BAND ALLOCATIONS	4
III.	SATELLITE FAMILY TREE	5
IV.	SATELLITE STABILIZATION TYPES	6
V.	LAUNCH SITES	7
VI.	LAUNCH DATES	8

TABLE OF CONTENTS—(Continued)

PART B: SATELLITE DESCRIPTIONS

COMMUNICATION AND BROADCAST SATELLITES

Satellite	*System*	*Page*
I. INTERNATIONAL		
INTELSAT	International Telecommunications Satellite Organization (Includes "Early Bird")	21
MARECS	Maritime Satellite (European)	47
MARISAT	Maritime Satellite (COMSAT)	51
MAROTS	Maritime Satellite (European)	55
MCS	Maritime Communications Subsystem (INTELSAT V)	41
II. INTERNATIONAL/DOMESTIC		
AEROSAT	Aeronautical Satellite	61
LOUTCH	USSR Satellite (Communication and TV Broadcast); "Beam" or "Ray"	65
MOLNIYA	USSR Satellite (Elliptical orbit); "Lightning" or "News Flash"	69
STATSIONAR	USSR Satellite	75
Raduga	USSR Satellite; "Rainbow"	79
Gorizont*	USSR Satellite; "Horizon"	84
VOLNA	USSR Satellite (Aeronautical, Maritime, and Land); "Wave"	87
III. REGIONAL		
ECS (EUROSAT)	European Communications Satellite	93

*Gorizont contains both 6/4 GHz and 8/7 GHz subsystems. Its description has been combined in Section II with Statsionar.

PART B: SATELLITE DESCRIPTIONS

COMMUNICATION AND BROADCAST SATELLITES

Satellite	*System*	*Page*
IV. DOMESTIC		
COMSTAR	United States Satellite (COMSAT, AT&T, and GTE)	99
INSAT	India Satellite (Communications and TV Broadcast)	103
PALAPA	Indonesia Satellite; "National Unity"	107
SATCOM	United States Satellite (RCA)	115
SBS	United States Satellite (Satellite Business Systems)	119
STATSIONAR-T*	USSR Satellite (TV Broadcast); "Ekran" or "Screen"	82
TELESAT	Canada Satellite; "Anik" or "Brother"	123
WESTAR	United States Satellite (Western Union)	135
ADV. WESTAR/TDRSS	United States Satellite (Western Union and Tracking and Data Relay Satellite System)	139
V. MILITARY		
DSCS	Defense Satellite Communications System (USA)	151
FLTSATCOM	Navy Fleet Satellite Communications System (USA)	159
GALS	USSR Satellite; "Tack"	163
GORIZONT**	USSR Satellite, "Horizon"	84
IDCSP	Interim Defense Communications Satellite (USA)	167
LEASAT	Leased Satellite (USA)	171
LES	Lincoln Experimental Satellite (USA)	175
NATO	North Atlantic Treaty Organization Satellite	207
SKYNET	United Kingdom Satellite	217
TACSAT	Tactical Satellite (USA)	225

*Statsionar—T (Ekran) is described in Section II with other Statsionar satellites.
**Gorizont contains both 6/4 GHz and 8/7 GHz subsystems. Its description has been combined in Section II with Statsionar.

TABLE OF CONTENTS—(Continued)

PART B: SATELLITE DESCRIPTIONS
COMMUNICATION AND BROADCAST SATELLITES

Satellite	*System*	*Page*
VI. EXPERIMENTAL		
ATS	Applications Technology Satellite (USA)	231
BS (also BSE)	Broadcast Satellite (Japan); "Yuri" or "Lily"	241
COURIER	Early Low-Medium Orbit Satellite (USA)	245
CS	Communications Satellite (Japan); "Sakura" or "Cherry Blossom"	249
CTS	Communications Technology Satellite (Canada/USA); "Hermes"	253
ECHO	Early Low-Medium Orbit Passive Satellite	257
ECS (Japanese)	Experimental Communications Satellite (Japan); "Ayame" or "Iris"	263
ETS II	Engineering Test Satellite (Japan); "Kiku II" or "Chrysanthemum"	267
OTS	Orbital Test Satellite (European)	271
RELAY	Early Low-Medium Orbit Satellite (USA)	275
SCORE	Early Low-Medium Orbit Satellite (USA)	279
SIRIO	Propagation and Communication Satellite (Italy)	283
SYMPHONIE	French-German Satellite	287
SYNCOM	Early Geostationary Satellite (USA)	291
TELSTAR	Early Low-Medium Orbit Satellite (USA)	295
WESTFORD	Early Low-Medium Orbit Satellite (USA)	299
VII. AMATEUR		
OSCAR (also AMSAT)	Amateur Satellite (USA and Australia)	305
RS	Amateur Satellite (USSR); "Radio" or "Radio Satellite"	327
VIII. REFERENCES		329

PART A

OVERVIEW
COMMUNICATION AND BROADCAST SATELLITES

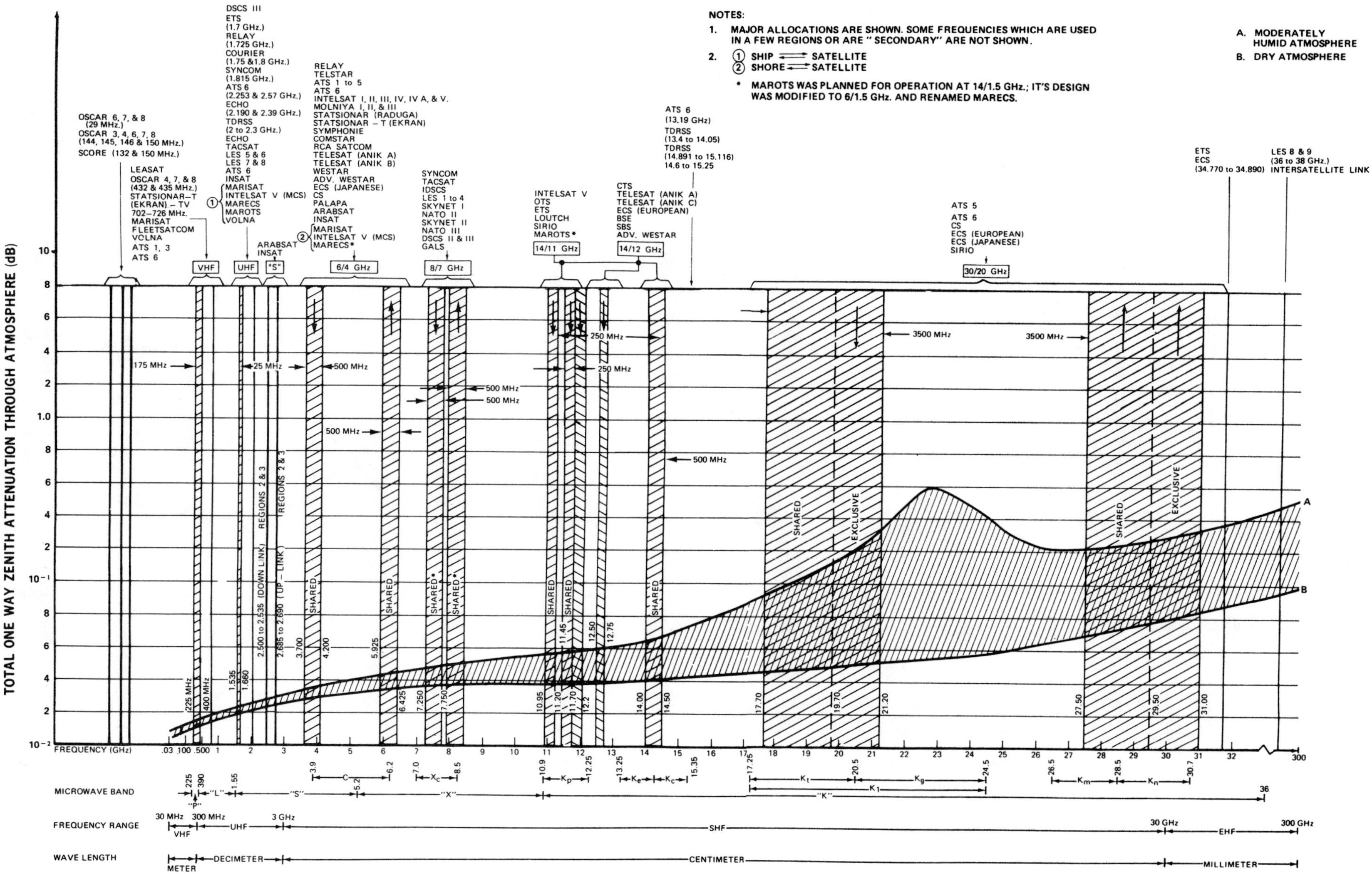

BEFORE GWARC-79
FREQUENCY ALLOCATIONS FOR COMMUNICATION & BROADCAST SATELLITES

Martin P. Brown, Jr.
IEEE Compendium

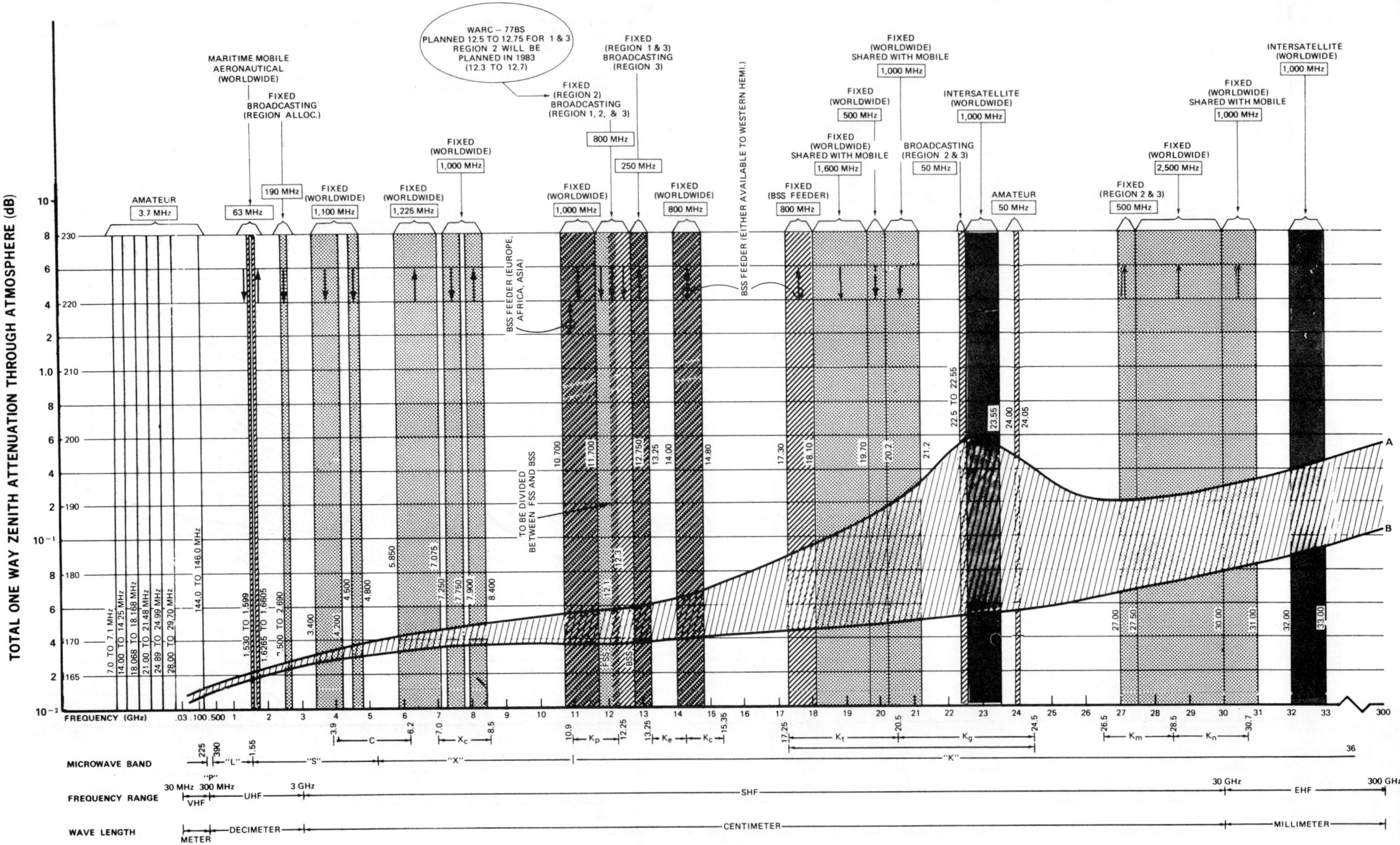

A. MODERATELY HUMID ATMOSPHERE
B. DRY ATMOSPHERE

AFTER GWARC-79
FREQUENCY ALLOCATIONS FOR COMMUNICATION & BROADCAST SATELLITES

Martin P. Brown, Jr.
IEEE Compendium

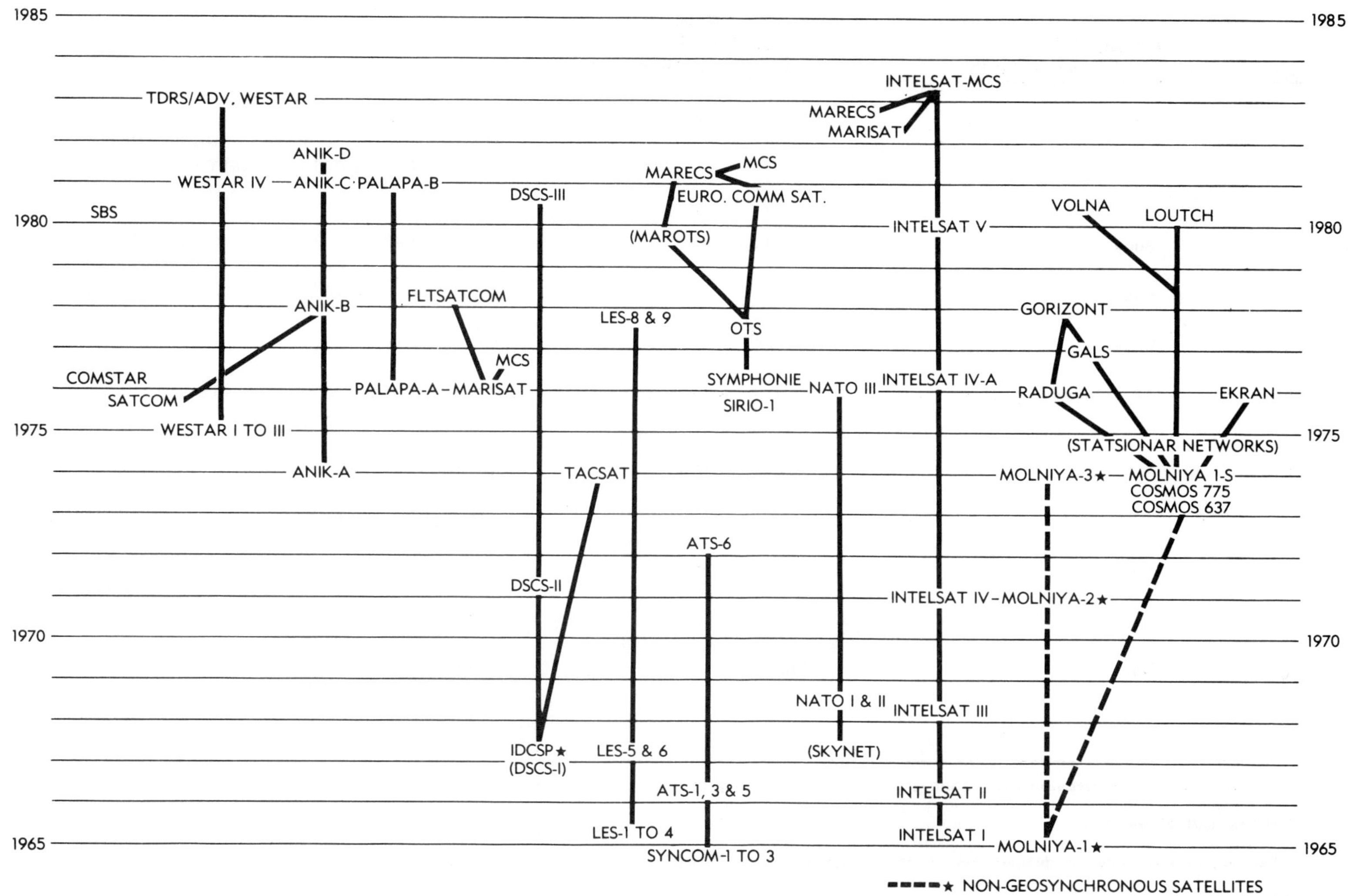

FAMILY TREE
COMMUNICATION AND BROADCAST SATELLITES

Walter L. Morgan
IEEE Compendium

Type	Portion of Satellite Oriented towards Earth	Portion of Satellite Spinning for Inertial Reference	Satellites
1	Body	Momentum Wheel [1]	SATCOM, ANIK B, SYMPHONIE, OTS, CTS, INSAT MARECS, INTELSAT V, ECS (EUROPEAN), FLTSATCOM, TDRSS/ADV. WESTAR; LES–8 & –9,
2	Body	Reaction Wheel [1]	ATS–6, DSCS III, STATSIONAR FAMILY, BSE
3	Body	None	MOLNIYA[2], SCORE[2], OSCAR 5 & 6[3], OSCAR–9[4]
4	Body	Body [5]	OSCAR–7 & –8[3]
5	RF Subsystem Platform	Solar Array & Propulsion	TACSAT, INTELSAT IV, IVA, COMSTAR, SBS, ANIK C, ANIK D, PALAPA B.
6	Antenna	Rest of Satellite	INTELSAT III, DCSC–II, ANIK A, WESTAR, PALAPA–A, SIRIO, NATO II, III, CS, MARISAT, ECS (JAPAN), SKYNET I, II, ETS II, ATS–3
7	None	All of Satellite	LES–1, to –6[6], COURIER[6], TELSTAR[4], RELAY[4], SYNCOM, INTELSAT I & II, ATS–1[6], –2[6], –4[6], & –5, IDCSP, OSCAR 4
8	None	None	ECHO, WESTFORD, OSCAR 1–3, RS–1 & –2.

[1] The demarkation line between momentum and reaction wheels may be blurred in some cases.

[2] During part of the orbit.

[3] Bar magnet and hysteresis damper used to orient spacecraft toward earth.

[4] Electromagnetic torque coils.

[5] A solar pressure vane was used to spin the satellite at a 1 revolution per 10 minutes rate for thermal control.

[6] Electronically despun antenna.

STABILIZATION OF COMMUNICATION SATELLITES

Walter L. Morgan
IEEE Compendium

LAUNCH SITES

Communication and Broadcast Satellites

	Site	Country
1.	Baikonur	USSR
2.	Cape Canaveral* (Eastern Test Range)	USA
3.	Kourou	French Guiana
4.	Plesetsk	USSR
5.	Tanegashima	Japan
6.	Vandenberg (Western Test Range)	USA

*Cape Canaveral is the geographic location of the Cape Kennedy launch facility in the State of Florida, U.S.A.

LAUNCH DATES

Communication and Broadcast Satellites

I. INTERNATIONAL SATELLITES

Satellite		Launch Date	Out-of-Service	Launch Site
INTELSAT I (Early Bird)		6 Apr 1965	May 1977*	Cape Canaveral
INTELSAT II	F-1	26 Oct 1966	Launch Failure	"
	F-2	11 Jan 1967	May 1977*	"
	F-3	23 Mar 1967	May 1977*	"
	F-4	28 Sep 1967	May 1977*	"
INTELSAT III	F-1	18 Sep 1968	Launch Failure	"
	F-2	18 Dec 1968	May 1977*	"
	F-3	5 Feb 1969	Dec 1979*	"
	F-4	21 May 1969	May 1977*	"
	F-5	26 Jul 1969	Launch Failure	"
	F-6	14 Jan 1970	May 1977*	"
	F-7	22 Apr 1970	Mar 1972	"
	F-8	23 Jul 1970	Launch Failure	"
INTELSAT IV	F-1	22 May 1975	-	"
	F-2	25 Jan 1971	-	"
	F-3	19 Dec 1971	-	"
	F-4	22 Jan 1972	-	"
	F-5	13 Jun 1972	-	"
	F-6	20 Feb 1975	Launch Failure	"
	F-7	23 Aug 1973	-	"
	F-8	21 Nov 1974	-	"
INTELSAT IVA	F-1	25 Sep 1975	-	"
	F-2	29 Jan 1976	-	"
	F-3	7 Jan 1978	-	"
	F-4	26 May 1977	-	"
	F-5	29 Sep 1977	Launch Failure	"
	F-6	31 Mar 1978	-	"
INTELSAT V	F-1	PLANNED	-	Cape Canaveral/Kourou
	F-2	6 Dec 1980	-	"
	F-3	PLANNED	-	"
	F-4	"	-	"
	F-5	"	-	"
	F-6	"	-	"
	F-7	"	-	"
	F-8	"	-	"
	F-9	"	-	"

* Removed from synchronous orbit, but still in space.

LAUNCH DATES (continued)

Communication and Broadcast Satellites

I. INTERNATIONAL SATELLITES (Con't.)

Satellite		Launch Date	Out-of-Service	Launch Site
MARISAT	F-1	19 Feb 1976	-	Cape Canaveral
	F-2	9 Jun 1976	-	"
	F-3	14 Oct 1976	-	"
MARECS**	F-1	PLANNED	-	Kourou
	F-2	"	-	"
	F-3	"	-	"
INTELSAT V	F-5	PLANNED	-	
(MCS)***	F-6	"	-	Cape Canaveral/Kourou
	F-7	"	-	"
	F-8	"	-	"
	F-9	"	-	"

II. INTERNATIONAL/DOMESTIC SATELLITES

Satellite		Launch Date	Out-of-Service	Launch Site
AEROSAT		PLANNED	-	-
LOUTCH		PLANNED	-	USSR
MOLNIYA I	F-1	23 Apr 1965	Unknown	Baikonur
	F-2	13 Oct 1965	17 Mar 1967	"
	F-3	25 Apr 1966	11 Jun 1973	"
	F-4	20 Oct 1966	11 Sep 1968	"
	F-5	25 May 1967	26 Nov 1971	"
	F-6	3 Oct 1967	4 Mar 1969	"
	F-7	22 Oct 1967	31 Dec 1969	"
	F-8	21 Apr 1968	29 Jan 1974	"
	F-9	5 Jul 1968	15 May 1971	"
	F-10	5 Oct 1968	16 Jul 1976	"
	F-11	11 Apr 1969	17 Apr 1974	"
	F-12	22 Jul 1969	18 Jun 1971	"
	F-13	19 Feb 1970	29 Sep 1975	Plesetsk
	F-14	26 Jun 1970	16 Feb 1976	"
	F-15	29 Sep 1970	20 Mar 1976	"
	F-16	27 Nov 1970	25 Nov 1975	"

** Formerly known as MAROTS when it carried an 11 GHz band.
*** MCS = Maritime Communication Subsystem. To be carried on INTELSAT V.

LAUNCH DATES (continued)

Communication and Broadcast Satellites

II. INTERNATIONAL/DOMESTIC SATELLITES (Con't.)

Satellite		Launch Date	Out-of-Service	Launch Site
MOLNIYA I	F-17	25 Dec 1970	22 Dec 1975	Baikonur
	F-18	28 Jul 1971	19 Jul 1977	Plesetsk
	F-19	20 Dec 1971	13 Apr 1977	"
	F-20	4 Apr 1972	30 Jan 1974	Baikonur
	F-21	14 Oct 1972	1 Nov 1977	Plesetsk
	F-22	2 Dec 1972	11 Feb 1976	Baikonur
	F-23	3 Feb 1973	23 Oct 1977	Baikonur
	F-24	30 Aug 1973	Unknown	Plesetsk
	F-25	14 Nov 1973	"	Baikonur
	F-26	30 Nov 1973	"	Plesetsk
	F-27	20 Apr 1974	"	"
	F-28	24 Oct 1974	"	"
	F-29	29 Apr 1975	"	"
	F-30	5 Jun 1975	"	"
	F-31	2 Sep 1975	"	"
	F-32	22 Jan 1976	"	"
	F-33	21 Mar 1976	"	"
	F-34	19 Mar 1976	"	"
	F-35	23 Jul 1976	"	"
	F-36	24 Mar 1977	"	"
	F-37	24 Jun 1977	"	"
	F-38	30 Aug 1977	"	"
	F-39	3 Mar 1978	"	"
	F-40	2 Jun 1978	"	"
	F-41	14 Jul 1978	"	"
	F-42	22 Aug 1978	"	"
	F-43	12 Apr 1979	"	"
	F-44	31 Jul 1979	"	"
	F-45	20 Oct 1979	"	"
	F-46	11 Jan 1980	"	"
MOLNIYA II	F-1	24 Nov 1971	10 May 1976	Plesetsk
	F-2	19 May 1972	22 Mar 1977	"
	F-3	30 Sep 1972	12 Jan 1978	"
	F-4	12 Dec 1972	22 Jan 1978	"
	F-5	5 Apr 1973	6 Jan 1979	Baikonur
	F-6	11 Jul 1973	Unknown	Plesetsk
	F-7	19 Oct 1973	"	"
	F-8	25 Dec 1973	"	"
	F-9	26 Apr 1974	"	"
	F-10	23 Jul 1974	"	"
	F-11	21 Dec 1974	"	"
	F-12	6 Feb 1975	"	"
	F-13	8 Jul 1975	"	"
	F-14	9 Sep 1975	"	"

LAUNCH DATES (continued)

Communication and Broadcast Satellites

II. INTERNATIONAL/DOMESTIC SATELLITES (Con't.)

Satellite		Launch Date	Out-of-Service	Launch Site
	F-15	17 Dec 1975	"	"
	F-16	2 Dec 1976	"	"
	F-17	11 Feb 1977	"	"
MOLNIYA III	F-1	21 Nov 1974	Unknown	Plesetsk
	F-2	14 Apr 1975	"	"
	F-3	14 Nov 1975	"	"
	F-4	27 Dec 1975	"	"
	F-5	12 May 1976	"	"
	F-6	28 Dec 1976	"	"
	F-7	28 Apr 1977	Aug 1977	"
	F-8	28 Oct 1977	Unknown	"
	F-9	24 Jan 1978	"	"
	F-10	13 Oct 1978	"	"
	F-11	18 Jan 1979		
	F-12	6 Jun 1979	"	"
MOLNIYA 1-S (1st Sync.)		29 Jul 1974	Unknown	Baikonur
STATSIONAR 1	(Raduga 1)	22 Dec 1975	Unknown	Baikonur
2	(Raduga 2)	11 Sep 1976	"	"
3	(Raduga 3)	23 Jul 1977	Dec 1979	"
4	(Gorizant 2)	5 Jul 1979	Unknown	"
5	(Gorizant 3)	28 Dec 1979	"	"
7	?	PLANNED	-	-
8	?	"	-	-
9	?	"	-	-
10	?	"	-	-
STATSIONAR ?	(Raduga 4)	19 Jul 1978	Unknown	Baikonur
?	(Gorizant 1)	19 Dec 1978	"	"
?	(Raduga 5)	25 Apr 1979	"	"
?	(Raduga 6?)	20 Feb 1980	"	"
STATSIONAR-T	(Ekran-1)	26 Oct 1976	Unknown	Baikonur
-T	(Ekran-2)	20 Sep 1977	"	"
-T	(Ekran-3)	21 Feb 1979	"	"
-T	(Ekran-4)	3 Oct 1979	"	"
VOLNA		PLANNED	-	-

LAUNCH DATES (continued)

Communication and Broadcast Satellites

III. REGIONAL SATELLITES

Satellite		Launch Date	Out-of-Service	Launch Site
ECS (European)		PLANNED	-	-

IV. DOMESTIC SATELLITES

Satellite		Launch Date	Out-of-Service	Launch Site
COMSTAR	D-1	13 May 1976	-	Cape Canaveral
	D-2	22 Jul 1976	-	"
	D-3	29 Jun 1978	-	"
	D-4	PLANNED	-	"
INSAT	F-1	PLANNED	-	Unknown
PALAPA A	F-1	8 Jul 1976	-	Cape Canaveral
	F-2	10 Mar 1977	-	"
PALAPA B	F-1	PLANNED	-	"
	F-2	"	-	"
SATCOM (RCA	F-1	12 Dec 1975	-	"
	F-2	26 Mar 1976	-	"
	F-3	6 Dec 1979	Launch Failure	"
	F-4	PLANNED	-	"
SBS	F-1	15 Nov 1980	-	"
	F-2	PLANNED	-	"
	F-3	"	-	"
STATSIONAR-T		(See Statsionar Series (International/Domestic)		
TELESAT (Anik A)	F-1	9 Nov 1972	-	"
	F-2	20 Apr 1973	-	"
	F-3	7 May 1975	-	"
TELESAT (Anik B)	F-1	15 Dec 1978	-	"
TELESAT (Anik C)	F-1	PLANNED	-	"
	F-2	"	-	"
	F-3	"	-	"
WESTAR	F-1	13 Apr 1974	-	"
	F-2	10 Oct 1974	-	"
	F-3	9 Aug 1979	-	"

LAUNCH DATES (continued)

Communication and Broadcast Satellites

IV. DOMESTIC SATELLITES

Satellite		Launch Date	Out-of-Service	Launch Site
ADV WESTAR/ TDRSS	F-1	PLANNED	-	Cape Canaveral
	F-2	"	-	"
	F-3	"	-	"
	F-4	"	-	"

V. MILITARY SATELLITES

Satellite		Launch Date	Out-of-Service	Launch Site
DSCS II	F-1	3 Nov 1971	15 Sep 1973	Cape Canaveral
	F-2	3 Nov 1971	8 Sep 1972	"
	F-3	13 Dec 1973	Unknown	"
	F-4	13 Dec 1973		"
	F-5	20 May 1975	Launch Failure	"
	F-6	20 May 1975	Launch Failure	"
	F-7	12 May 1977	Unknown	"
	F-8	12 May 1977	"	"
	F-9	13 Dec 1978	"	"
	F-10	13 Dec 1978	"	"
	F-11	20 Nov 1979	"	"
	F-12	20 Nov 1979	"	"
DSCS III	F-1 to F-12	PLANNED	-	"
FLTSATCOM	F-1	9 Feb 1978	-	"
	F-2	4 May 1979	-	"
	F-3	18 Jan 1980	-	"
	F-4	28 Oct 1980	-	"
GALS	F-1	PLANNED	-	USSR
	F-2	"	-	"
	F-3	"	-	"
	F-4	"	-	"
IDCSP	F-1	16 Jun 1966	Unknown	Cape Canaveral
	F-2	"	"	"
	F-3	"	"	"
	F-4	"	"	"
	F-5	"	"	"
	F-6	"	"	"
	F-7	"	"	"
	F-8	"	"	"
	F-9	18 Jan 1967	"	"
	F-10	"	"	"
	F-11	"	"	"
	F-12	"	"	"
	F-13	"	"	"
	F-14	"	"	"
	F-15	"	"	"
	F-16	1 Jul 1967	"	"
	F-17	"	"	"
	F-18	"	"	"
	F-19	13 Jun 1968	"	"
	F-20	"	"	"

LAUNCH DATES (continued)

Communication and Broadcast Satellites

V. MILITARY SATELLITES

Satellite		Launch Date	Out-of-Service	Launch Site
IDCSP	F-21	13 Jun 1968	Unknown	Cape Canaveral
	F-22	"	"	"
	F-23	"	"	"
	F-24	"	"	"
	F-25	"	"	"
	F-26	"	"	"
LEASAT	F-1	PLANNED	-	"
	F-2	"	-	"
	F-3	"	-	"
	F-4	"	-	"
LES	F-1	11 Feb 1965	-	"
	F-2	6 May 1965	-	"
	F-3	21 Dec 1965	6 Apr 1968	"
	F-4	21 Dec 1965	-	"
	F-5	1 Jul 1967	-	"
	F-6	26 Sep 1968	-	"
	F-7	Not Built	-	"
	F-8	14 Mar 1976	-	"
	F-9	14 Mar 1976	-	"
NATO I			(See IDCSP)	
NATO II	F-1	20 Mar 1970	Unknown	"
	F-2	2 Feb 1971	"	"
NATO III	F-1	22 Apr 1976	-	"
	F-2	27 Jan 1977	-	"
	F-3	19 Nov 1978	-	"
SKYNET I	F-1	22 Nov 1969	-	"
	F-2	19 Aug 1970	Launch Failure	"
SKYNET II	F-1	18 Jan 1974	Launch Failure	"
	F-2	22 Nov 1974	-	"
TACSAT	F-1	9 Feb 1969	4 Dec 1972	"
WESTFORD	F-1	21 Oct 1961	Unknown	"
	F-2	8 May 1963	"	

LAUNCH DATES (continued)

Communication and Broadcast Satellites

VI. EXPERIMENTAL

Satellite		Launch Date	Out-of-Service	Launch Site
ATS-1		6 Dec 1966	Unknown	Cape Canaveral
-2		5 Apr 1967	2 Sep 1969	"
-3		5 Nov 1967	Unknown	"
-4		10 Aug 1968	Launch Failure	"
-5		12 Aug 1969	Unknown	"
-6		30 May 1974	-	"
BS (or BSE)	F-1	7 Apr 1978	-	"
COURIER	F-1	18 Aug 1960	Launch Failure	"
	F-2	4 Oct 1960	Operated for 17 days	"
CS	F-1	14 Dec 1977	-	"
CTS	F-1	17 Jan 1976	-	"
ECHO	A-10	13 May 1960	Launch Failure	"
	F-1	12 Aug 1960	24 May 1968	"
	F-2	25 Jan 1964	7 Jan 1969	"
ECS	F-1	6 Feb 1979	Launch Failure	Tanegashima
(Japanese)	F-2	22 Feb 1980	Launch Failure	"
ETS-II		23 Feb 1977	Unknown	Tanegashima
OTS	F-1	13 Sep 1977	Launch Failure	Cape Canaveral
	F-2	11 May 1978	-	"
RELAY	F-1	13 Dec 1962	Feb 1965	"
	F-2	21 Jan 1964	26 Sep 1965	"
SCORE	F-1	18 Dec 1958	21 Jan 1959	"
SIRIO	F-1	25 Aug 1977	-	"
SYMPHONIE	F-1	19 Dec 1974	-	"
	F-2	27 Aug 1975	-	"
SYNCOM	F-1	14 Feb 1963	Launch Failure	"
	F-2	26 Jul 1963	Unknown	"
	F-3	19 Aug 1964	"	"
TELSTAR	F-1	10 Jul 1962	21 Feb 1963	"
	F-2	21 Jan 1964	Unknown	"

LAUNCH DATES (continued)

Communication and Broadcast Satellites

VII. AMATEUR

Satellite	Launch Date	Out-of-Service	Launch Site
OSCAR-1	12 Dec 1961	31 Jan 1962	Vandenberg
-2	2 Jun 1962	21 Jun 1962	"
-3	9 Mar 1965	25 Mar 1965	"
-4	21 Dec 1965	-	"
-5	23 Jan 1970	-	"
-6	15 Oct 1972	-	Cape Canaveral
-7	15 Nov 1974	-	Vandenberg
-8	5 Mar 1979	-	"
-9	23 May 1980	Launch Failure	Kourou
RS 1 and 2	26 Oct 1978	-	Plesetsk

PART B

SATELLITE DESCRIPTIONS

SECTION I

INTERNATIONAL SATELLITES

- INTELSAT I (Early Bird)
- INTELSAT II
- INTELSAT III
- INTELSAT IV
- INTELSAT IVA
- INTELSAT V
- MARECS
- MARISAT
- MAROTS
- MCS*

*See INTELSAT V, the Maritime Communications Subsystem (MCS) is carried as a separate payload on INTELSAT V

INTELSAT I (EARLY BIRD)

PHOTO COURTESY OF COMSAT

Michael J. Robusto
IEEE Compendium

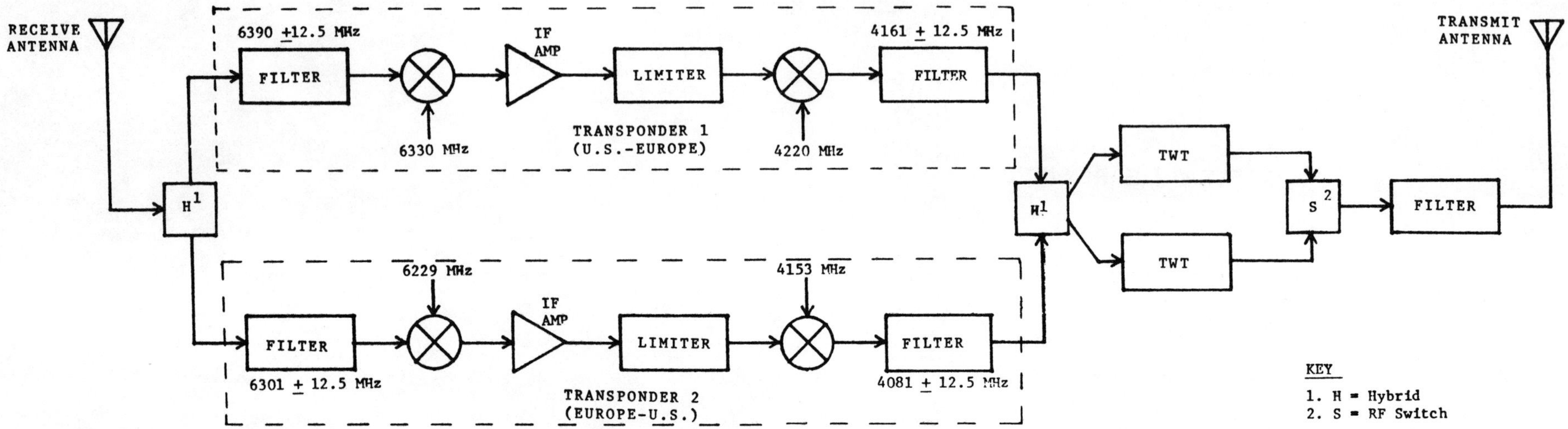

INTELSAT I COMMUNICATIONS BLOCK DIAGRAM (EARLY BIRD)

Michael J. Robusto
IEEE Compendium

General

No. satellites built	2 (1 launched)
Design life	18 months
Orbit	Synchronous equatorial
Launch vehicle	McDonnell/Douglas 3 stage thrust augmented Delta
Apogee injection motor	JPL Starfinder
Major contractor	Hughes Aircraft Company
Launch date	1 April 1965
Operational dates	28 June 1965 - 20 January 1968 29 June 1969 - 21 August 1969

Physical Description

Cylinder diameter	72 cm (28 in)
Cylinder height	59 cm (23 in)
Overall height	120 cm (47 in)
Weight	
Launch	68 kg (150 lb)
Orbit (initial)	39 kg (85 lb)

Control System

Stabilization	Spin (120-180 rpm)
Configuration	2 independent H_2O_2 systems
Capability	3 yr station keeping life

Electrical Power

Solar array	
Power	45W max, 33W min after 3 yrs
No. and size	6000 solar cells, 1.3m^2 area
Batteries	2 Ni-Cad 0.8 amp-hr each No eclipse operation

Telemetry

Configuration	2 1.8W, 136 MHz transmitters 2 4 GHz beacons 4 VHF whip antennas
Encoders	2 PAM/FM

Command

Configuration	Commands entered through repeater receivers
Decoders	1 per receiver
No. commands	12

Communications Antennas

Receive	
Type	3 element cloverleaf array
Gain	4 dBi
Pattern	40° x 360°
Polarization	Linear in plane perpendicular to spin axis
Transmit	
Type	6 element colinear slot array
Gain	9 dBi
Pattern	11° x 360°, tilted 7° above equatorial plane for Northern Hemispheric coverage
Polarization	Linear Orthogonal to receive polarization

Communications Capability

Total useable bandwidth	50 MHz
Frequency range	
Uplink (receive)	6301 ± 12.5 MHz, 6390 ± 12.5 MHz
Downlink (transmit)	4081 ± 12.5 MHz, 4161 ± 12.5 MHz
Repeaters	
Type	2 independent double conversion repeaters with limiters
Bandwidth	25 MHz each
Power amplifier	2 6W TWT's 1 operating, 1 spare
Saturation flux density	-74 dBW/m^2
Receiver noise figure	9 dB
Available e.i.r.p.	10 dBW (nominal)
Capacity	240 2-way voice circuits or 1 TV channel

Significant Features

Experimental and operational satellite
Provided regular commercial service
Based on Syncom design
No multiple access capability
Coverage confined to Northern Hemisphere

INTELSAT I TECHNICAL PARAMETERS

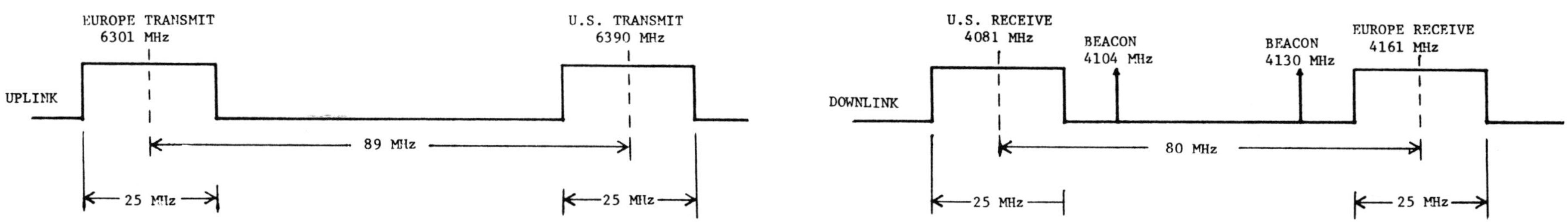

TYPICAL INTELSAT I FREQUENCY PLAN (EARLY BIRD)

Michael J. Robusto
IEEE Compendium

INTELSAT II

PHOTO COURTESY OF COMSAT

Michael J. Robusto
IEEE Compendium

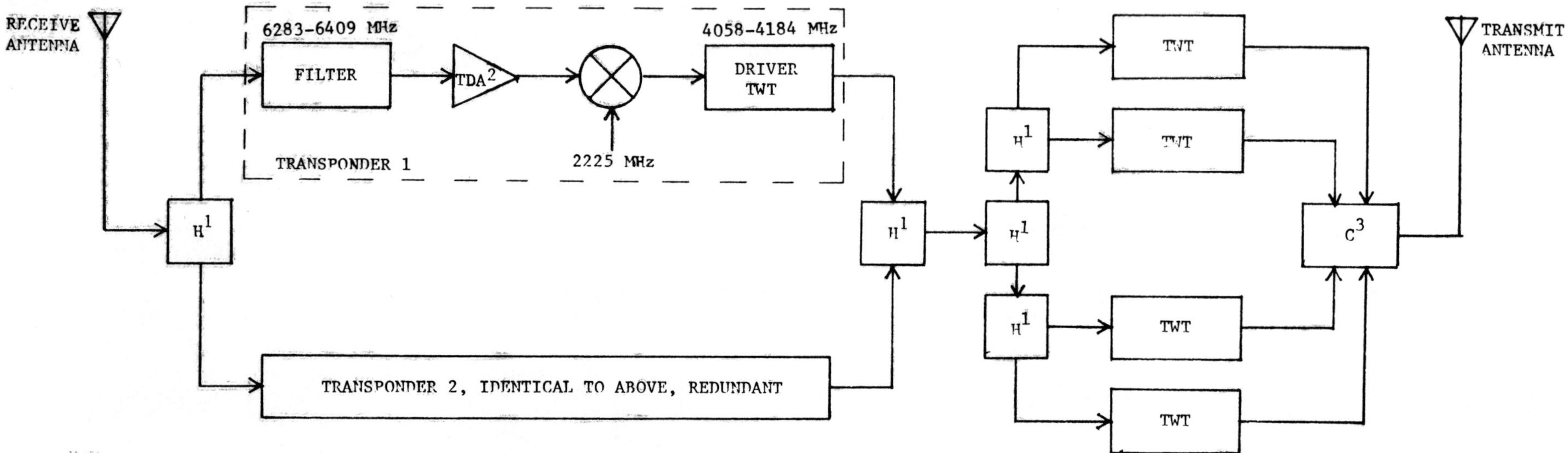

KEY
1. H = Hybrid
2. TDA = Tunnel Diode Amplifier
3. C = Combiner Switch, able to select 1,2, 3 or 4 TWT's simultaneously.

INTELSAT II COMMUNICATION BLOCK DIAGRAM

Michael J. Robusto
IEEE Compendium

General

No. satellites built	5 (4 launched)
Design life	3 yrs
Orbit	Synchronous equatorial
Launch vehicle	McDonnell/Douglas 3 stage, thrust augmented improved Delta
Apogee injection motor	Aerojet SVM-1
Major contractor	Hughes Aircraft Company
Date of 1st launch	26 October 1966

Physical Description

Cylinder diameter	142 cm (56 in)
Cylinder height	67 cm (26 in)
Overall height	130 cm (51 in)
Weight	
Launch	162 kg (357 lb)
Orbit (initial)	86 kg (190 lb)

Control System

Stabilization	Spin (80-140 rpm)
Configuration	2 independent H_2O_2 systems
Capability	5 yr station keeping life

Electrical Power

Solar array	
Power	100W max, 85W min after 5 yrs
No. and size	12756 solar cells, $3m^2$ area
Batteries	2 Ni-Cad, 4.5 amp-hr each Eclipse operation

Telemetry

Configuration	2 1.8W, 136 MHz transmitters 8 VHF whip antennas 2 4 GHz beacons
Encoders	2 PAM/FM

Command

Configuration	Commands entered through repeater receivers
Decoders	1 per receiver
No. commands	15

Communications Antennas

Receiver	
Type	1 cloverleaf array 1 slot array
Gain	2.5 dBi each
Pattern	$40^\circ \times 360^\circ$
Polarization	Linear
Transmit	
Type	4 stacked biconical horns
Gain	4.5 dBi
Pattern	$12^\circ \times 360^\circ$ centered at equator
Polarization	Linear Orthogonal to receive polarization

Communications Capability

Total usable bandwidth	126 MHz
Frequency range	
Uplink (receive)	6283-6409 MHz
Downlink (transmit)	4058-4184 MHz
Repeaters	
Type	2 redundant quasi-linear single conversion repeaters
Bandwidth	126 MHz each
Power amplifier	4 6W TWT's 1, 2, 3, or 4 operated in parallel
Saturation flux density	-68 dBW/m^2
Receiver noise figure	6 dB
Available e.i.r.p.	15.5 dBW (2 TWT's on) over $\pm 36^\circ$ latitude
Capacity	240 2-way voice circuits or 1 TV channel

Significant Features

Designed to satisfy NASA requirements and have additional capacity for other commercial traffic
Extension of INTELSAT I in short timeframe
Wideband quasi-linear repeater
Multiple access capability
Both Northern and Southern Hemispheric coverage

INTELSAT II TECHNICAL PARAMETERS

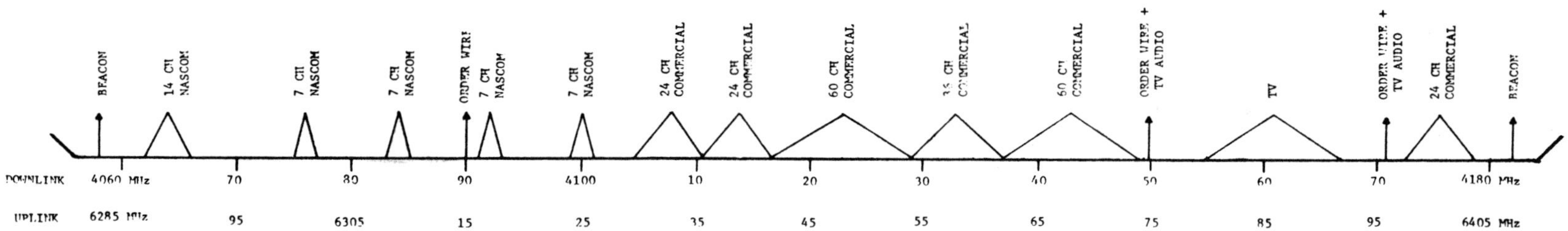

TYPICAL INTELSAT II FREQUENCY PLAN

Michael J. Robusto
IEEE Compendium

INTELSAT III

PHOTO COURTESY OF COMSAT

Michael J. Robusto
IEEE Compendium

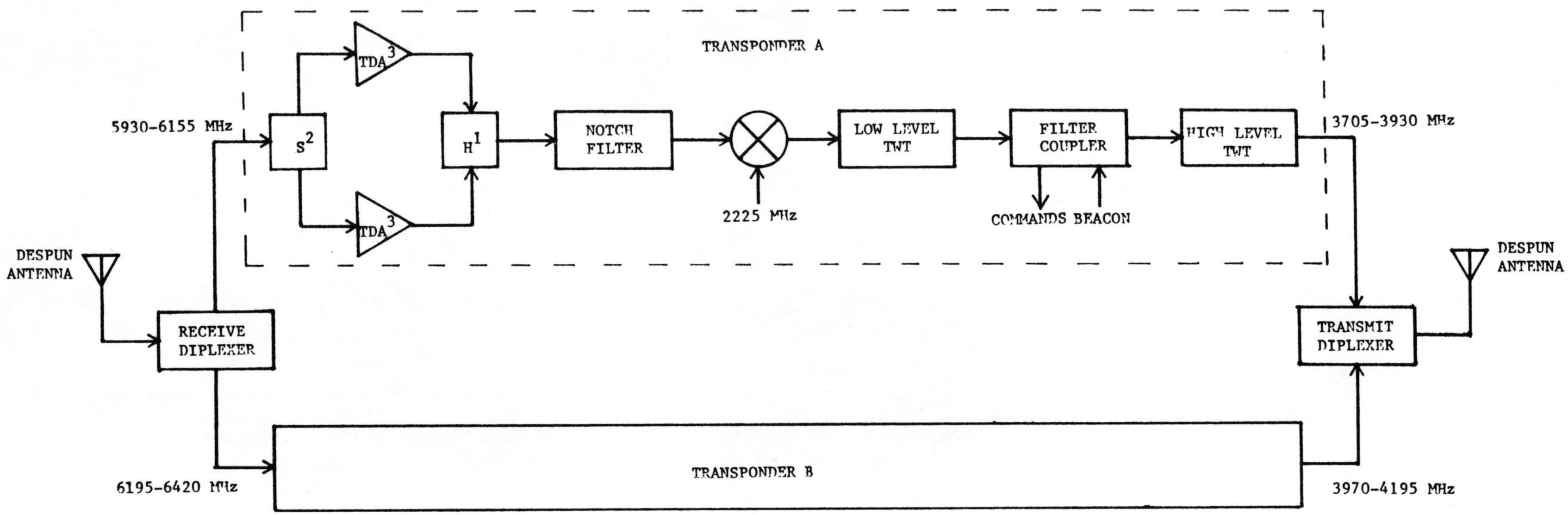

KEY
1. H = HYBRID
2. S = SWITCH
3. TDA = TUNNEL DIODE AMPLIFIER

INTELSAT III COMMUNICATION BLOCK DIAGRAM

Michael J. Robusto
IEEE Compendium

General

No. satellites built	8 (all launched)
Design life	5 yrs
Orbit	Synchronous equatorial
Launch vehicle	McDonnell/Douglas 3 stage, long-tank Delta
Apogee injection motor	Aerojet SVM-2
Major contractor	TRW Systems
Date of 1st launch	18 September 1968

Physical Description

Cylinder diameter	142 cm (56 in)
Cylinder height	104 cm (41 in)
Overall height	198 cm (78 in)
Weight	
Launch	294 kg (647 lb)
Orbit (initial)	152 kg (334 lb)

Control System

Stabilization	Dual spin (65-120 rpm) Despun earth oriented antenn spinning drum
Configuration	2 independent N_2H_4 systems
Capability	5 yr station keeping life

Electrical Power

Solar array	
Power	178W max, 130W min after 5 yrs
No. and size	10720 solar cells, 4.5m^2 area
Batteries	1 Ni-Cad, 10 amp-hr Eclipse operation

Command

Configuration	Single command frequency in guard band between transponders Omnidirectional antenna
Decoders	2 redundant
No. commands	28

Communications Antennas

Type	Mechanically despun conical horn with flat reflector 45° to horn axis
Gain	
Receive	21 dBi peak, 15 dBi over 19.3°
Transmit	18 dBi peak, 14 dBi over 19.3°
Polarization	
Receive	Left hand circular
Transmit	Right hand circular

Telemetry

Configuration	2 4 GHz beacons Omnidirectional antenna

Communications Capability

Total usable bandwidth	450 MHz
Frequency range	
Uplink (receive)	3705-4195 MHz
Downlink (transmit)	5930-6420 MHz
Repeaters	
Type	2 independent quasi-linear single conversion repeaters
Bandwidth	225 MHz each
Power amplifier	2 10W TWT's 1 for each repeater
Saturation flux density	-74 dBW/m^2
Receiver noise figure	< 7 dB
Available e.i.r.p.	27 dBW per repeater (peak) 22 dBW per repeater (beam edge)
Capacity	1500 2-way voice circuits or 4 TV or combinations thereof

Significant Features

First satellite series to provide global services with satellites over each of the three ocean regions
Despun antenna
Five times the capacity of INTELSAT II

INTELSAT III TECHNICAL PARAMETERS

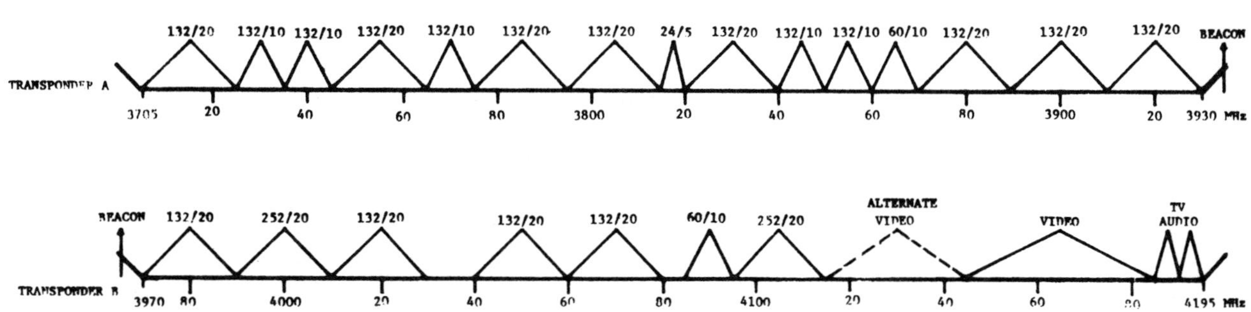

132/20 = 132 CHANNELS / 20 MHz BANDWIDTH FDM/FM CARRIER

TYPICAL INTELSAT III FREQUENCY PLAN

Michael J. Robusto
IEEE Compendium

INTELSAT IV

PHOTO COURTESY OF COMSAT

Martin P. Brown, Jr.
IEEE Compendium

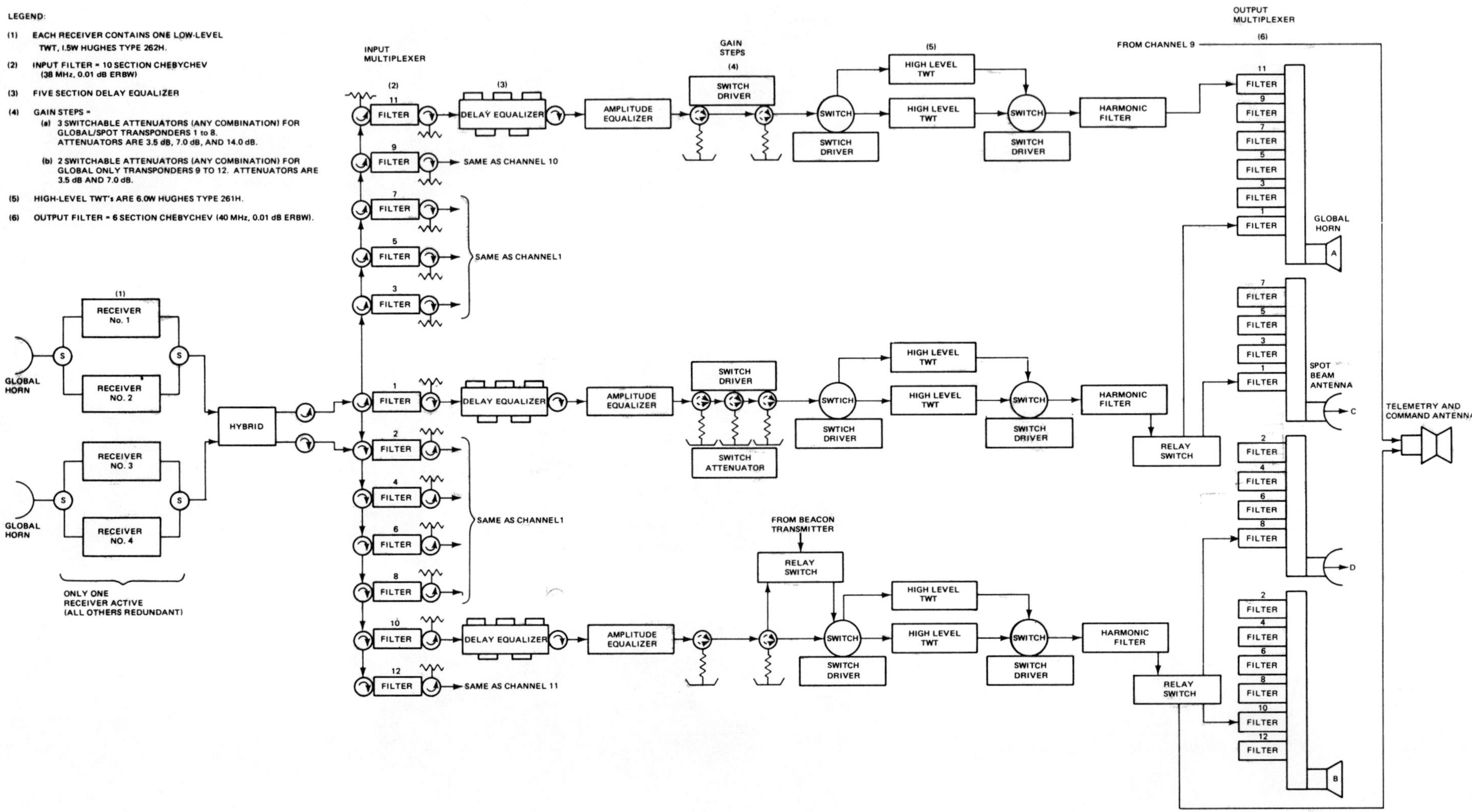

INTELSAT IV COMMUNICATION BLOCK DIAGRAM

Martin P. Brown, Jr.
IEEE Compendium

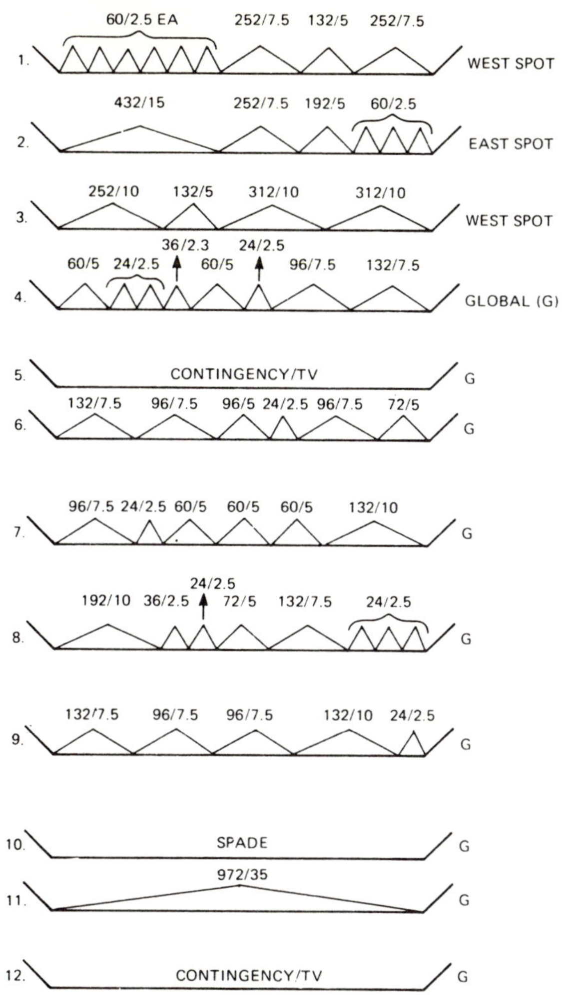

TYPICAL INTELSAT IV ATLANTIC PRIMARY FREQUENCY PLAN

60/5 = 60 CHANNEL/5.0 MHz BANDWIDTH CARRIER

General	
No. Satellites Built	8 (F-1 thru F-8)
Design Life	7 years
Launch Vehicle	Atlas Centaur
Major Contractor	Hughes
Date of 1st Launch	25 January 1971(IV F-2)
Physical Description	
Stabilization	Dual Spin (Despun Antenna Earth Oriented, Spinning Drum)
Cylinder Diameter	238.1 cm (93.75 in)
Cylinder Height	281.9 cm (112.8 in)
Height w/antenna mast	528.3 cm (208 in)
Weight	
Launch	1370.0 kg (3048 lbs)
Orbit (initial)	696.0 kg (1547 lbs)
Antenna Configuration	
Global	2 Transmit 2 Receive (one redundant)
Spot	1 West Transmit 1 East Transmit
Polarization	
Receive Global	LHC
Transmit Global	RHC
Transmit Spot	RHC

Communication Capability	
Total Usable Bandwidth	432 MHz
Assigned Frequency Range	
Up-Link (Receive)	5932-6418 MHz
Down-Link (Transmit)	3707-4193 MHz
Repeaters	
No. on Line	12
No. Redundant	12
Type	Hughes 261-H
Bandwidth	36 MHz
RF power	6.3 Watts (8dBW)
Available e.i.r.p. (Nominal)	
Global	
Beam Edge	22.0 dBW
Beam Center	25.5 dBW
Spot	
Beam Edge	33.7 dBW
Beam Center	37.0 dBW
Receive G/T (Nominal)	
Beam Edge	-18.6 dB/K
Beam Center	-15.0 dB/K
Saturation Flux Density at Beam Edge	
Min Gain Step Attn setting	-73.7 dBW/m^2
Max " " " "	-55.7 dBW/m^2
Communication Capability	
Per Global Repeater	
FDM/FM/FDMA	336-1092 channels
SPADE	800 channels (Demand Assigned)
1 TV	S/N=54 dB
2 TV	S/N=49 dB
Per Spot Repeater	
FDMA/FM/FDMA	800-1872 channels
Total Satellite Capacity	
10 FDM/FM/FDMA Repeaters	7500 one way channels
1 SPADE Repeater	800 channels
1 TV Repeater	1 TV

INTELSAT IV TECHNICAL CHARACTERISTICS

Martin P. Brown, Jr.
IEEE Compendium

INTELSAT IVA

PHOTO COURTESY OF INTELSAT

Martin P. Brown, Jr.
IEEE Compendium

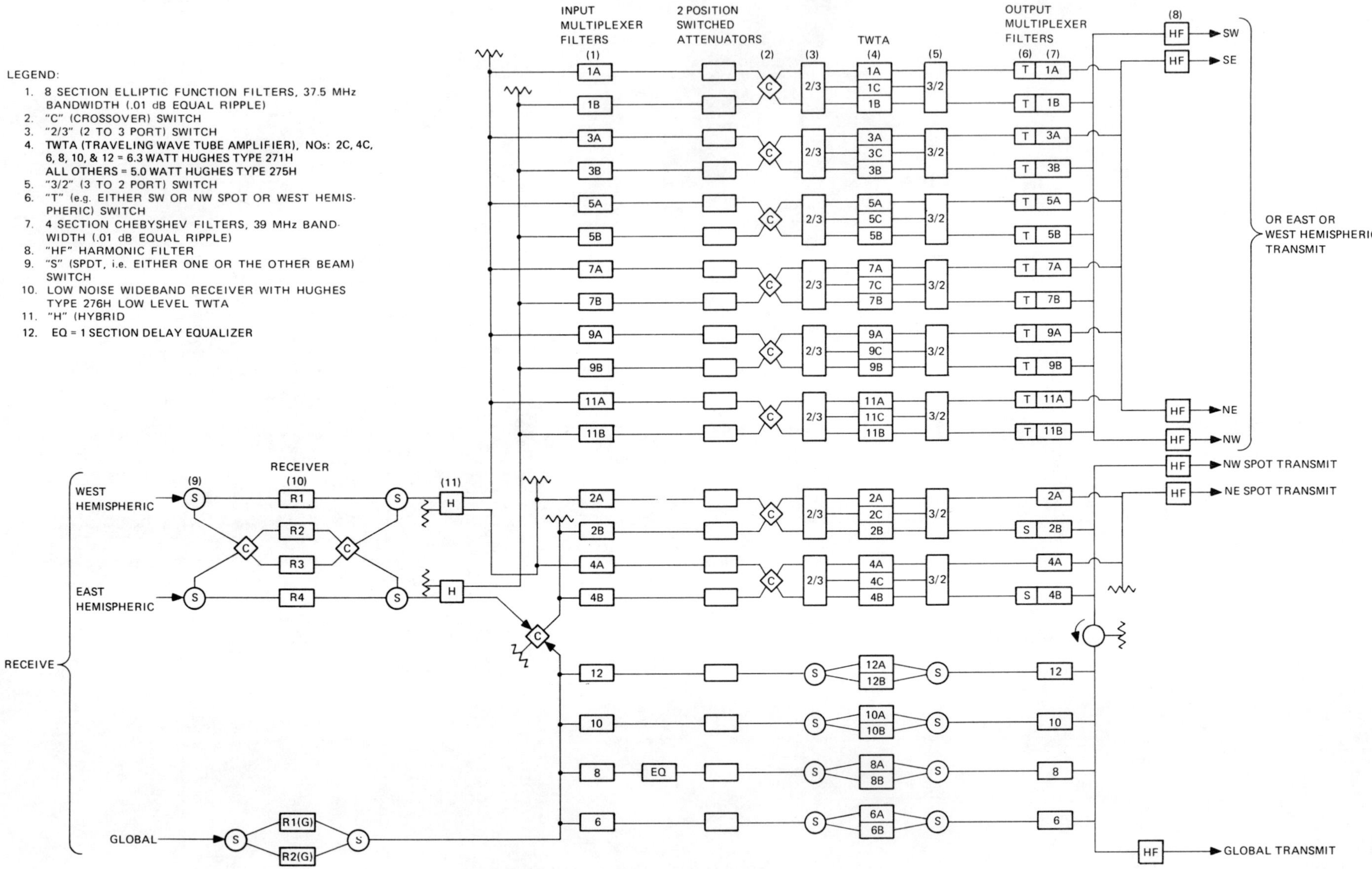

INTELSAT IVA COMMUNICATION BLOCK DIAGRAM

Martin P. Brown, Jr.
IEEE Compendium

INTELSAT IV–A (20 TRANSPONDERS)

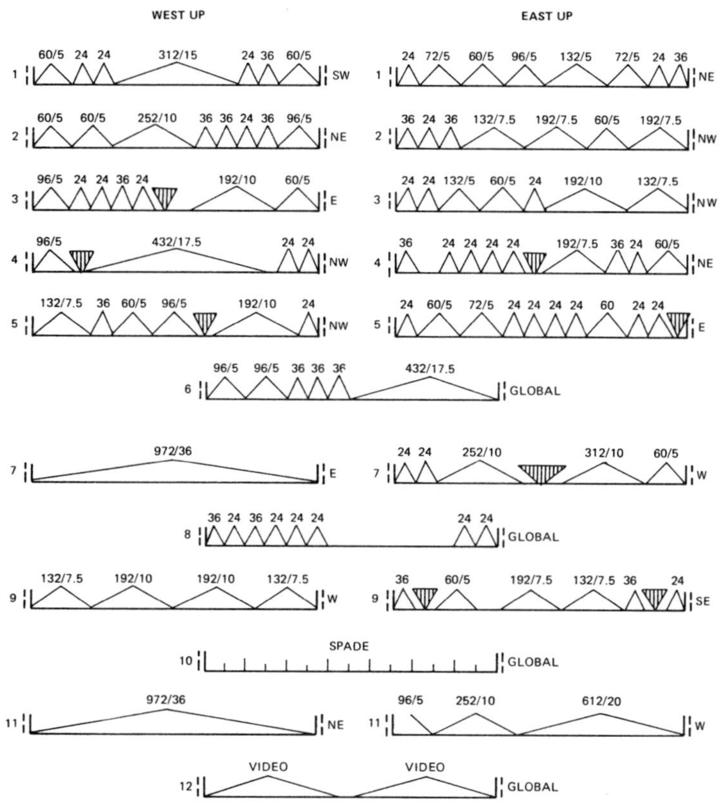

1. 60/5 = 60 CHANNEL/5.0 MHz BANDWIDTH CARRIER
2. [symbol] BANDWIDTH NOT AVAILABLE DUE TO CO-CHANNEL INTERFERENCE.

TYPICAL INTELSAT IVA FREQUENCY PLAN

General

No. Satellites Built	6 (F-1 thru F-6)
Design Life	7 years
Launch Vehicle	Atlas Centaur
Major Contractor	Hughes
Date of 1st launch	27 September 1975 (F-1)

Physical Description

Stabilization	Dual Spin (Despun Antenna Earth Oriented, Spinning Drum)
Cylinder Diameter	238.1 cm (93.75 in)
Cylinder Height	281.9 cm (111 in)
Height w/antenna mast	698.4 cm (275 in)
Weight	
Launch	1501 kg (3,340 lbs)
Orbit (initial)	818 kg (1,820 lbs)

Antenna Configuration

Global Horns	1 Transmit chan 2,4,6, 1 Receive 8,10,12
Hemispheric/Spot	2 Transmit (one for chans 2&4, one for chans 1,3,5,7,9&11)
Hemispheric	1 Receive (chan 1,2,3,4,5,7, 9,11)

Polarization

Receive Global	LHC
Transmit Global	RHC
Receive Hemi/Spot	LHC
Transmit Hemi/Spot	RHC

Communication Capability

Total Usable Bandwidth	2xfreq reuse (800 MHz)*
Assigned Frequency	
Up-link (receive)	5925-6425 MHz
Down-link (Transmit)	3700-4200 MHz
Repeaters	
No. on line	20
No. redundant	12
Type	Hughes 271H and 275H
Bandwidth	36 MHz
RF power	6.3W and 5.0W (see block diagram)
Beam edge e.i.r.p. (nominal)	
Global	22.0 dBW
Hemi	26.0 dBW
Spot	29.0 dBW
Beam edge G/T (nominal)	
Global	-17.6 dB/K
Hemi	-11.6 dB/K
Saturation Flux Density at Beam Edge (nominal)	
min (0 dB gain step)	-75.0 dBW/m^2
max (7.5 dB gain step)	-67.5 dBW/m^2
Communication Capability	
Per global repeater	
FDM/FM/FDMA**	450
SPADE	800
1 TV (30 MHz)	(S/N=54 dB)
2 TV (17.5 MHz each)	(S/N=49 dB)
Per Hemi repeater	
FDM/FM/FDMA**	600
Per Spot repeater	
FDM/FM/FDMA**	700

Total Satellite Capability

18 FDM/FM/FDMA repeaters	12,500 channels
1 SPADE repeater	800 channels
1 TV repeater	2,17.5 MHz each

*2xfrequency reuse through spatial isolation of the East and West hemispheric beams. Approx 500 MHz East to West plus 500 MHz West to East. Not all of potential 1,000 MHz is usable due to the requirement for global transponders.
**Multicarrier mode.

INTELSAT IVA TECHNICAL CHARACTERISTICS

Martin P. Brown, Jr.
IEEE Compendium

INTELSAT V

PHOTO COURTESY OF INTELSAT

Martin P. Brown, Jr.
IEEE Compendium

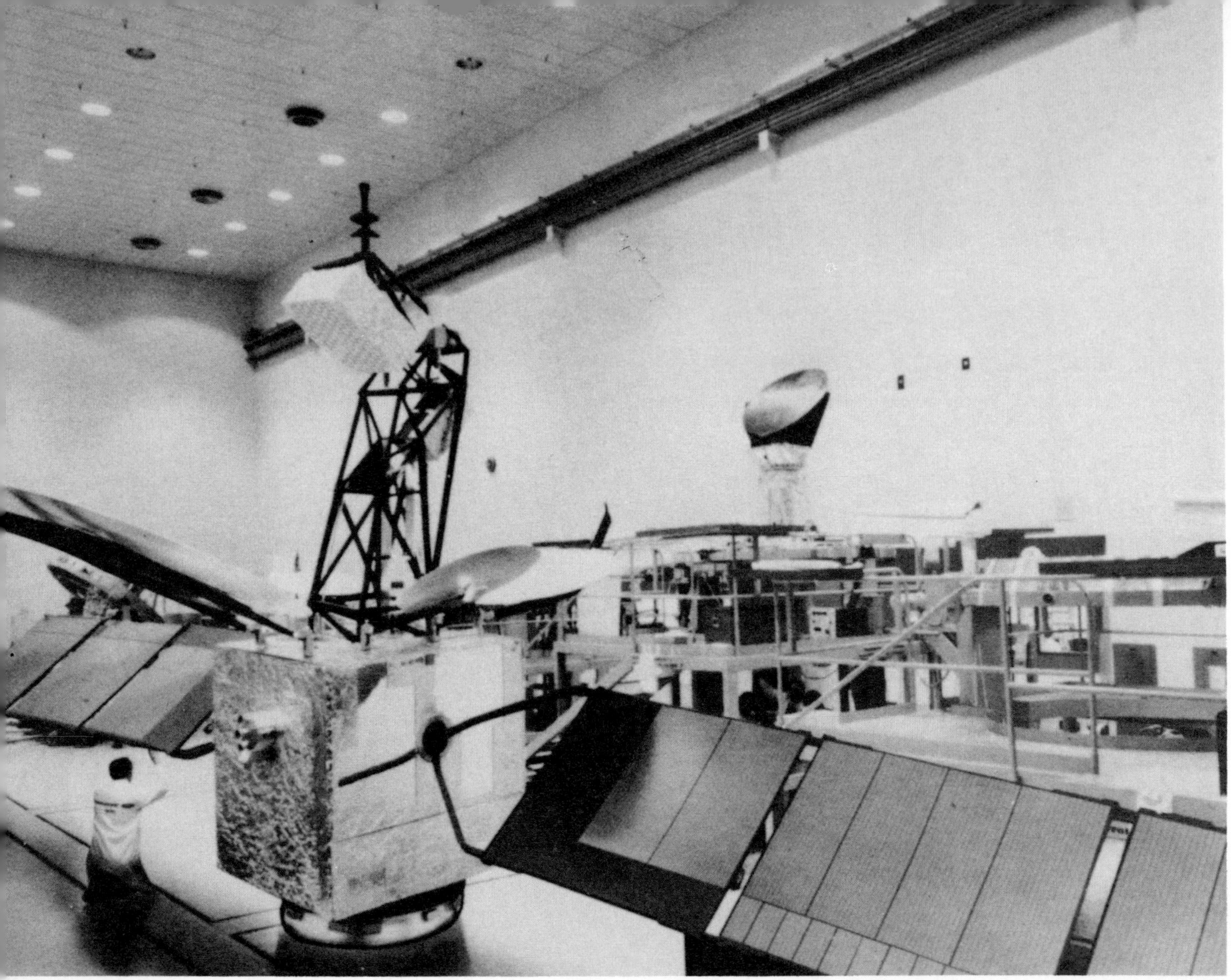

INTELSAT V SCALE MODEL

PHOTO COURTESY OF INTELSAT
MCS not shown
To be carried as a separate
pay-load on Flights F-5 thru F-8

Martin P. Brown, Jr.
IEEE Compendium

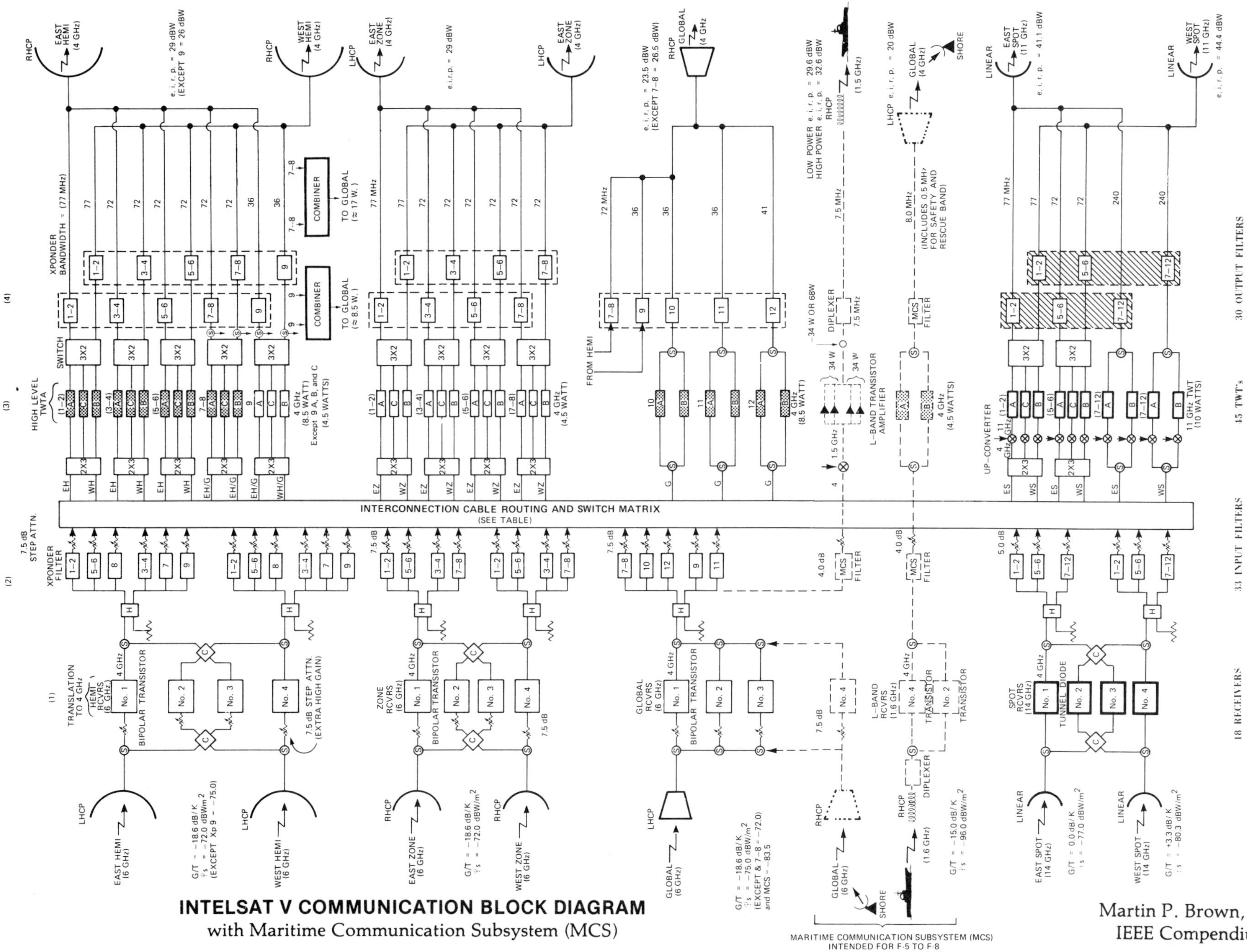

INTELSAT V COMMUNICATION BLOCK DIAGRAM
with Maritime Communication Subsystem (MCS)

Martin P. Brown, Jr.
IEEE Compendium

General

No. Satellites Built	**9 (F-1 Thru F-9)**
Design Life	7 years
Launch Vehicle	Atlas Centaur, Shuttle or Ariane
Major Contractor	Ford Aerospace and Comm. Corp
Date of 1st launch	6 Dec 1980

Physical Description

Stabilization	3 axis - momentum wheel
Length (with solar panels deployed)	15.9m (52 ft)
Height w/antenna mast	6.4m (21 ft)
Weight	
Launch	1847 kg (4,100 lbs)
Orbit (initial)	1003 kg (2,231 lbs)

Antenna Configuration

Global horns	1 transmit (4 GHz)
	1 receive (6 GHz)
Hemispheric/Zone	1 transmit (4 GHz)
	1 receive (6 GHz)
Spot (East)	1 Tx/Rx (14/11 GHz)
Spot (West)	1 Tx/Rx (14/11 GHz)

Polarization

Receive Global	LHC	
Transmit Global	RHC	
Receive Hemi	LHC	Xpol
Receive Zone	RHC	
Transmit Hemi	RHC	Xpol
Transmit Zone	LHC	
Receive Spot	Linear	orth.
Transmit Spot	Linear	

Communication Capability

Total Usable Bandwidth	4xfreq reuse at 6/4 GHz 2xfreq reuse at 14/11 GHz	2,300 MHz*
Assigned frequency Range		
Up-link (Receive)	5925-6425 MHz 14000-14500 MHz	
Down-link (Transmit)	3700-4200 MHz 10950-11200 MHz 11450-11700 MHz	
Repeaters		
No. on line	27	
No. Redundant	16	
Type	See block diagram	
Bandwidth	36 MHz global, 80 MHz hemi/zone/spot, 241 Spot	
RF power	See block diagram	

Beam Edge e.i.r.p. (nominal)	
Global	23.5 dBW
Hemi	29 dBW
Zone	29 dBW
Spot	41.1 dBW (East) 44.4 dBW (West)

Beam Edge G/T (nominal)	
Global	-18.6 dB/K
Hemi	-11.6 dB/K
Zone	-11.6 dB/K
Spot	0.0 dB/K (East) + 3.3 dB/K (West)

Saturation Flux Density at beam edge (minimum)	
Global	$-75.0\ dBW/m^2$
Hemi	$-72.0\ dBW/m^2$
Zone	$-72.0\ dBW/m^2$
Spot	$-77.0\ dBW/m^2$ (East) $-80.3\ dBW/m^2$ (West)

Total Satellite Capacity	
25 FDM/FM/FDMA repeaters	24,500 channels
1 SPADE repeater	800 channels
1 TV repeater	2 video chane (17.5 MHz ea.)

*Total potential bandwidth thru 4 x frequency reuse at 6/4 GHz and 2 x frequency reuse at 14/11 GHz is 3000 MHz. Not all of the 3,000 MHz is usable due to the requirement to provide global transponders.

INTELSAT V (MCS), Maritime Communication Subsystem
To be carried on F-5 to F-8
See next page

INTELSAT V TECHNICAL CHARACTERISTICS

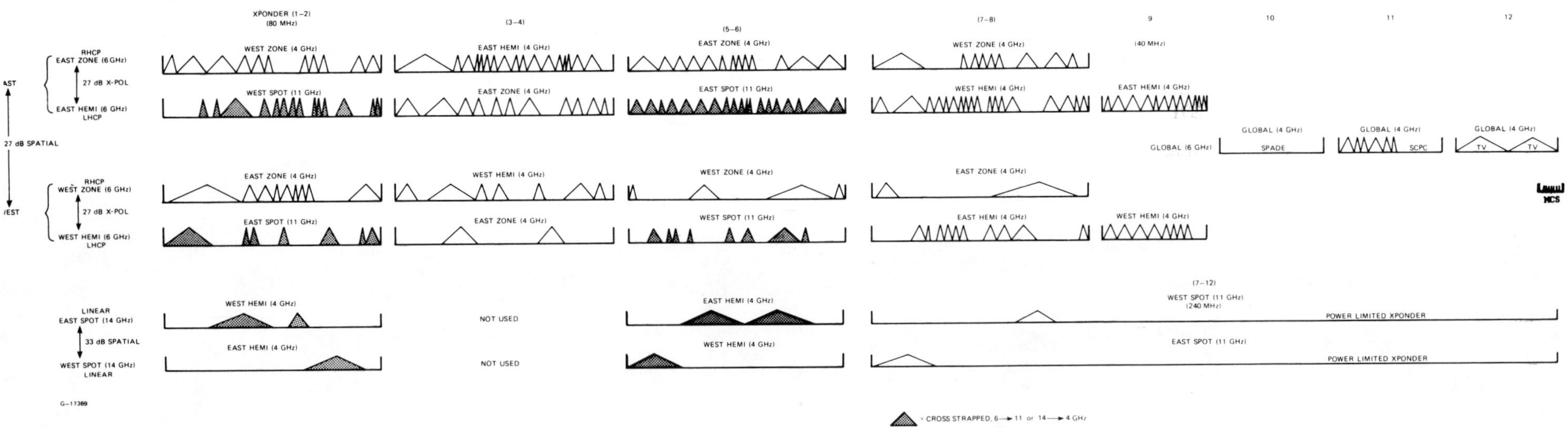

TYPICAL INTELSAT V FREQUENCY PLAN

Martin P. Brown, Jr.
IEEE Compendium

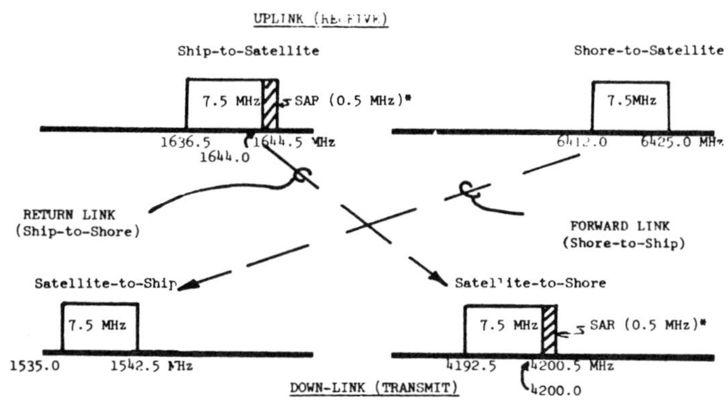

*SAR = Safety and Rescue band for ship distress signals.

Assigned Frequency Band

Up-Link (Receive)

·Shore-to-Satellite	6417.5 to 6425.0 MHz (7.5 MHz)
·Ship-to-Satellite	1636.5 to 1644.5 MHz (8.0 MHz) (includes 0.5 MHz for Safety and Rescue)

Down-Link (Transmit)

·Satellite-to-Shore	4192.5 to 4200.5 MHz (8.0 MHz)
·Satellite-to-Ship	1535.0 to 1542.5 MHz (7.5 MHz)

INTELSAT V (MCS) FREQUENCY PLAN

INTELSAT V (MCS) FREQUENCY PLAN

General

No. Satellites — To be carried on INTELSAT V F-5 to **F-9**

MCS system will be compatible with MARISAT and MARECS systems and may support the INMARSAT system in conjunction with one or both of these networks.

Physical Description

MCS will be added to the INTELSAT V antenna farm by upgrading the existing global hours for dual polarized operation and adding helix antennas for "L" band operation.

Polarization

Shore-to-Satellite (6 GHz)	RHCP
Satellite-to-Ship (1.5 GHz)	RHCP
Ship-to-Satellite (1.6 GHz)	RHCP
Satellite-to-Shore (4 GHz)	LHCP

Transmission Parameters

Beam edge e.i.r.p.	
4 GHz (to shore)	20.0 dBW
1.5 GHz (to ship)	32.6 dBW
G/T	
6 GHz (from shore)	-18.6 dB/K
1.6 GHz (from ship)	-15.0 dB/K

Modulation Techniques

SCPC/FDMA: with compandors for speech

: without compandors for data

TDM, 2ØCPSK

Channel Capacity

Forward Link (Shore-to-Ship)
30 voice or equivalent TDM

Return Link (Ship-to-Shore)
30 to 35 voice
·plus SAR
·plus approx 80 data

INTELSAT V (MCS) TECHNICAL PARAMETERS

Martin P. Brown, Jr.
IEEE Compendium

MARECS

PHOTO COURTESY OF
EUROPEAN SPACE AGENCY

European Space Agency
IEEE Compendium

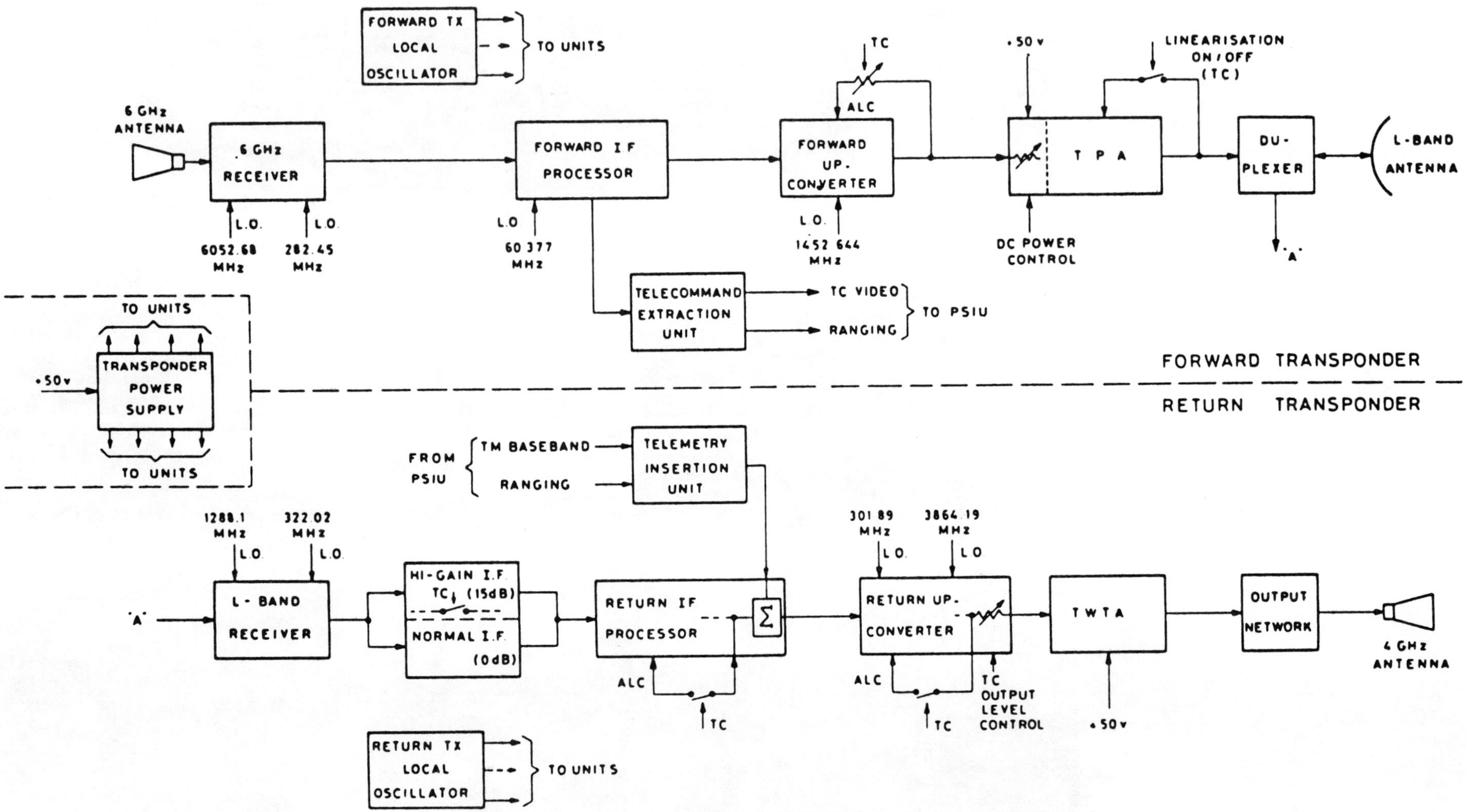

MARECS COMMUNICATION BLOCK DIAGRAM

European Space Agency
IEEE Compendium

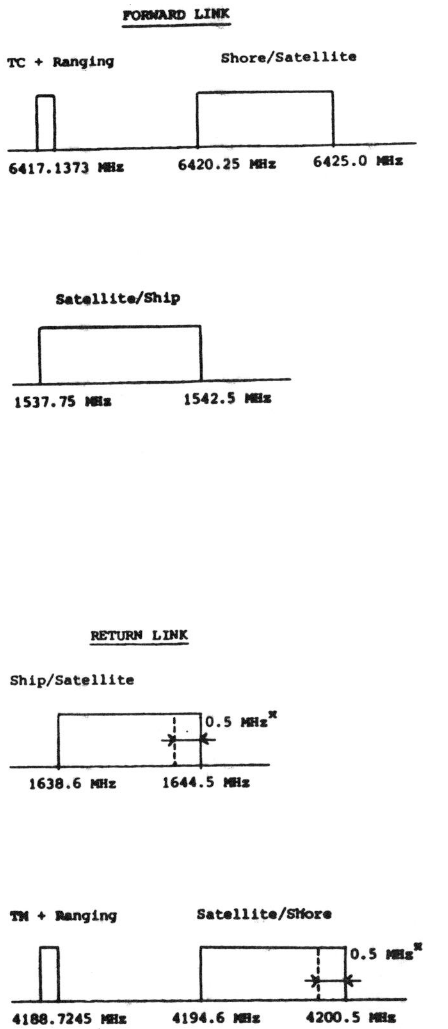

* 0.5 MHz will be used for a distress channel

MARECS FREQUENCY PLAN

Discussions at international level associated with the creation of the INMARSAT organisation led to a greater emphasis on the operational use of the maritime satellite, which was developed by the European Space Agency under the initial designation of MAROTS. In 1976 it was decided to make the spacecraft compatible with the existing MARISAT ground segment.

GENERAL

Number of satellites :	at least 2
Design life :	7 years
Launch Vehicle :	ARIANE
Major contractors :	
- Platform/spacecraft integration :	BADG (UK)
- Communications payload :	MSDS (UK)
Dates of launches :	mid February 1981 mid August 1981

PHYSICAL DESCRIPTION

Stabilisation :	3-axis momentum wheel
Spacecraft size :	
- diameter envelope :	2.20 m
- height :	2.50 m
- deployed arrays :	8.70 m
Weight at launch :	960 kg
" in orbit :	504 kg

ANTENNA CONFIGURATION

L-band :	1 global shaped-beam transmit/receive RHC polarisation
4 GHz :	1 global transmit- LHC polarisation
6 GHz :	1 global receive- RHC polarisation

COMMUNICATION PAYLOAD CHARACTERISTICS

Frequencies :	see frequency plan
Repeater chains :	
- Number :	2 redundant chains
- Bandwidth (1.4 dB peak to peak)	
Shore-to-ship :	4.75 MHz
Ship-to-shore :	5.9 MHz
- RF power output	
L-band :	63 W (linear)
4 GHz :	8.8 W(saturation)
- Available e.i.r.p. (nominal)	
L-band :	34.4 dBW (edge)
4 GHz :	14.5 dBW (edge)
- Receive G/T	
6 GHz :	-17 dB/K
L-band :	-12.1 dB/K

COMMUNICATION CAPABILITY

Shore-to-ship :	34.5 channels (SCPC 53 dBHz) or 55 channels (SCPC 50 dBHz)
Ship-to-shore :	50 channels (SCPC 53 dBHz)

MARECS TECHNICAL PARAMETERS

Consequently, the frequencies for transmission between shore and satellite were changed from 11 and 14 GHz to 4 and 6 GHz. At the same time advantage was taken of the newly developed ECS spacecraft platform to incorporate full eclipse capability and an extended lifetime. The satellite's name was changed to MARECS to reflect this reorientation. In early 1977 the scope of the programme was also extended to cover the manufacture of two satellites.

MARISAT

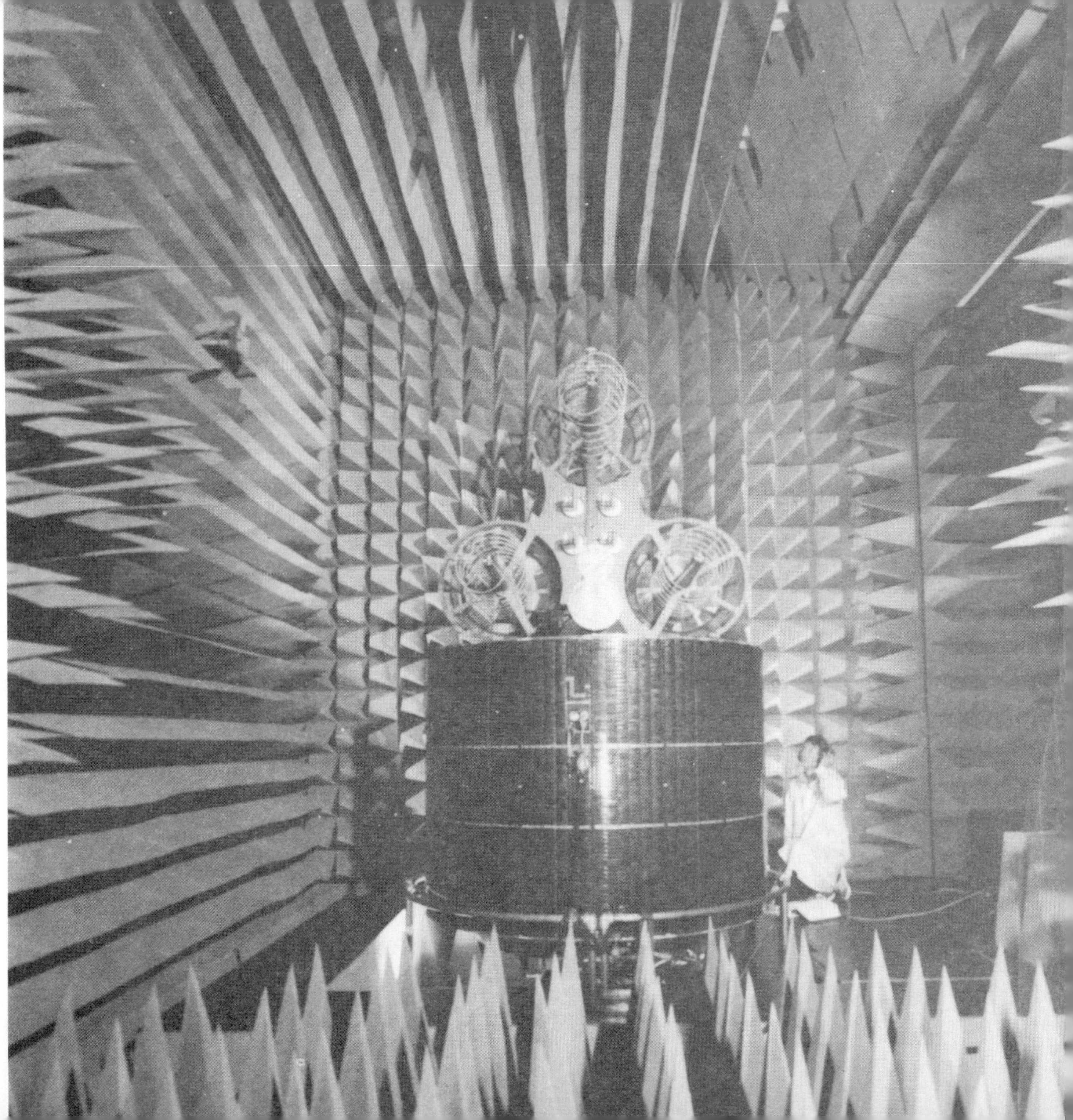

PHOTO COURTESY OF COMSAT

Hans Dodel
IEEE Compendium

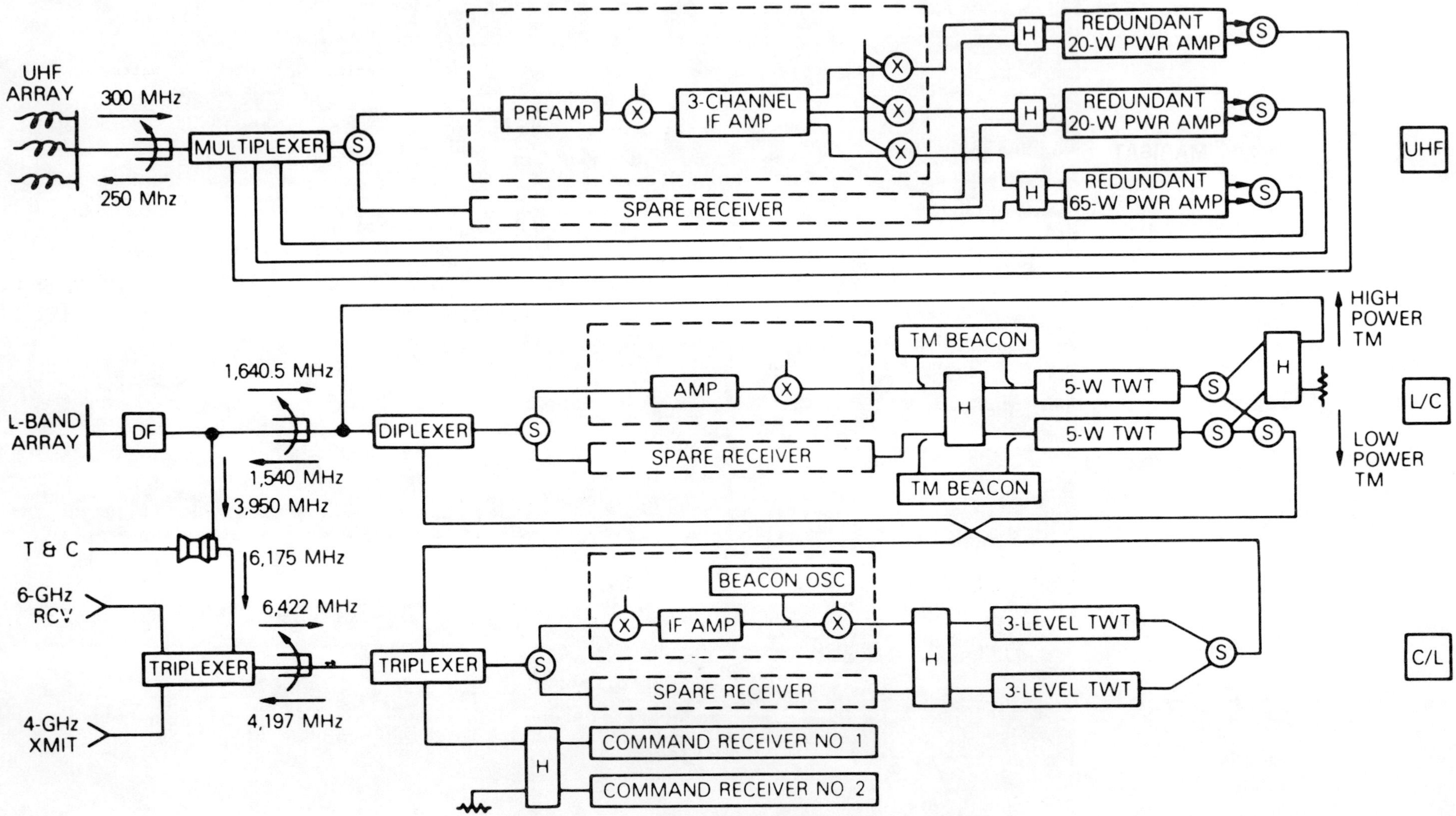

MARISAT COMMUNICATIONS BLOCK DIAGRAM

Hans Dodel
IEEE Compendium

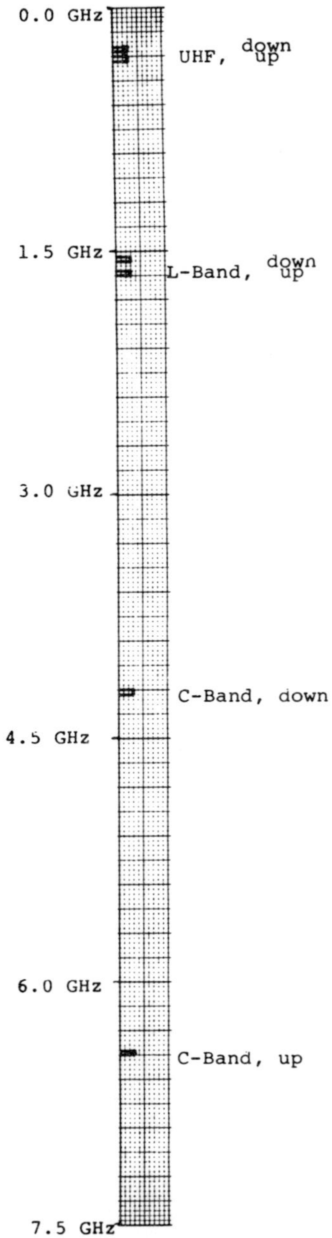

MARISAT FREQUENCY PLAN

General

No. Satellites Built	3 (F-1 thru F-3)
Design Life	5 years
Launch Vehicle	Delta 2914
Major Contractor	Hughes
Date of 1st Launch	19 Feburary 1976 (F-1)

Physical Description

Stabilization	Dual Spin (Despun Antenna Earth Oriented, Spinning Drum)
Cylinder Diameter	216 cm (85 inches)
Cylinder Height	163 cm (64 inches)
Height w/antenna mast	381 cm (150 inches)
Weight	
Launch	654.6 kg (1445 lbs.)
Orbit (initial)	326.2 kg (720 lbs.)

Antenna Configuration

UHF	3 helices 122 cm (48 in.) mounted 122 cm apart
L-Band	4 helices 38 cm (15 in.) mounted in center of UHF array
4-GHz	circular horn antenna for receive
6-GHz	circular horn antenna for transmit

Polarization

Receive Global	RHC
Transmit Global	LHC

Communication Capability

Total Usable Bandwidth	8 MHz
Assigned Frequency Range, C-Band	
Up-Link (Receive)	6420-6424 MHz
Down-Link (Transmit)	4195-4199 MHz
Assigned Frequency Range, L-Band	
Up-Link (Recieve)	1638.5-1642.5 MHz
Down-Link (Transmit)	1537.0-1541.0 MHz

Repeaters

No. on Line	2
No. Redundant	2
Bandwidth	4 MHz
RF power, shore-to-ship	
High Power Mode	60 Watts
Medium Power Mode	27 Watts
Low-Power Mode	7 Watts
Ship-to-shore	5 Watts

Available e.i.r.p. (Nominal) at Beam Edge

Shore-to-ship,	
High Power Mode	29.5 dBW
Medium Power Mode	26.0 dBW
Low-power mode	20.0 dBW
Ship-to-shore	18.8 dBW

Receive G/T (Nominal) at Beam Edge

Shore-to-ship	-25.4 dBi/K
Ship-to-shore	-17.0 dBi/K

Saturation Flux Density at Beam Edge

Shore-to-Ship	-87.0 dBW/m^2
Ship-to-Shore,	
Low gain setting	-99.5 dBW/m^2
High gain setting	-96.0 dBW/m^2

Communication Capability

Shore-to-ship	
High-Power Mode	16 voice, 88 telex channels
Medium Power Mode	8 voice, 44 telex channels
Low Power Mode	1 voice, 44 telex channels
Ship-to-shore	16 voice, 88 telex channels

MARISAT TECHNICAL CHARACTERISTICS

Hans Dodel
IEEE Compendium

MAROTS

PHOTO COURTESY OF
EUROPEAN SPACE AGENCY

Marcel R. E. Bichara
IEEE Compendium

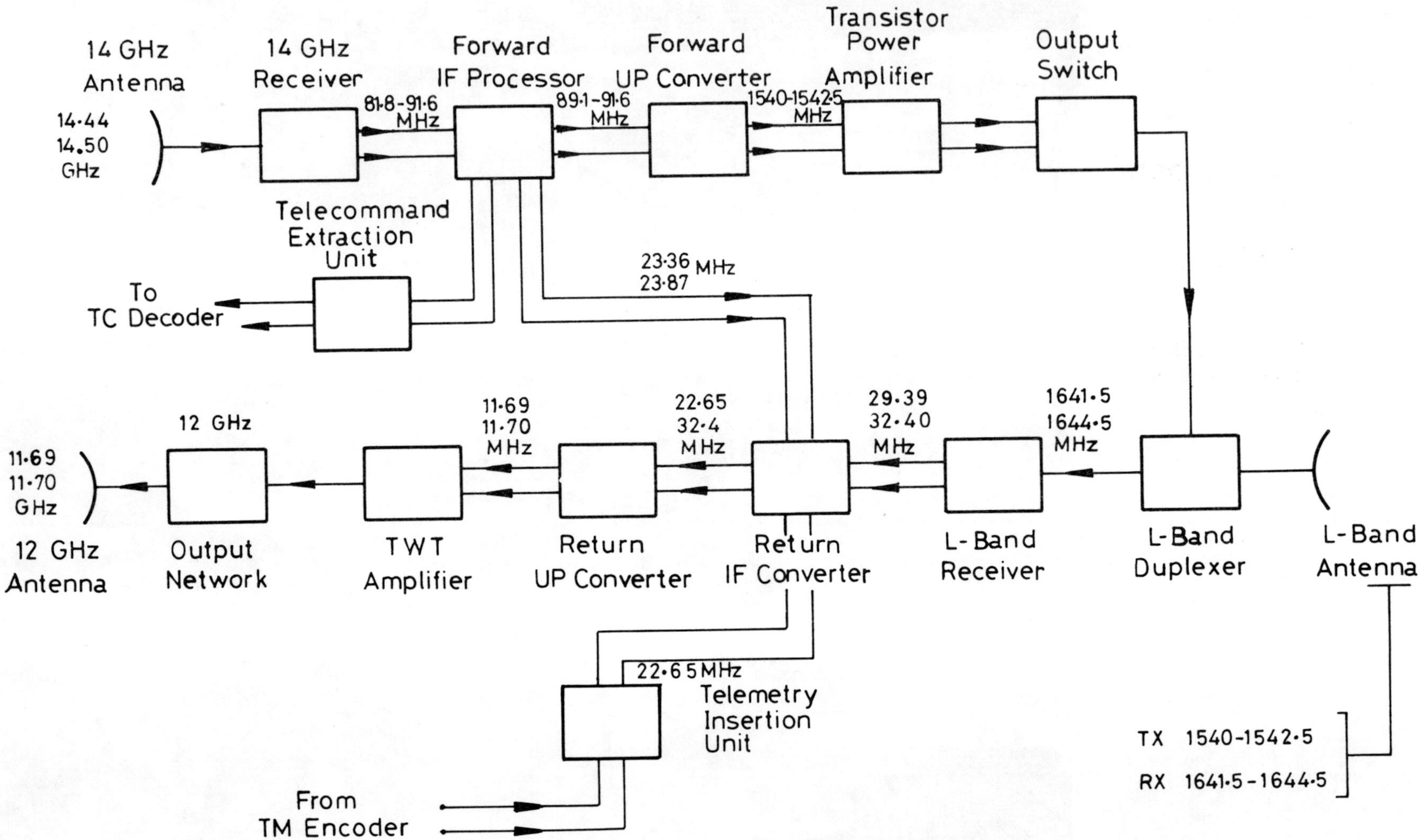

MAROTS COMMUNICATION BLOCK DIAGRAM

Marcel R. E. Bichara
IEEE Compendium

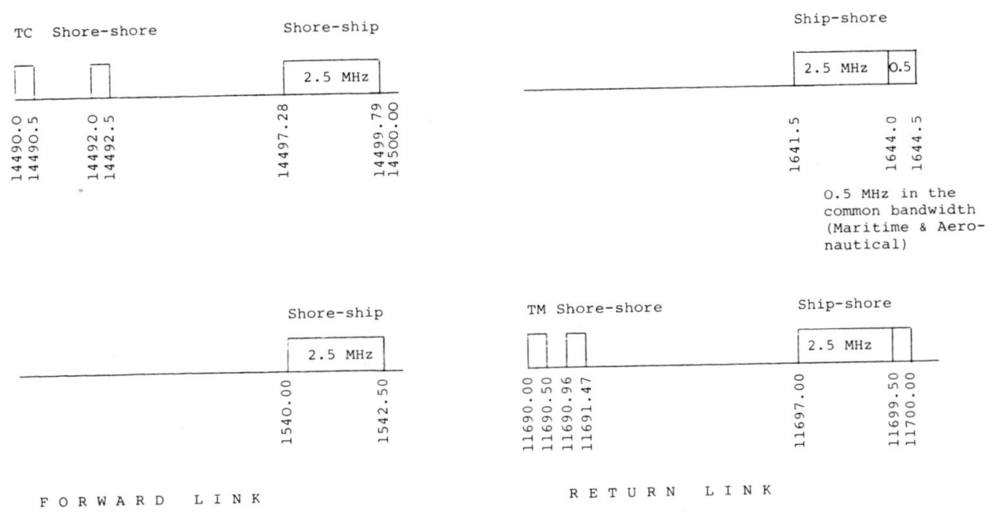

MAROTS FREQUENCY PLAN

*MAROTS FREQUENCY PLAN

* A decision to change the frequency bands of MAROTS was made after this section was prepared. The new satellite is to be called MARECS and is shown as another contribution. The MAROTS section has been retained since it has been widely referred to in literature.

General

No of satellites	one
Design life	7 years
Launch vehicle	Delta 3914
Major contractors:	
- Platform & space-craft integration	HSD (UK)
- Communications payload	MSDS (UK)

Physical description

Stabilization	3-axis momentum wheel
Spacecraft size:	
diameter envelope	2.18 m.
height	2.42 m.
deployed arrays	8.62 m.
Weight at launch	906 kg
in orbit	443 kg (initial)

Antenna configuration

L - Band	1 global shaped-beam transmit/receive RHC polarization
11 GHz	1 global transmit LHC polarization
14 GHz	1 global receive RHC polarization

Communication Payload Characteristics

Frequencies	See frequency plan
Repeater chains	
- Number	3 redundant chains
- Bandwidth	
Shore to ship	2.5 MHz
Ship to shore	3.0 MHz
Shore to shore	0.5 MHz
- RF power output	
L-band	60 W
11 GHz	20 W
Available EIRP (nominal)	
L-band	34 dBW (edge)
11 GHz	26.9 dBW (edge)
Receive G/T	
14 GHz	-16.5 dB/K
L-band	-12 dB/K

Communication Capability

Shore to ship	36 channels (DA/SCPC 51 dBHz)
Ship to shore	50 channels (DA/SCPC 52 dBHz)
Ship to ship	4 channels (SCPC 53 dBHz)

MAROTS TECHNICAL CHARACTERISTICS

Marcel R. E. Bichara
IEEE Compendium

SECTION II

INTERNATIONAL / DOMESTIC SATELLITES

- AEROSAT
- LOUTCH and LOUTCH-P
- MOLNIYA I
- MOLNIYA II
- MOLNIYA III
- STATSIONAR
    - RADUGA
    - EKRAN*
    - GORIZONT**
- VOLNA

---

*EKRAN is a dedicated domestic TV satellite which has been placed in this section due to its similarity to the STATSIONAR series.

**GORIZONT carries both 6/4 GHz and 7/8 GHz subsystems and has been placed in this section due to its similarity to the STATSIONAR series.

AEROSAT

Hans Dodel
IEEE Compendium

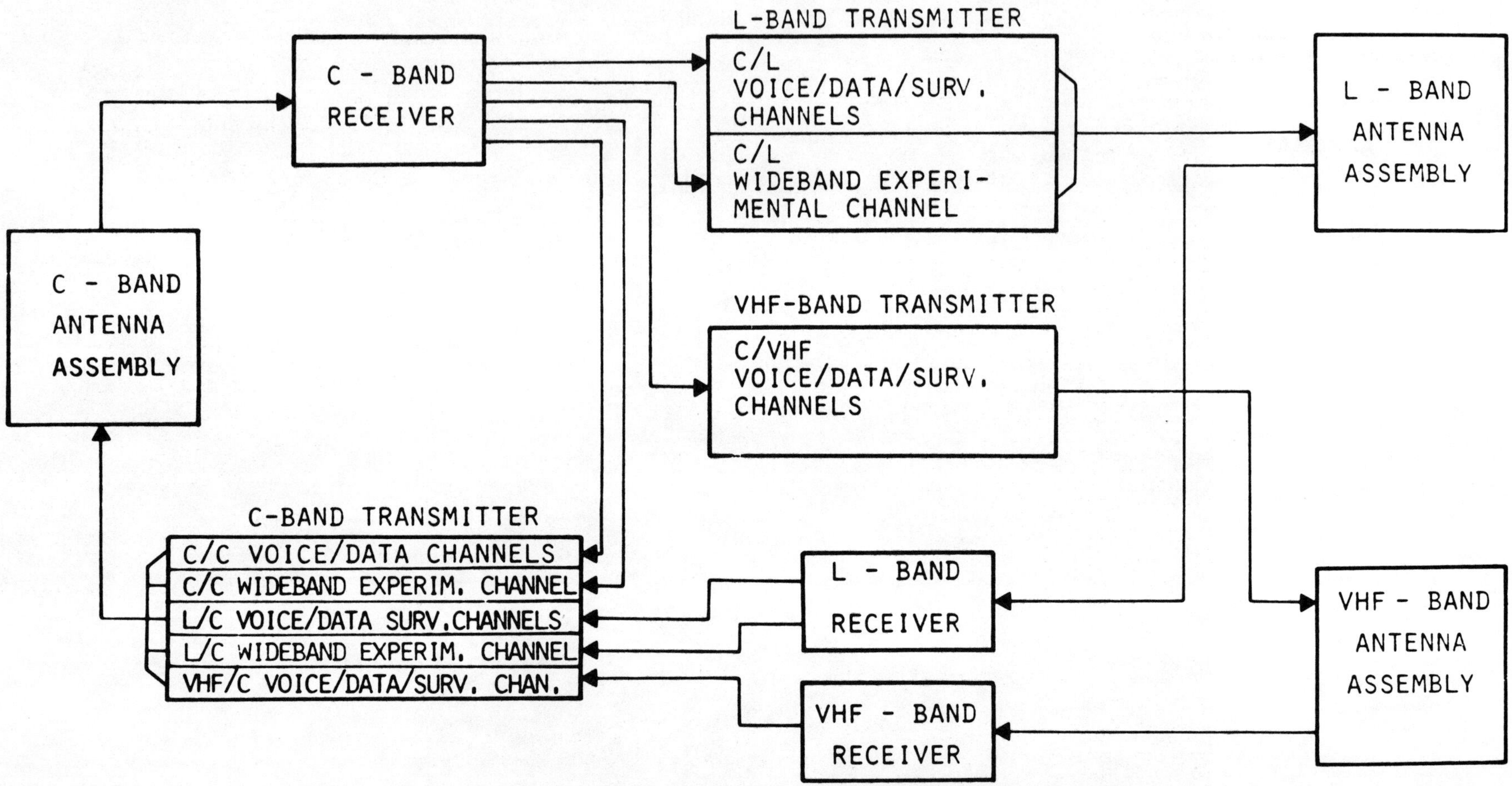

AEROSAT COMMUNICATION BLOCK DIAGRAM

Hans Dodel
IEEE Compendium

Purpose: Air Traffic Control

Launch Vehicle: Delta 3914

Design Life: Seven Years

Mass: 925.9 kg (1703 lbs)

Stabilization: 3-axis body stabilized

Repeaters: Five repeaters are required,
- from C to L-band
- from L to C-band
- from C to C-band
- from C to VHF-band
- from VHF to C-band

Six Types of Channels:
- narrowband voice/data or surv. channels for Ground Station to Aircraft transmissions (forward channels) through the C to L-band and C to VHF-band repeaters;
- narrowband voice/data or surv. channels for Aircraft to Ground Station transmissions (return channels) through the L to C-band and VHF to C-band repeaters;
- narrowband voice/data channels for Ground Station to Ground Station transmission through the C to C-band repeaters;
- a wideband experimental channel for Ground Station to Aircraft transmission through the C to L-band repeaters;
- a wideband experimental channel for Aircraft to Ground Station transmission through the L to C-band repeaters;
- a wideband experimental channel for Aircraft or Ground Station transmission through the C to C-band repeaters;

Channel Modulation:

Type of Repeater	Channel Type	Tone Frequency	Peak Deviation
C to L	Voice/data/surv.	1 kHz	39 kHz
	Wideband Experimental	50 kHz	2.450 MHz
L to C	Voice/data	1 kHz	19 kHz
	Surveillance	1 kHz	39 kHz
	Wideband Experimental	50 kHz	2.450 MHz
C to C	Voice/data	1 kHz	39 kHz
	Wideband Experimental	50 kHz	2.450 MHz
C to VHF	Voice/data	1 kHz	11 kHz
VHF to C	Voice/data	1 kHz	11 kHz

Radio Frequencies:

i) Ground to Spacecraft (C-band)
- a. Voice/data/surveillance channels: 5000 to 5030 MHz
- b. Experimental Channels: 5000 to 5040 MHz
- c. Telecommand and Tracking channels: 5000 to 5125 MHz

ii) Spacecraft to Ground (C-band)
- a. Voice/data/surveillance channels: 5220 to 5250 MHz
- b. Experimental Channels: 5197 to 5250 MHz
- c. Telemetry channel: 5125 to 5250 MHz

iii) Aircraft to Spacecraft (C-band)
- Experimental Channel: 5000 to 5040 MHz

iv) Spacecraft to Aircraft (L-band)
- a. Voice/data/surveillance channels: 1543.5 to 1558.5 MHz
- b. Experimental Channel: 1543.5 to 1570.5 MHz

v) Aircraft to Spacecraft (L-band)
- a. Voice/data/surveillance channels: 1645.0 to 1660.0 MHz
- b. Experimental channel: 1622.5 to 1633.5 MHz or 1649.0 to 1660.0 MHz

vi) Spacecraft to Aircraft (VHF-band)
- Voice/data/surveillance channels: 125.425 to 125.975 MHz

vii) Aircraft to Spacecraft (VHF-band)
- Voice/data/surveillance channels: 131.425 to 131.975 MHz

C/L: The frequencies for the wideband experimental channel: C-Band Receive, 5028.1 MHz; L-Band Tran., 1565.5 MHz

G/T-Value: -16.5 dBi/K at edge of C coverage;

Input Signal Levels: C-Band Receive, -102 dBW/m^2 for voice/data/surveillance; - 90 dBW/m^2 wideband experimental

E.I.R.P.: L-Band voice/data/surv., 34.5 dBW (37.5 dBW) for 10° elevation
Wideband experimental 37.5 dBW Zone A (34.5 dBW Zone B)

L/C: The frequencies for the wideband experimental channel: L-Band Receive 1627.5 MHz; C-Band Transmit 5214.3 MHz

G/T-Value: -13.3 dBi/K at edge of L coverage;

Input Signal Levels: L-Band Receive, Voice/Data/Surv. - 129 dBW/m^2; Wideband Experimental - 106 dBW/m^2

E.I.R.P.: Voice/Data/Surveillance 1.2 dBW
Wideband Experimental 16.9 dBW

Bandwidth: Voice/Data 40 MHz
Surveillance 80 MHz
Wideband Experimental 5 MHz

C/C: The frequencies for the wideband experimental channel: C-Band Receive 5028.1 MHz; C-Band Transmit 5225.7 MHz

G/T-Value: -16.5 dBi/K at edge of C coverage;

Input Signal Levels: Voice/Data - 100 dBW/m^2; Wideband Exp. -99 dBW/m^2

E.I.R.P.: Voice/Data 4.2 dBW
Wideband Exp. 12.0 dBW

Bandwidth: Voice/Data 80 MHz
Wideband/Exp. 5 MHz

C/VHF: G/T-Value: -16.5 dBi/K at edge of C coverage;

Input Signal Levels: as C-Band immediately above

E.I.R.P.: Voice/Data/Surv. normal power mode 25.4 dBW (28.4 dBW High Power Mode)

Bandwidth: 25 MHz

VHF/C: G/T-Value: -22.2 dBi/K at edge of VHF coverage

Input Signal Levels: -135.0 dBW/m^2 for voice/data/surveillance

E.I.R.P.: 3.0 dBW at C-Band

Bandwidth: 25 kHz

AEROSAT TECHNICAL PARAMETERS

Hans Dodel
IEEE Compendium

LOUTCH and LOUTCH-P SERIES*
("Beam" or "Ray")

*To the author's knowledge, a picture of this satellite has not been released.

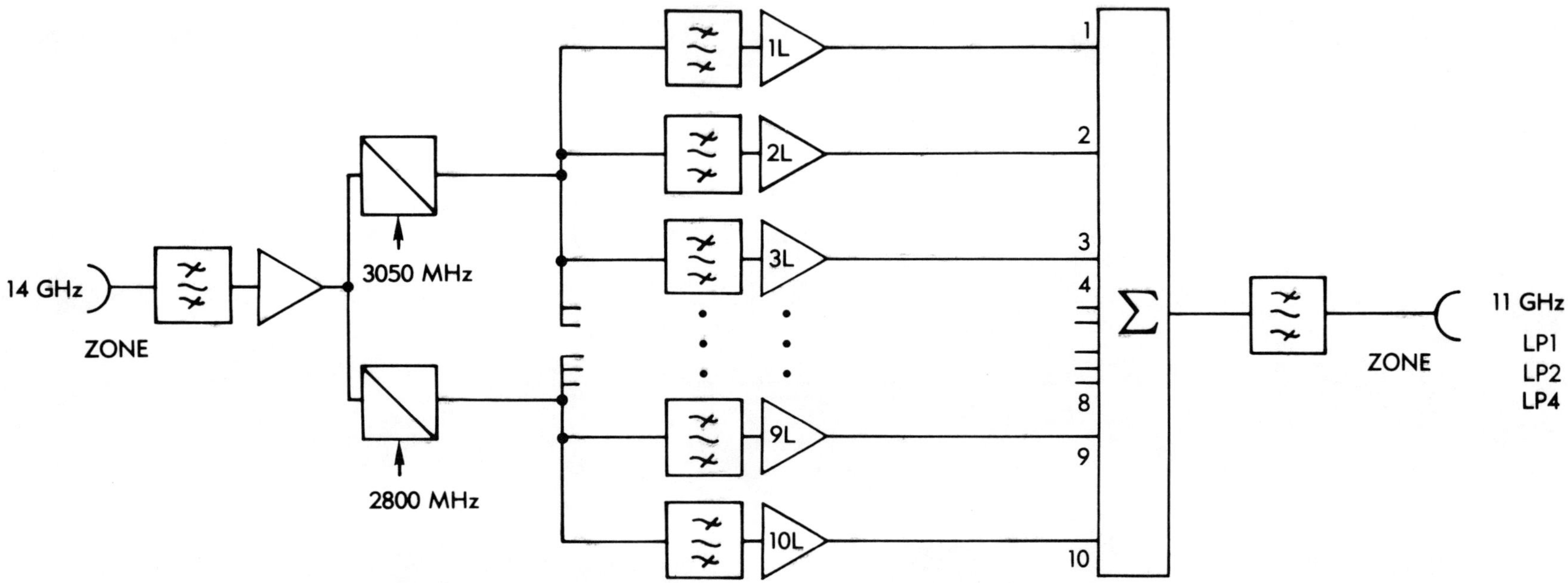

LOUTCH 1 THROUGH 4 and LOUTCH P1 THROUGH P4 BLOCK DIAGRAM

Walter L. Morgan
IEEE Compendium

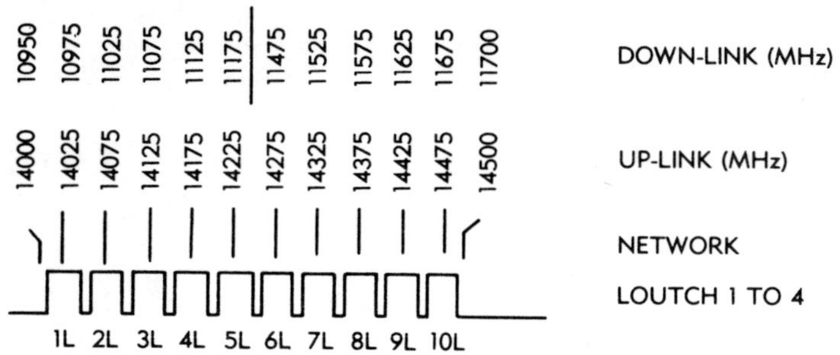

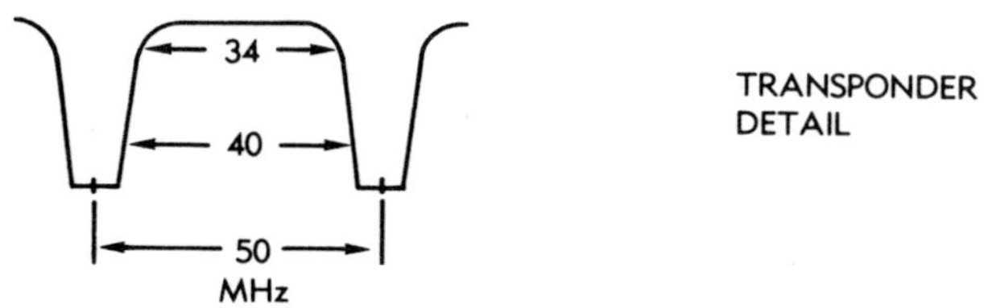

General (to 1/1/79)

This system is designed for telephone, telegraph, and phototelegraph services; transmission of television and sound broadcasting programs; and purposes of international cooperation.

No. Launched to Date	0
Network Location	
Loutch 1	346.0°E ± 0.5°
Loutch 2	53.0°E ± 0.5°
Later Loutch 2	55.5°E ± 0.5°
Loutch 3	90.0°E ± 0.5°
Loutch 4	140.0°E ± 0.5°
Loutch P1	335.0°E
Loutch P2	45.0°E
Loutch P3	85.0°E
Loutch P4	190.0°E
Inclination	±0.5°

Antenna Configuration	Zone
Caribbean (Havana)	30 dBi (Loutch 1)
Europe (Belgrade)	30 dBi (Loutch 1)
USSR (Moscow)	30 dBi (Loutch 2)
USSR (Tashkent)	30 dBi (Loutch 3)
Mongolia	30 dBi (Loutch 4)
Receive Polarization	LHC
Transmit Polarization	RHC

Communications Capability

Total Usable Bandwidth	400 MHz
Assigned Frequency Range	
Up-Link (receive)	14.00-14.50 GHz
Down-Link (transmit)	10.95-11.20 & 11.45-11.70 GHz
Repeaters	
No. on Line	10
Bandwidth	34 MHz
RF Power	5 W (TV: 10 W)
Available e.i.r.p. (nom.)	
Beam Center (total)	36.8-39.8 dBW
Per Carrier	16.5 (SCPC/FM)-39.8 (TV) dBW
Receive G/T (nom.)	
Beam Edge	-9 dB/K
Beam Center	-6 dB/K
System Noise Temperature	4000 K
Satellite Capacity per Transponder	
SCPC/FM	200 channels
FDM/FM	
2 carriers @ 60 chan.	120 channels
6 carriers @ 12 chan.	72 channels
TV/FM	1 channel

LOUTCH 1 THROUGH 4 and LOUTCH P1 THROUGH P4
FREQUENCY PLAN and TECHNICAL PARAMETERS

Walter L. Morgan
IEEE Compendium

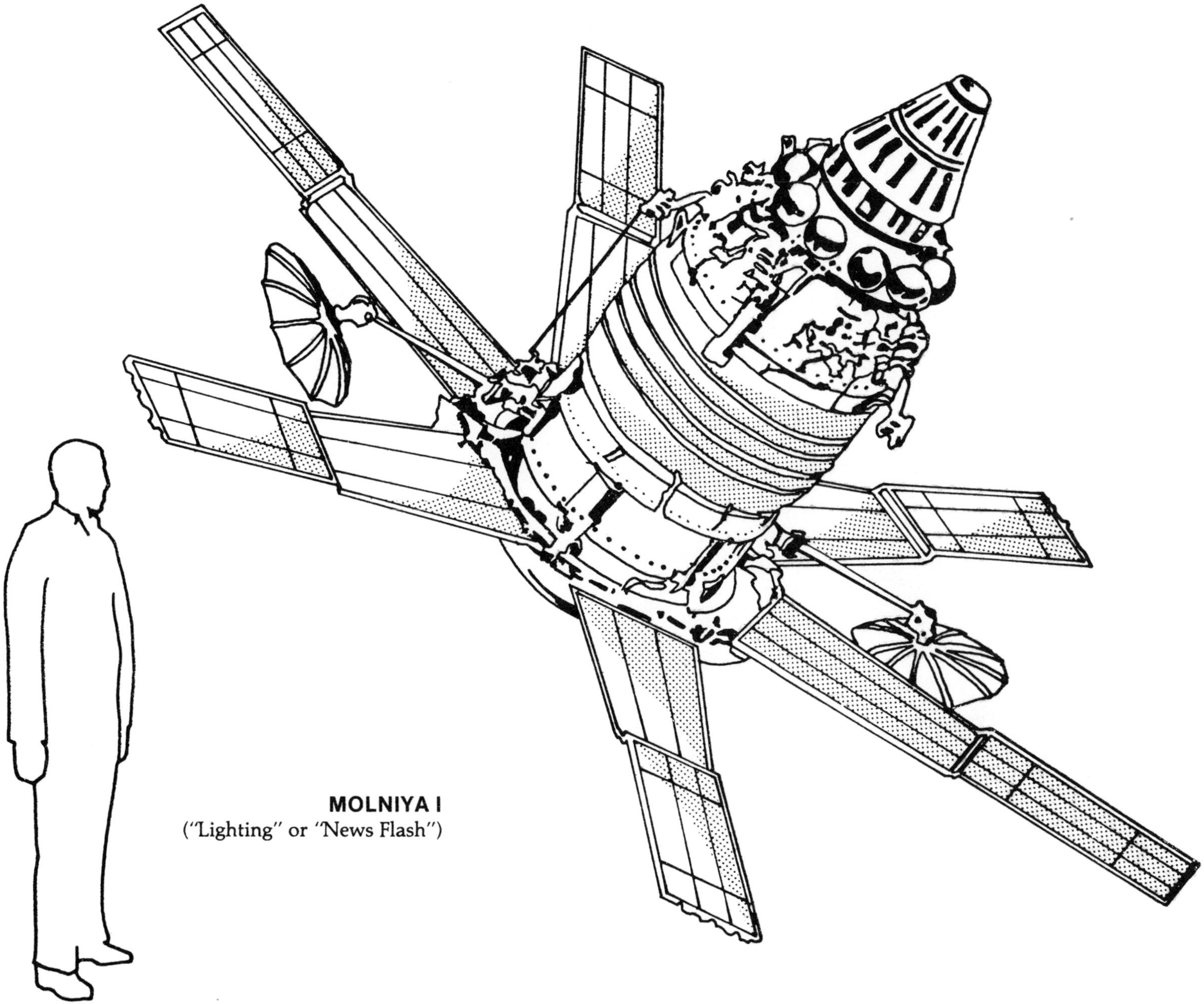

MOLNIYA I
("Lighting" or "News Flash")

Martin P. Brown, Jr.
IEEE Compendium

MOLNIYA II and III
("Lighting" or "News Flash")

PHOTO COURTESY OF AVIATION WEEK

Martin P. Brown, Jr.
IEEE Compendium

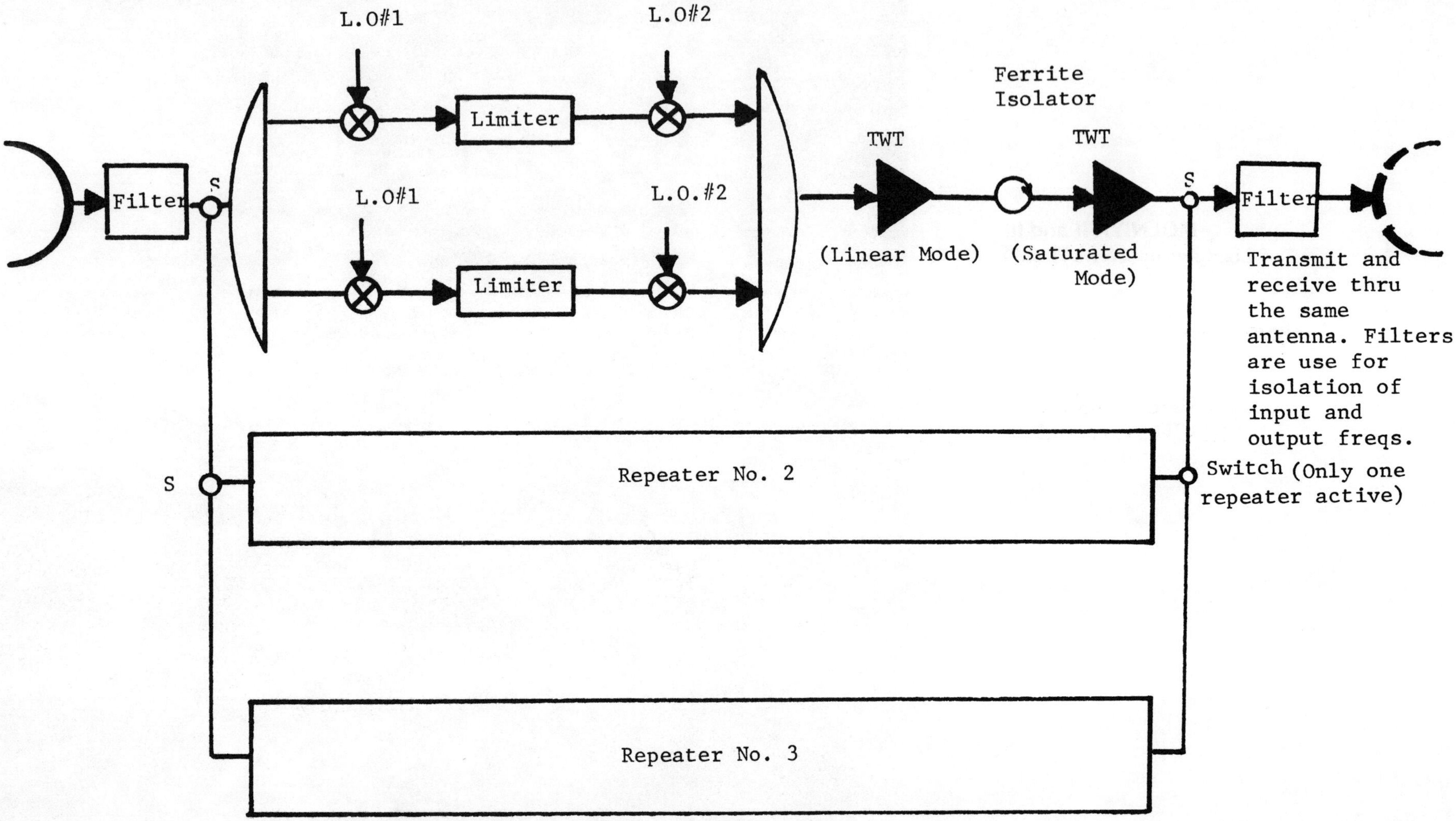

Note: 14 Watts for telephony; 40 Watts available for TV

MOLNIYA I COMMUNICATION BLOCK DIAGRAM*

Martin P. Brown, Jr.
IEEE Compendium

*To the authors knowledge, block diagrams for Molniya II and III are not available.

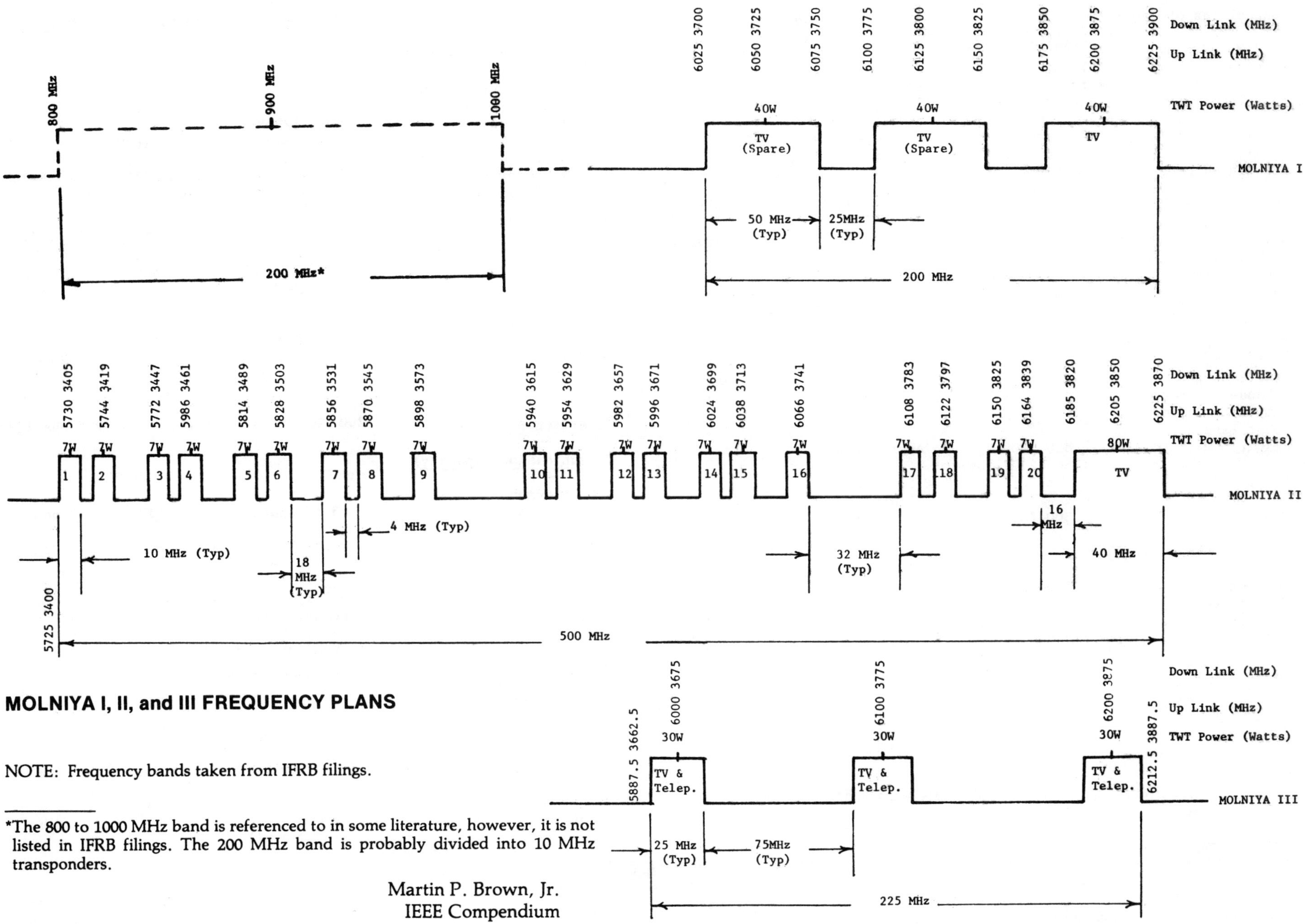

MOLNIYA I, II, and III FREQUENCY PLANS

NOTE: Frequency bands taken from IFRB filings.

---

*The 800 to 1000 MHz band is referenced to in some literature, however, it is not listed in IFRB filings. The 200 MHz band is probably divided into 10 MHz transponders.

Martin P. Brown, Jr.
IEEE Compendium

General (as of Jan 79)

Molniya means "lightning" in Russian (a colloquial meaning is "news flash"). Also spelled Molnya or Molnia. The satellites transmit voice, video, and data to a large network of ground antennas (>50) called "Orbita" earth stations. Molniya II satellites are upgraded versions of their predecessor with expanded capacity at 6/4 GHz. Molniya III is the version used with the hot line between Washington D.C. and Moscow.

	Molniya I	Molniya II	Molniya III
No. Satellites Built (as of 1976)	31	11	2 = (total 44)
Design Life (approx)	2 years	Same	?
Launch Vehicle	A-2-e	A-2-e	A-2-e
Date of 1st launch	23 April 1965	24 November 1971	21 November 1974
Physical Description			
Stabilization	Body	Same	Same
Cylinder Height	3.8m (11.3ft)	Same	?
Number of Silicon Solar Panels	12	24	?
Weight	820-1000 kg (1800-2200 lbs)	Approx Same	?
Positioning fuel	Nitrogen	Nitrogen (extra containers)	
Power	500-700W	700W+	?
Antenna Configuration			
Coverage	Global & Spot	Global	?
No. reflectors	2 Spot (one redundant)	Global horns	?
Gain	18 dBi Global	18 dBi	?

	Molniya I	Molniya II	Molniya III
Communications Capability			
Assigned Frequency Range			
·Up-Link (receive)			
1) Television	6025-6225MHz	6185-6225MHz	5987.5-6212.5
2) Telephony	Unknown**	5725-6169MHz	" "
·Down-Link (transmit)			
1) Television	3700-3900MHz	3820-3870MHz	3662.5-3887.5
2) Telephony	800-1000MHz	3400-3844MHz	" "
Number TWT active			
80 watt	N/A	1	N/A
30 watt	N/A	N/A	3
20 watt/40 watt*	3 (2 spare)	N/A	N/A
7 watt	10 ?	20	N/A
TWT Bandwidth			
80 watt	N/A	40 MHz	N/A
30 watt	N/A	N/A	25 MHz
20 watt/40 watt*	50 MHz	N/A	N/A
7 watt		10 MHz	N/A
Satellite Bandwidth	200 MHz (6/3)	500 MHz (6/3)	225 MHz (6/3)
Communication Capacity)			
No TV Channels (B&W or color)	3 (2 spare)	1	3 possible
Audio	Multiplexed with Video	Same	Same
Other	Multiplexed telephony, telegraphy and photofax (FDM/FDMA and SCPC) 120 chan/10 MHz	Same	Hot Line and FDM/FDMA
Orbit			
Inclination	63° to 65°	Same	Same
Orbit Period	12 hours	Same	Same
Apogee (Northern Hemisphere over USSR)	40,000 km (25,000 s mi)	Same	Same
Perigee	480 km (300 s mi)	Same	Same

MOLNIYA I, II, and III TECHNICAL PARAMETERS

*Two TWT's are probably used, one with 20 watts for use with good geographical advantage between satellite and earth station and 40 watts as required for poor advantage.

**Possibly 800 MHz to 1000 MHz in 10 MHz segments.

Martin P. Brown, Jr.
IEEE Compendium

STATSIONAR SATELLITE SERIES*
(RADUGA and EKRAN)
("Rainbow" and "Horizon")

*To the author's knowledge, with the exception of Statsionar-T, pictures of these satellites have not been released.

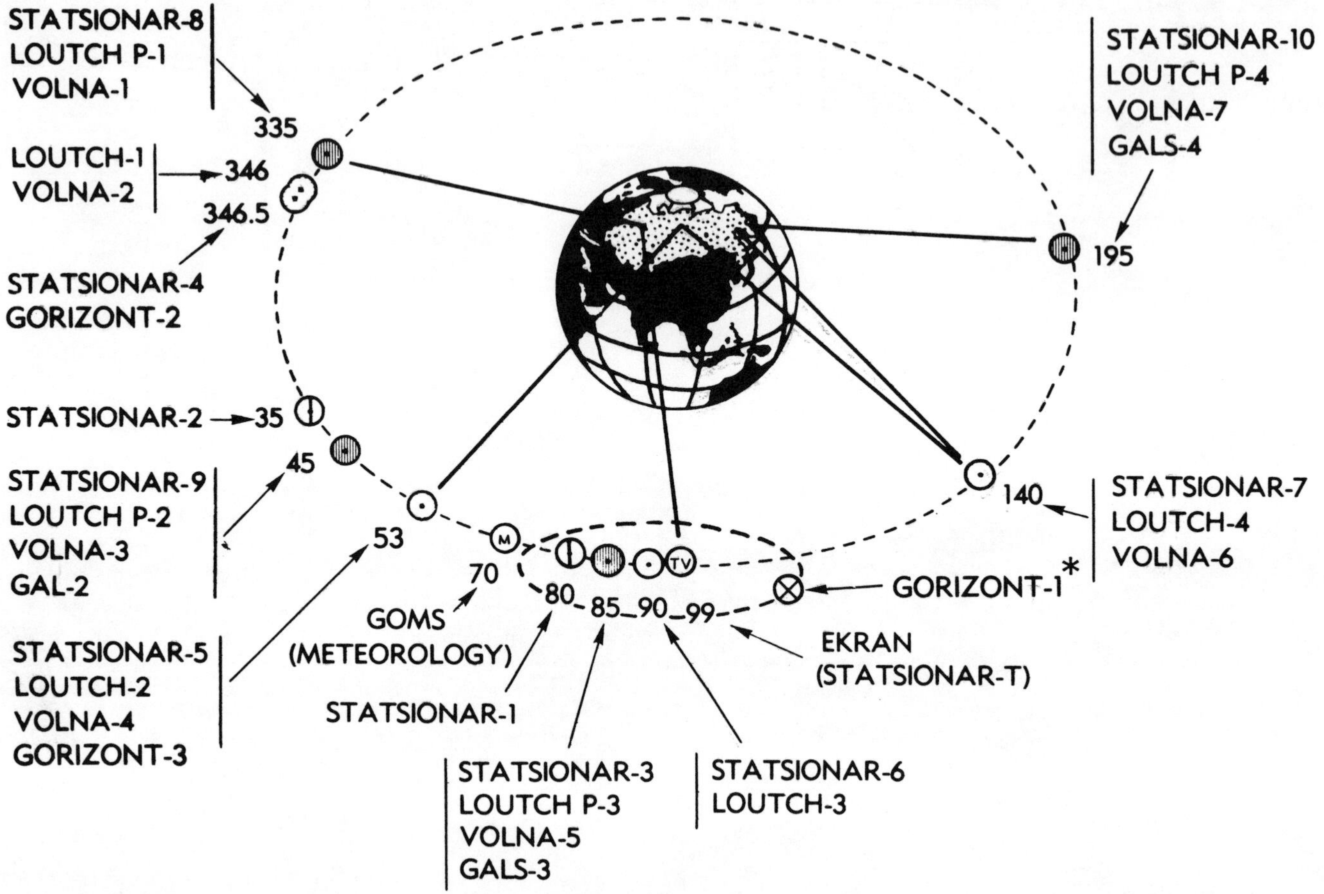

* GORIZONT-1 IS A NON-STATIONARY GEOSYNCHRONOUS SATELLITE THAT HAS A DIURNAL MOTION

USSR SATELLITE NETWORK LOCATIONS

Walter L. Morgan
IEEE Compendium

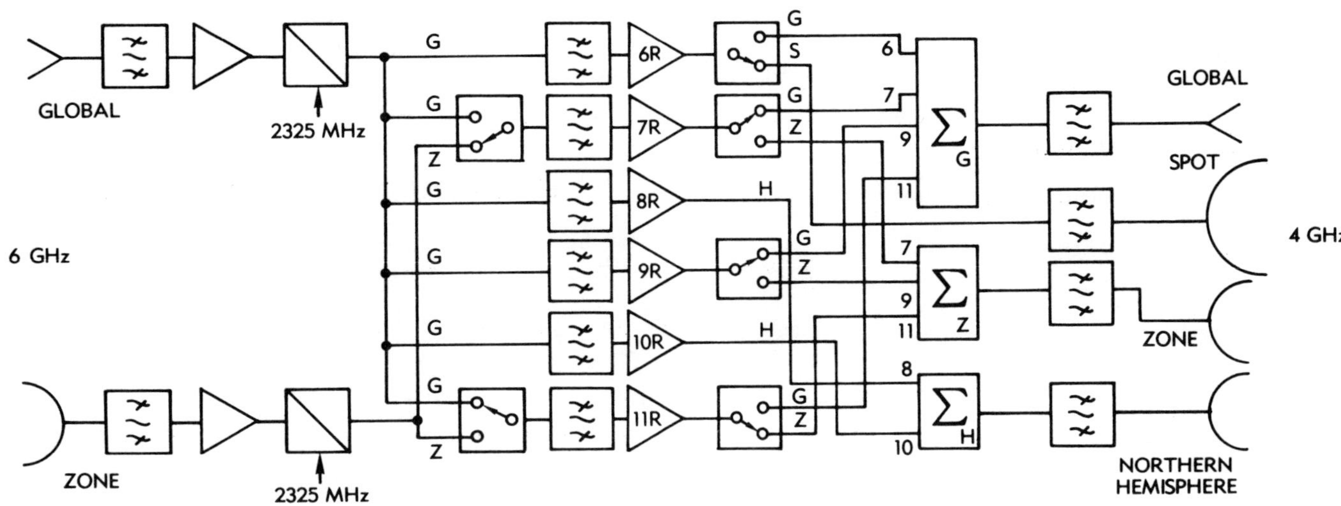

STATSIONAR 4 THROUGH 7 COMMUNICATION BLOCK DIAGRAM

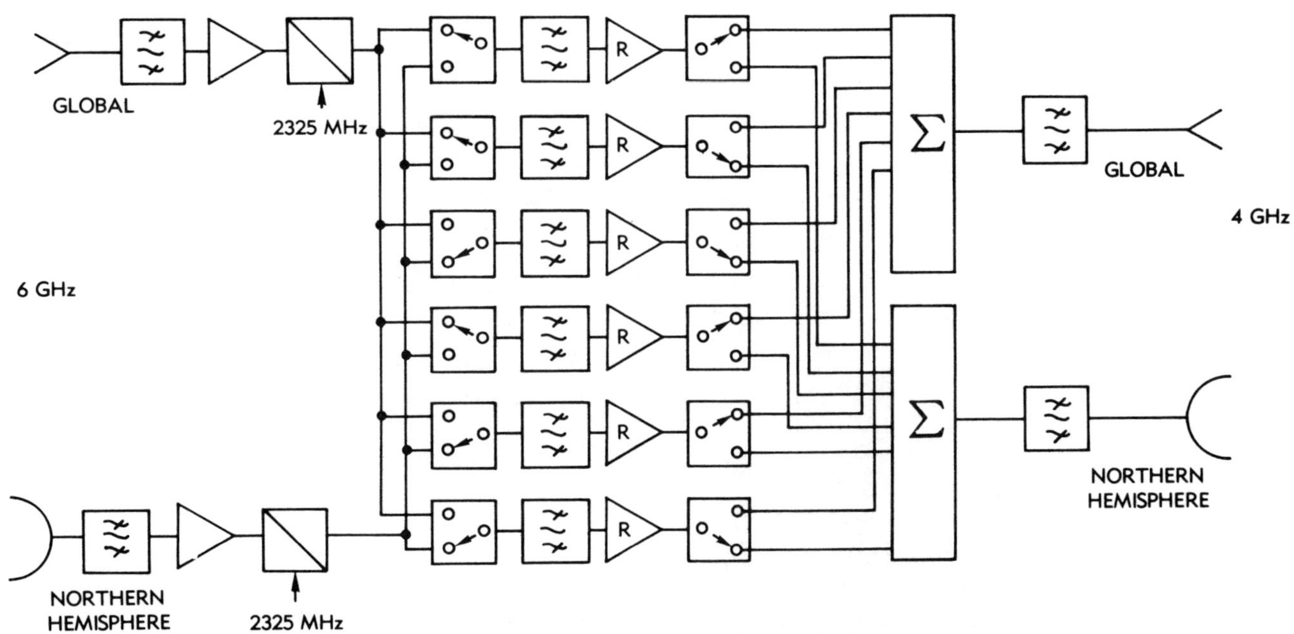

STATSIONAR 3 AND 8 THROUGH 10 COMMUNICATION BLOCK DIAGRAM

Walter L. Morgan
IEEE Compendium

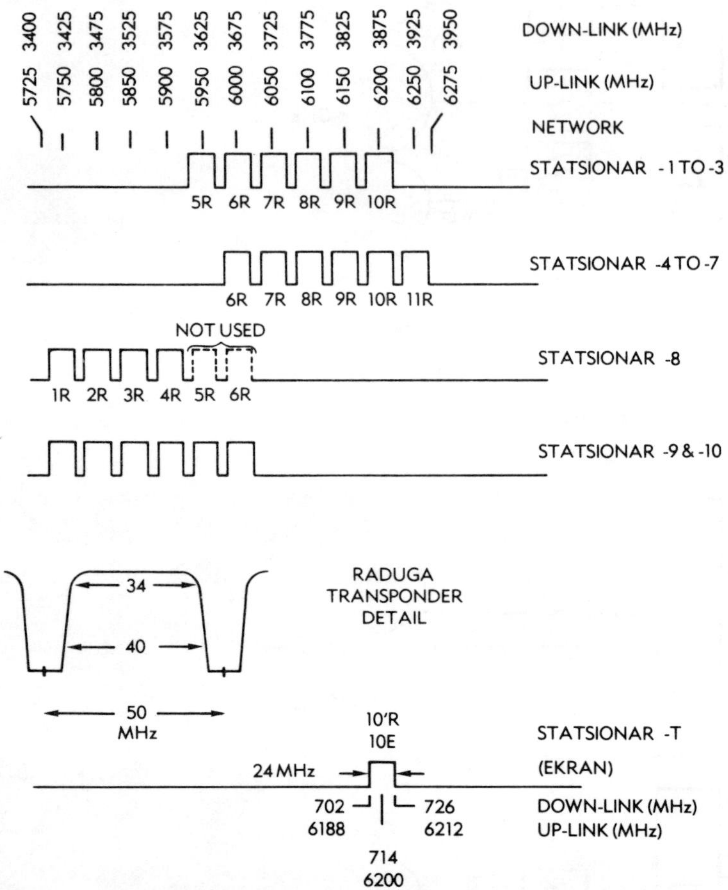

STATSIONAR SERIES FREQUENCY PLAN

STATSIONAR TRANSPONDER USE[a]

Statsionar Network	Transponder										
	1R	2R	3R	4R	5R	6R	7R	8R	9R	10R	11R
1, 2, 3	--	--	--	--	A or B	B	A or B	A	A or B	C	--
4	--	--	--	--	--	C	D	A	D	A or C	D
5	--	--	--	--	--	C	A	A	A	A or C	A
6	--	--	--	--	--	C	A or B	B	A or B	C	B
7	--	--	--	--	--	?	?	?	?	?	?
8	?	?	?	?	--[b]	--[b]	--	--	--	--	--
9, 10	?	?	?	?	A	A	--	--	--	--	--
T	--	--	--	--	--	--	--	--	--	E	--

[a]A = SCPC/FM, 400 channels.
B = FDM/FM, 10 carriers.
C = TV/FM (34 MHz), 1 carrier.
D = SCPC/FM, 200 channels.
E = TV/FM (24 MHz), 1 direct TV (down-link: UHF).
[b]Not used per intersystem coordination with INTELSAT.

Walter L. Morgan
IEEE Compendium

General (to 3/1/80)

The system is designed for telephone, telegraph, and phototelegraph communications and for sound broadcasting and television program transmission.

Launch Vehicle	Proton D (D-l-e or D-l-E-e)
Date of First Launch	December 22, 1975
Network Location	
Statsionar-1	80° ± 0.5°E
Statsionar-2	35° ± 0.5°E
Statsionar-3	85° ± 0.5°E
Inclination Tolerance	±3° (may be ±0.5°)

Physical Description (from *Spaceflight*)

Stabilization	3 axis
Approx. Dimensions	5 m long x 2-m-dia. cylinder + 2 solar panels + 2 antennas (?)
Initial Orbit Weight	5000 kg (?)

Antenna Configuration

Transmit and Receive, Northern Hemisphere	23 dBi

Communications Capability (see table)

Assigned Frequency Range	
Up-Link (receive)	5925-6225 MHz
Down-Link (transmit)	3600-3900 MHz
Repeaters	
No. on Line	6
Bandwidth	34 MHz on 50 MHz centers
RF Power	8, 10, 15 W
Group Delay	20 ns
Available e.i.r.p. (nominal)	
Beam Center	19-30 dBW (typical 29 dBW)
Receive G/T (nominal)	
Beam Edge, Statsionar-1	-12.8 dB/K
Beam Center, Statsionar-2 and -3	-14.7 dB/K
Receive Noise Temperature	3000 K
Satellite Capacity per Transponder	
SCPC/FM	200-400 channels
FDM/FM	10 carriers
TV/FM	1 channel

*The Statsionar, Loutch, Gals, and Volna satellites appear in groups as shown in the network location diagram.

STATSIONAR 1, 2, and 3 (RADUGA) TECHNICAL PARAMETERS

General (to 3/1/80)	
Network Location	
Gorizont-2 (Statsionar 4)	346.0° ± 0.2°E
Gorizont-3 (Statsionar 5)	53.0° ± 0.5°E
Statsionar-6	90.0° ± 0.5°E
Statsionar-7	140.0° ± 0.5°E

Antenna Configuration	
Receive, Global (G)	19 dBi
Transmit, Global (G)	19 dBi
Northern Hemisphere	22 dBi
Zone (Z)	30 dBi
Spot (S)	35 dBi

Up-Down Beams

Network Transponder[b]	Statsionar Network 4	5	6	7
6R	G-G	G-G	G-S	G-S
7R	G-G	Z-Z	Z-Z	Z-Z
8R	G-H	G-H	G-H	G-H
9R	G-G	Z-Z	Z-Z	Z-Z
10R	G-H	G-H	G-H	G-H
11R	G-G	Z-Z	Z-Z	Z-Z

Communications Capability	
Total Usable Bandwidth	300 MHz
Assigned Frequency Range	
Up-Link (receive)	5.975-6.275 GHz
Down-Link (transmit)	3.650-3.950 GHz
Repeaters	
No. on Line	6
Bandwidth	34 MHz on 50-MHz centers
Available e.i.r.p. (nominal)	
Beam Center (global)	23 dBW
Northern Hemisphere	30.5 dBW
Zone	
Capacity of Statsionar-4	
SCPC/FM Telephony	
2 Transponders (400 channels ea.)	800 channels
3 Transponders (200 channels ea.)	600 channels
	1400 channels
TV (1 transponder with 1 channel)	1 TV channel

[a]The Statsionar, Loutch, Gals, and Volna satellites appear in group as shown in the network location diagram.

[b]R = Raduga transponder number (see figure).

STATSIONAR 4 THROUGH 7 TECHNICAL PARAMETERS

Walter L. Morgan
IEEE Compendium

General (to 3/1/80)

Network Location	
Statsionar-8	335.0° ± 0.5°E
Statsionar-9	45.0° ± 0.5°E
Statsionar-10	190.0° ± 0.5°E

Antenna Configuration

Receive	
Global (G)	18 dBi
Northern Hemisphere (H)	23 dBi
Transmit	
Global (G)	18 dBi
Northern Hemisphere (H)	23 dBi

Antenna Beam

Network Transponder[b]	Statsionar Network		
1R	G or H	G or H	G or H
2R	G or H	G or H	G or H
3R	G or H	G or H	G or H
4R	G or H	G or H	G or H
5R	--[c]	G or H	G or H
6R	--[c]	G or H	G or H

Communications Capability

Total Usable Bandwidth	144-216 MHz
Assigned Frequency Range	5.725-6.025 GHz
Up-Link (receive)	5.725-6.025 GHz
Down-Link (transmit)	3.400-3.700 GHz
Repeaters	
No. on Line	6 (4 for Statsionar-8)
Bandwidth	34 MHz on 50-MHz centers

[a]The Statsionar, Loutch, Gals, and Volna satellites appear in groups as shown in the network location diagram.

[b]R = Raduga transponder number (see figure).

[c]Not used due to interference with INTELSAT.

STATSIONAR 8 THROUGH 10 TECHNICAL PARAMETERS

Walter L. Morgan
IEEE Compendium

EKRAN (STATSIONAR—T)

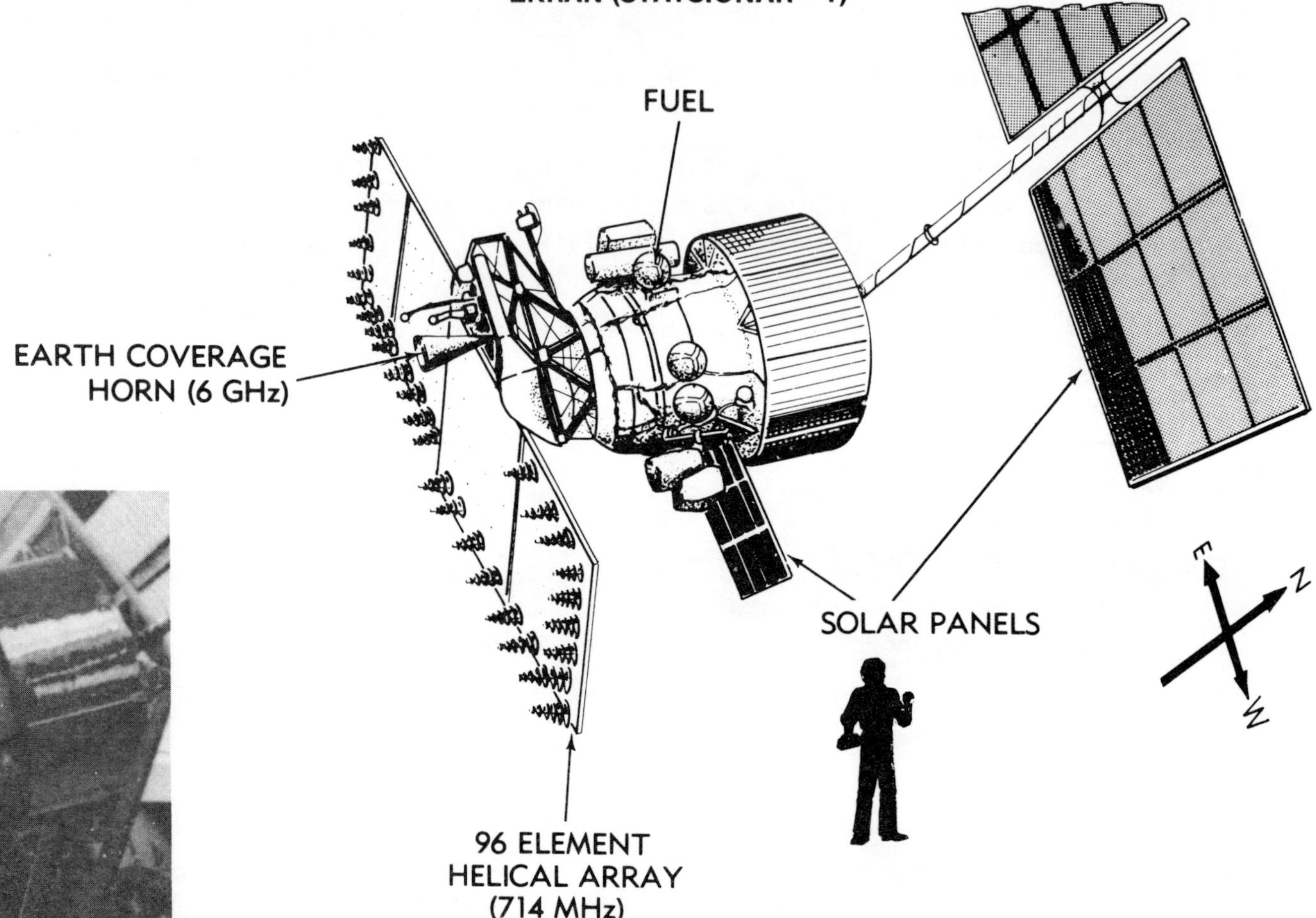

EKRAN
("Screen")

Walter L. Morgan
IEEE Compendium

General (to 3/1/80)

Statsionar-T is a satellite network intended for television broadcasting to regions of the USSR beyond the Urals, Siberia and the extreme north. The system uses a transmitting earth station at Gus-Khrustalnyi, a geostationary satellite, and a network of receiving earth stations intended for community and apartment house reception.

Launch Vehicle	D-1-E (Baikonur)
Date of First Launch	October 26, 1976
Longitude Location	99°E ± 1° and 53° E
Inclination Tolerance	±1°

Physical Description

Stabilization	3 axis with both wrap-around and deployed solar arrays; 96-element phased array antenna for TV transmit; 2 kW power.
Approx. Dimensions	5 m long, 2-m-dia. body plus solar paddles and phased array antennas.
Orbit Weight (initial)	approx. 5000 kg

Antenna Configuration

Up-Link Global Horn	19 dBi
Down-Link Siberia Phased Array	
On Axis	33.5 dBi
At Edge of Area	26 dBi
Transmit Polarization	RHC (ellipticity ≥0.7)

Communications Capacity

Total Usable Bandwidth	24 MHz
Assigned Frequency Range	
Up-Link (receive)	6188-6212 MHz
Down-Link (transmit)	702-726 MHz
Repeaters	
No. on Line	1
Bandwidth	24 MHz
RF Power	300 W (200 W ea. antenna)
Available e.i.r.p. (nominal)	
Coverage Zone Edge	49 dBW
Beam Center	56.5 dBW
Receive G/T (nominal)	
In Direction of Gus-Khrustalnyi	-18.6 dB/K
Beam Center	-15.8 dB/K
Total Satellite Capacity	1 TV repeater

EKRAN (STATSIONAR—T) TECHNICAL PARAMETERS

Walter L. Morgan
IEEE Compendium

GORIZONT
("Horizon")

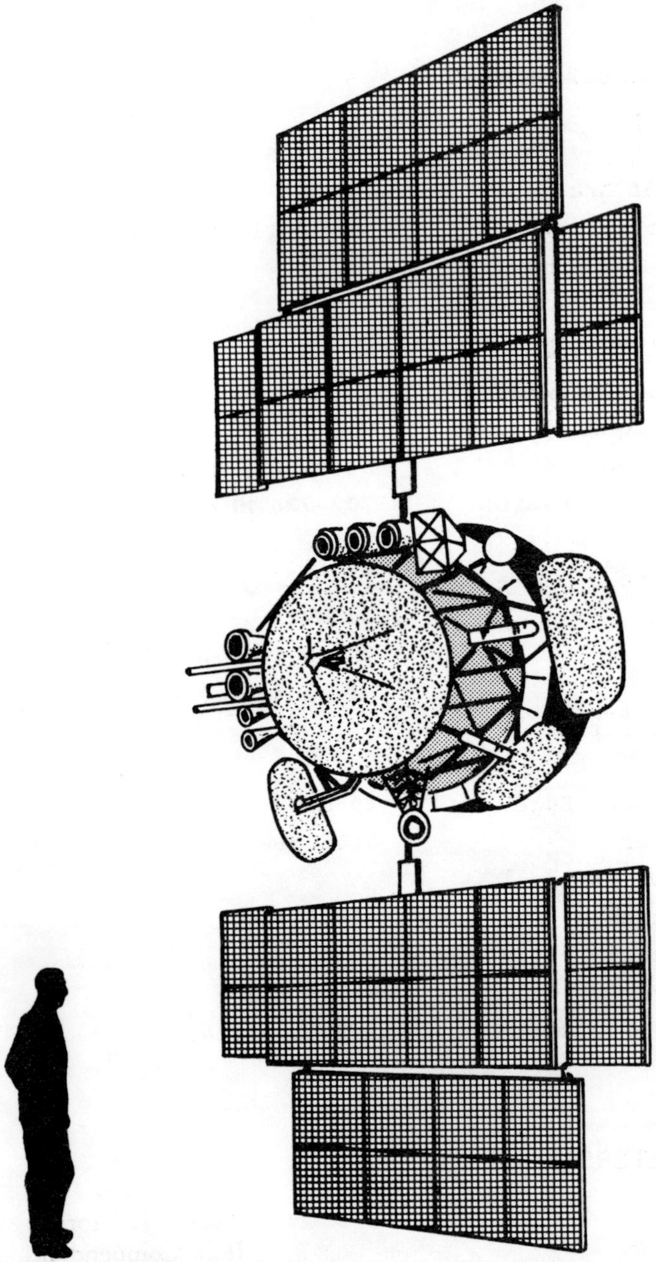

Walter L. Morgan
IEEE Compendium

General (3/1/80)

Gorizont has not been registered with the IFRB. This series apparently contains both 6/4-GHz (similar to Raduga) and 8/7-GHz (similar to Gals) equipment.

No. Launched to Date: 3

	Longitude	Network
Gorizont-1	*	
Gorizont-2	346.5°E	(Statsionar-4 & Gals-1)
Gorizont-3	53°E	(Statsionar-5)

Antenna Configuration (from Paris Air Show Model)

Global	6/4 GHz 8/7 GHz
Northern Hemisphere	8/7 GHz
Narrow Beam	8/7 GHz

Communications Capability (from Statsionar-4 and Gals-1)

Up-Link	5.975-6.275 & 7.9-8.4 GHz
Down-Link	3.65-3.95 & 7.25-7.75 GHz

Power Generation Capacity

1.0-1.5 kW (from Paris Air Show Model)

Source: "New Soviet Horizon Satellite Exhibited," Aviation Week & Space Technology, June 25, 1979.
*Gorizont-1 was placed in an orbit of 22,553 x 49,023 km with an inclination of 11.3°.
Gorizont-2 and -3 are both in geosynchronous, geostationary orbits.

GORIZONT SERIES TECHNICAL PARAMETERS

VOLNA 1 THROUGH 7 SERIES*

("Wave")

*To the author's knowledge, a picture of this satellite has not been released.

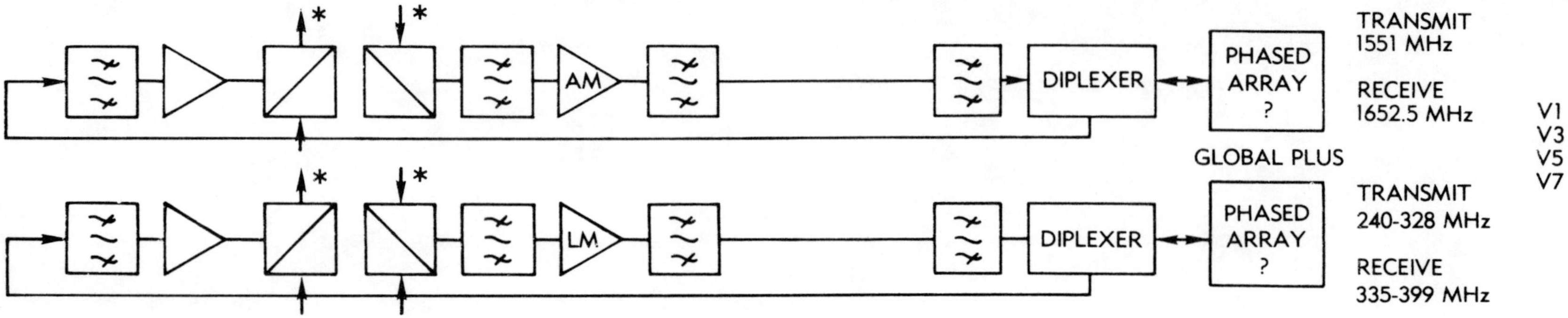

* TO MASTER EARTH STATION VIA A STATSIONAR OR LOUTCH TRANSMITTER
(SOME OF THE BAND MAY RETURN DIRECTLY TO THIS SERVICE)

VOLNA 1, 3, 5, and 7 COMMUNICATIONS BLOCK DIAGRAM

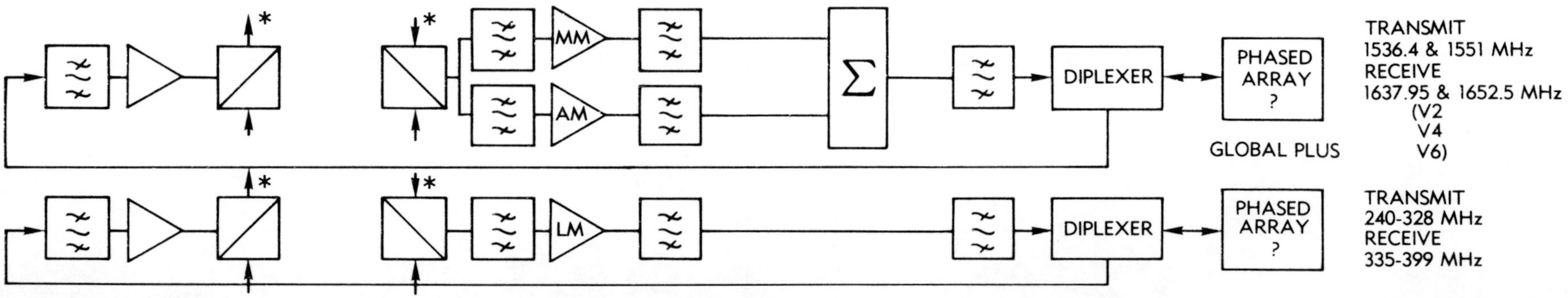

* TO MASTER EARTH STATION VIA A STATSIONAR OR LOUTCH TRANSPONDER.
(SOME OF THE BAND MAY RETURN DIRECTLY TO THIS SERVICE)

VOLNA 2, 4, and 6 COMMUNICATIONS BLOCK DIAGRAM

Walter L. Morgan
IEEE Compendium

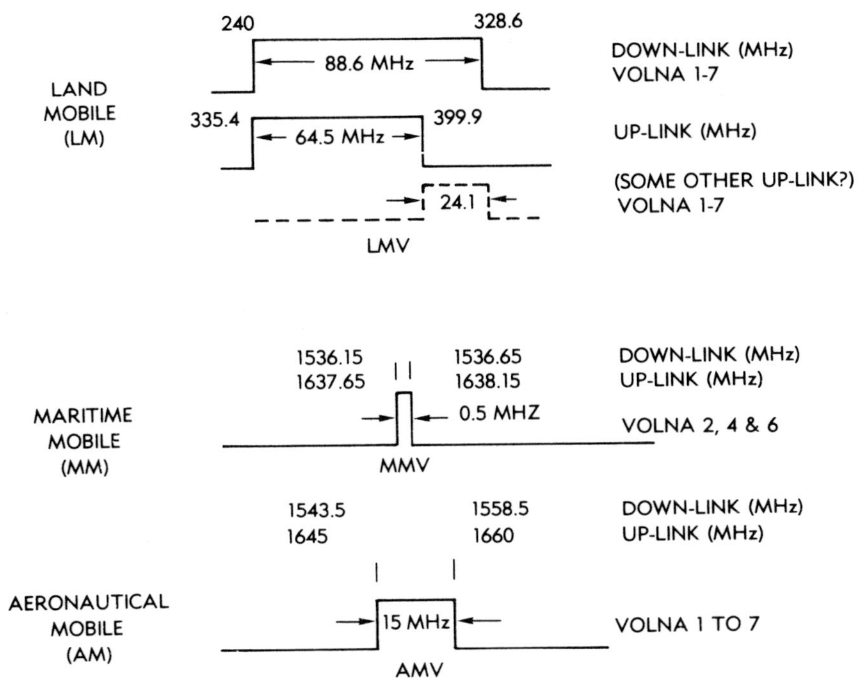

VOLNA FREQUENCY PLAN

General (to 3/1/80)

No. Launched to Date	0
Network Locations	
Volna 1	335.0° ± 0.5°E
Volna 2	316.5° ± 0.5°E
Volna 3	45.0° ± 0.5°E
Volna 4	58.0° ± 0.5°E
Volna 5	85.0° ± 0.5°E
Volna 6	140.0° ± 0.5°E
Volna 7	190.0° ± 0.5°E

Antenna Configuration

Global	12-18 dBi

Communications Capability

Assigned Frequency Ranges (MHz) Mobile Service	Land (LM)	Maritime (MM)	Aeronautical (AM)
Up-Link	335-399	1637.9 ± 0.25	1645-1660
Down-Link	240-328	1536.4 ± 0.25	1543.5-1558.5
Volna 1	yes	no	yes
Volna 2	yes	yes	yes
Volna 3	yes	no	yes
Volna 4	yes	yes	yes
Volna 5	yes	no	yes
Volna 6	yes	yes	yes
Volna 7	yes	no	yes
Receive G/T (nominal)			
Beam Center	-5 dB/K	-2 dB/K	-2 dB/K

VOLNA TECHNICAL PARAMETERS

Walter L. Morgan
IEEE Compendium

SECTION III

REGIONAL SATELLITES

ECS (EUROSAT)

ECS (EUROPEAN)

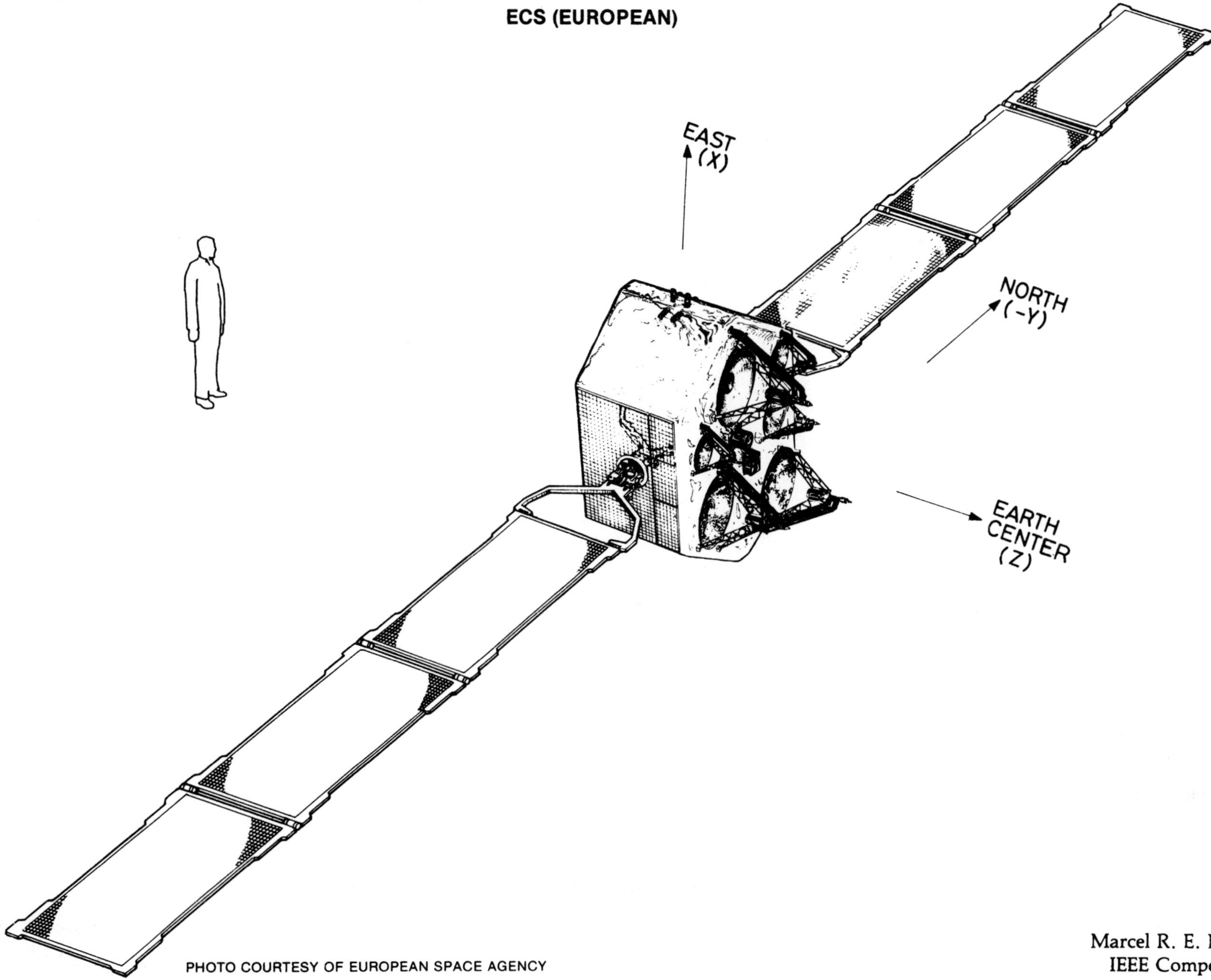

PHOTO COURTESY OF EUROPEAN SPACE AGENCY

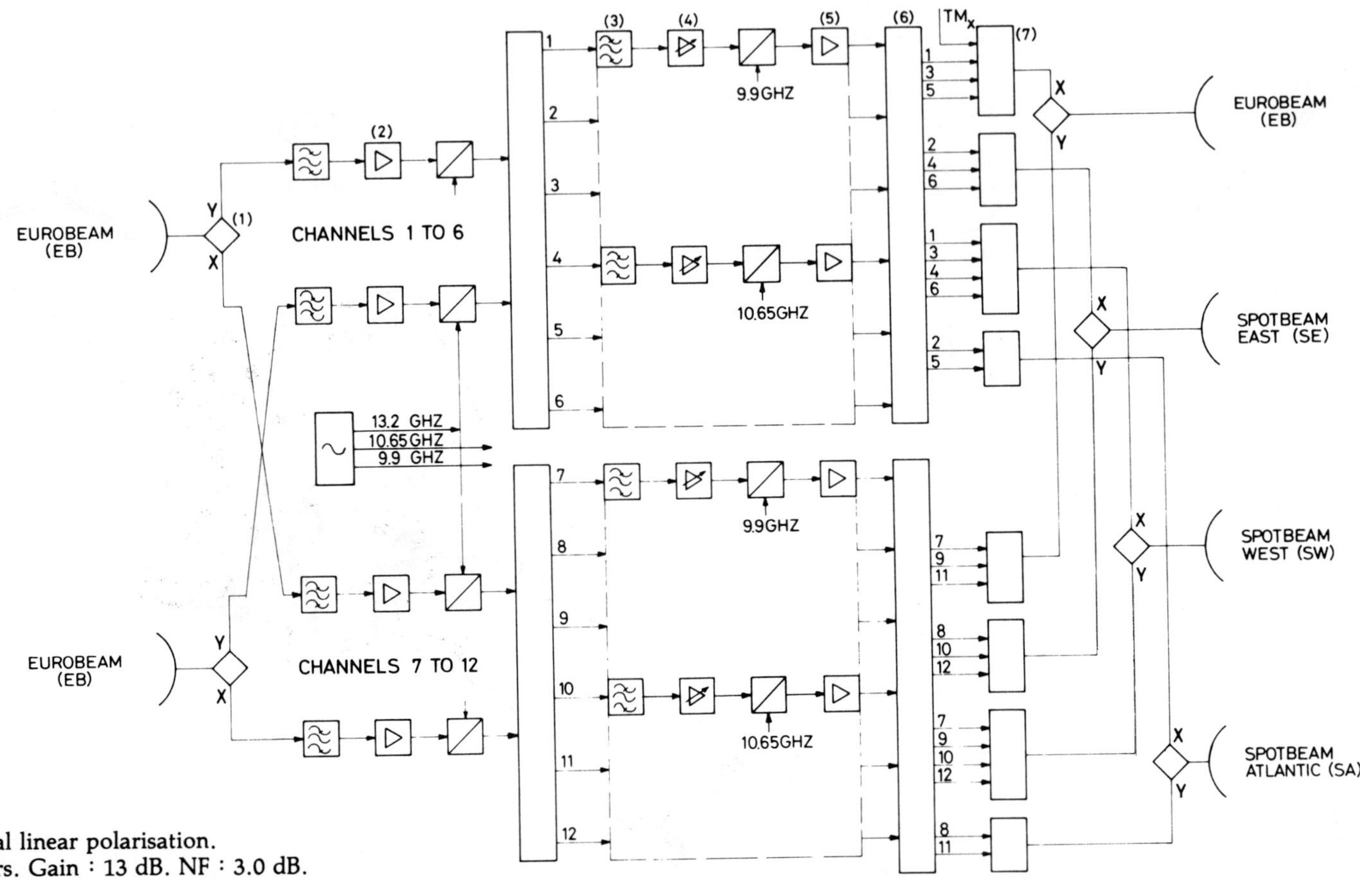

Legend

(1) Orthomode waveguide transitions for dual linear polarisation.
(2) GaAs varactor diode parametric amplifiers. Gain : 13 dB. NF : 3.0 dB.
(3) 14-cavity linear-phase IF filters. Bandwidth : 72 MHz.
(4) IF main amplifiers. Gain variable over a 15 dB range in 16 steps.
(5) Dual-collector TWTAs. Output power : 20 W.
(6) Waveguide circulator switches.
(7) Output filters. Bandwidth : 72 MHz.

ECS COMMUNICATION BLOCK DIAGRAM (EUROPEAN)

Marcel R. E. Bichara
IEEE Compendium

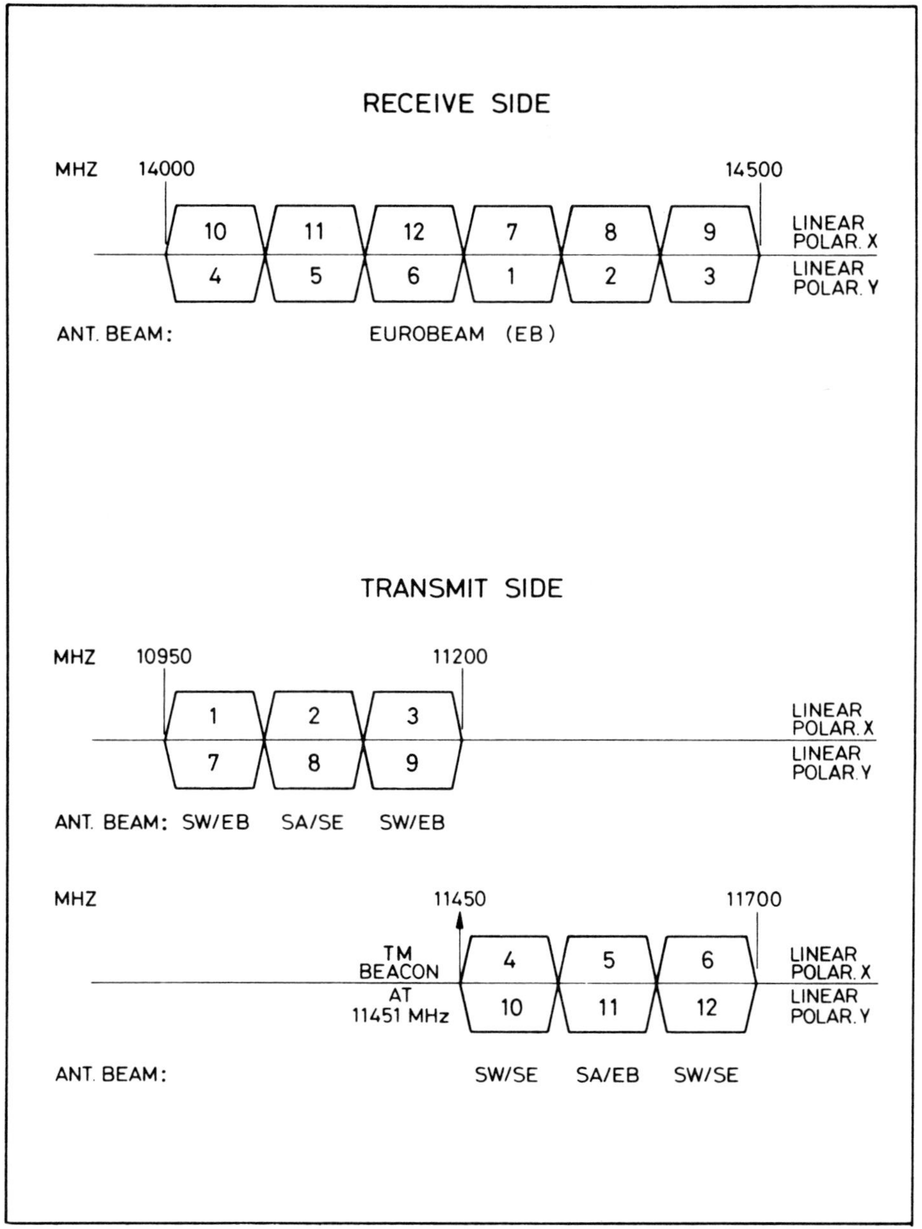

ECS FREQUENCY PLAN

General

No of satellites	Five
Design life	7 years
Launch vehicle	Ariane
Major contractor	MESH Consortium
Date of first launch	1981

Physical Description

Stabilisation	3-axis momentum wheel
Spacecraft size	
Diametrical envelope	2.18 m
Height	2.32 m
Deployed arrays	13.73 m
Weight at launch	957 kg
in orbit	544 kg (initial)

Antennas

Receive	2 Eurobeam (dual linear polar)
Transmit	1 Eurobeam and 3 Spotbeams (dual linear polar)

Communication Payload Characteristics

Total usable bandwidth	1000 MHz
Assigned freq. range	
Up-link (Receive)	14.0 - 14.5 GHz
Down-link (Transmit)	)10.95 - 11.2 GHz)11.45 - 11.7 GHz
Repeater chains	
Number/Bandwidth	12 x 80 MHz
RF power output	20 W
Available EIRP	
Eurobeam	42.7 dBW (centre) 34.8 dBW (edge)
Spotbeams	48.3 dBW (centre) 40.8 dBW (edge)
Saturation flux density at beam edge	
Min. gain setting	-61.5 dB (W/m^2)
Max. gain setting	-76.5 dB (W/m^2)

Communication Capability

Per Eurobeam chain	
1 TV	S/N = 55 dB
Per Spotbeam chain	
PSK/TDMA 120 Mbit/s	1800 channels
Total sat. capacity	18000 channels and 2 TV

ECS TECHNICAL CHARACTERISTICS (EUROPEAN)

Marcel R. E. Bichara
IEEE Compendium

SECTION IV

DOMESTIC SATELLITES

- COMSTAR
- INSAT
- PALAPA—A and PALAPA—B
- SATCOM
- SBS
- STATSIONAR—T (EKRAN)*
- TELESAT
    - ANIK A
    - ANIK B
    - ANIK C
- WESTAR
- WESTAR (ADVANCED)/TDRSS

---

*See STATSIONAR SERIES in International/Domestic Section.

COMSTAR

PHOTO COURTESY OF COMSAT

John F. Phiel, Jr.
IEEE Compendium

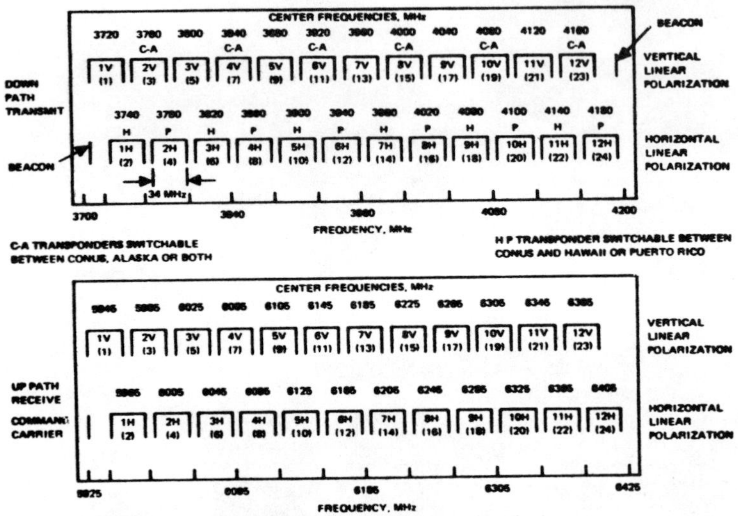

• ALL VERTICAL UP PATH RECEIVE TRANSPONDERS AVAILABLE TO CONUS AND ALASKA
• ALL HORIZONTAL UP PATH RECEIVE TRANSPONDERS AVAILABLE TO CONUS, HAWAII, AND PUERTO RICO

NOTE: NUMBERS BELOW FREQUENCIES REFER TO TRANSPONDER IDENTIFICATION AT HUGHES. NUMBERS IN PARENTHESES ARE USED IN CONTRACTUAL DOCUMENTS AND FCC FILING.

COMSTAR FREQUENCY PLAN

General

Number of Satellites	4 (D-1 thru D-4)
Design Life	7 years
Launch Vehicle	Atlas Centaur
Major Contractor	Hughes Aircraft Company
Date of Launch	D-1 - May 13, 1976
	D-2 - July 22, 1976
	D-3 - 29 June 1978
	D-4 - SPARE

Physical Description

Stabilization	Gyrostat
Solar Panel Diameter	238.0 cm (93.7 in.)
Solar Panel Height	281.0 cm (111.0 in.)
Height w/antenna mast	609.0 cm (239.7 in.)
Weight	
Launch (w/adapter)	1515.9 kg (3341.8 lbs.)
Orbit (initial)	862.7 kg (1901.8 lbs.)

Antenna Configuration

Polarization
Linear Dual Polarization for Frequency Reuse

Receive Coverage
Horizontal - CONUS, Hawaii, Puerto Rico
Vertical - CONUS and Alaska

Transmit Coverage
Horitzontal - CONUS, Hawaii, Puerto Rico (6 repeaters switchable to each offshore beam)
Vertical - CONUS, Alaska, CONUS/Alaska (6 repeaters switchable to Alaska or combined mode)

19/28 GHz Beacons

19.040 GHz	28.560 GHz
e.i.r.p. 21.0 dBW over CONUS	e.i.r.p. 21.0 dBW over CONUS
Vertical and horizontal Pol. switched at 1.0 kHz rate	Vertically polarized
	Phase modulated

Communication Capability

Total Usable Bandwidth	616 MHz
Frequency Bands	
Up-link (Receive)	
Horizontal	5948-6422 MHz
Vertical	5928-6398 MHz
Down-link (Transmit)	
Horizontal	3723-4197 MHz
Vertical	3703-4177 MHz
Repeaters	
No on-line	24
Bandwidth	36 MHz
Type and output power of TWTs	
Horizontal	Hughes 275H 5.0W
Vertical	Hughes 277H 5.5W
Available e.i.r.p. (nominal beam edge)	
CONUS, Hawaii, Alaska	33.0 dBW
Puerto Rico	
CONUS/Alaska combined	31.0 dBW
Receive G/T (nominal beam edge)	-8.8 dB/K
Saturation Flux Density (nominal beam edge)	
Gain Step 1	-81.7 dBW/m^2
Gain Step 2	-75.7 dBW/m^2
Gain Step 3	-72.7 dBW/m^2
Gain Step 4	-66.7 dBW/m^2
Cross Polarization Isolation (Minimum)	
CONUS, Hawaii, Alaska, Puerto Rico	33.0 dB
CONUS/Alaska combined	31.0 dB

Satellite Capacity

Design (per transponder)
1200 one-way channels at less than 4,800 pWOp
1800 one-way channels at less than 10,000 pWOp

Actual Capacity (Nominal total)
1500 channels per transponder
36000 channels per satellite

COMSTAR TECHNICAL CHARACTERISTICS

John F. Phiel, Jr.
IEEE Compendium

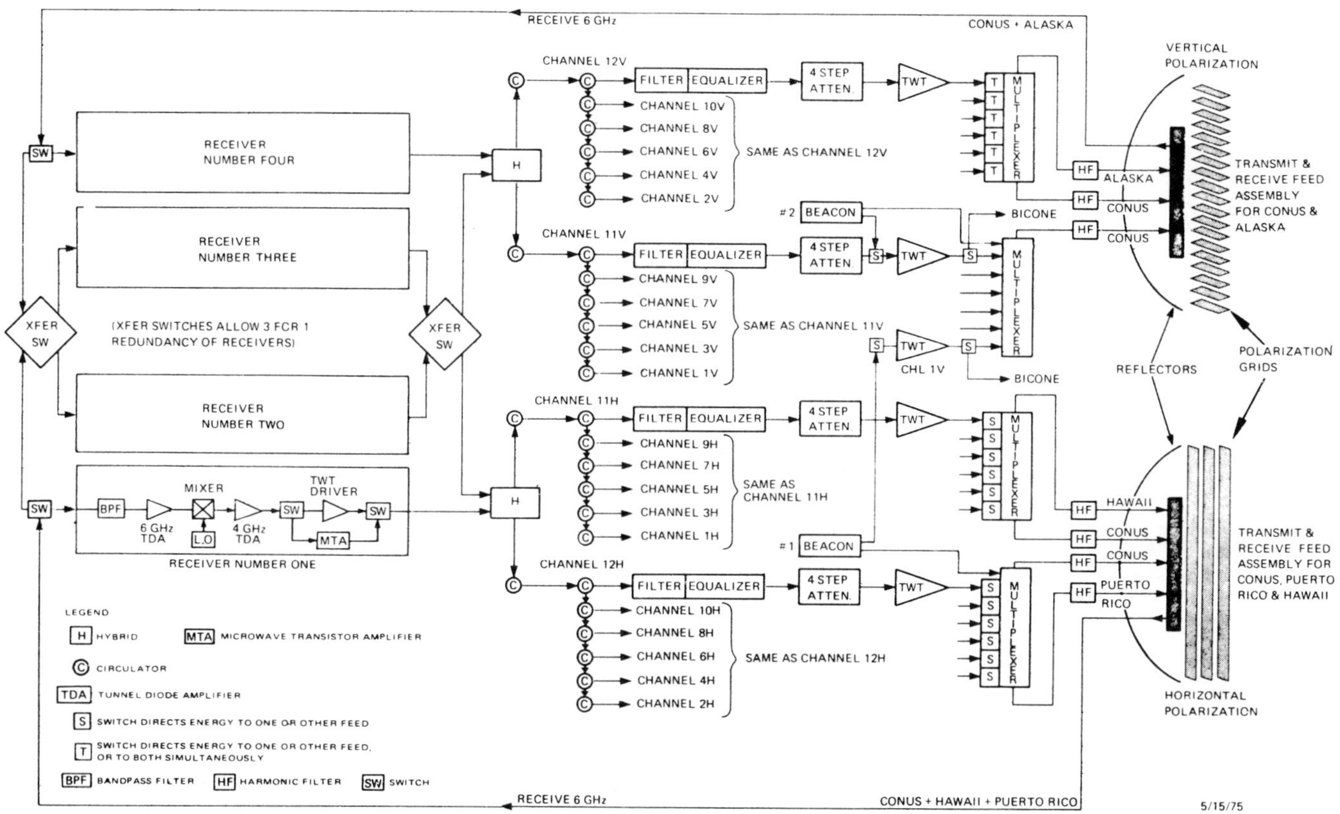

COMSTAR COMMUNICATION BLOCK DIAGRAM

INSAT

PHOTO COURTESY OF
FORD AEROSPACE AND
COMMUNICATIONS CORP.

Ramsawami Parthasarathy
IEEE Compendium

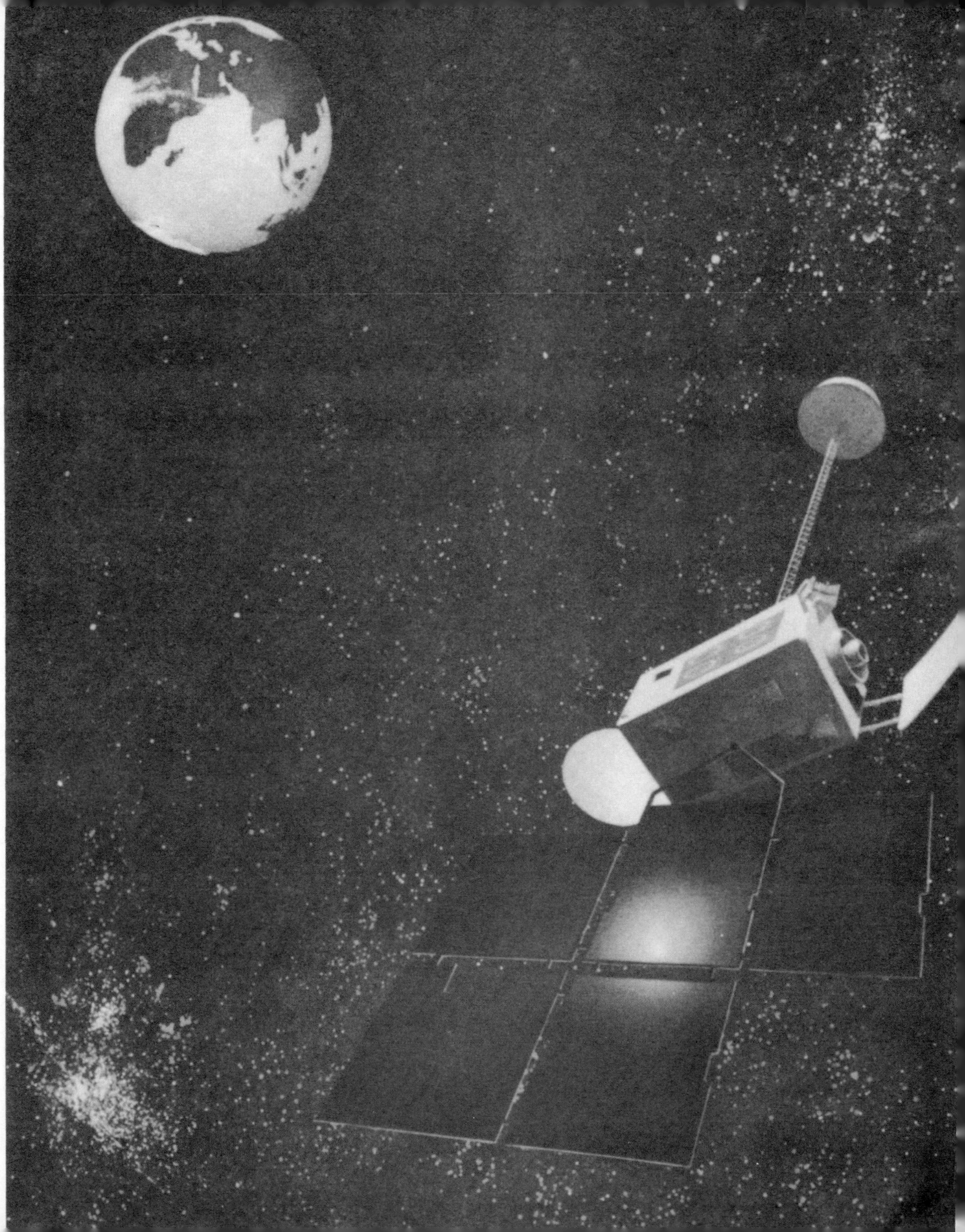

INSAT SYSTEM CONCEPT

Ramsawami Parthasarathy
IEEE Compendium

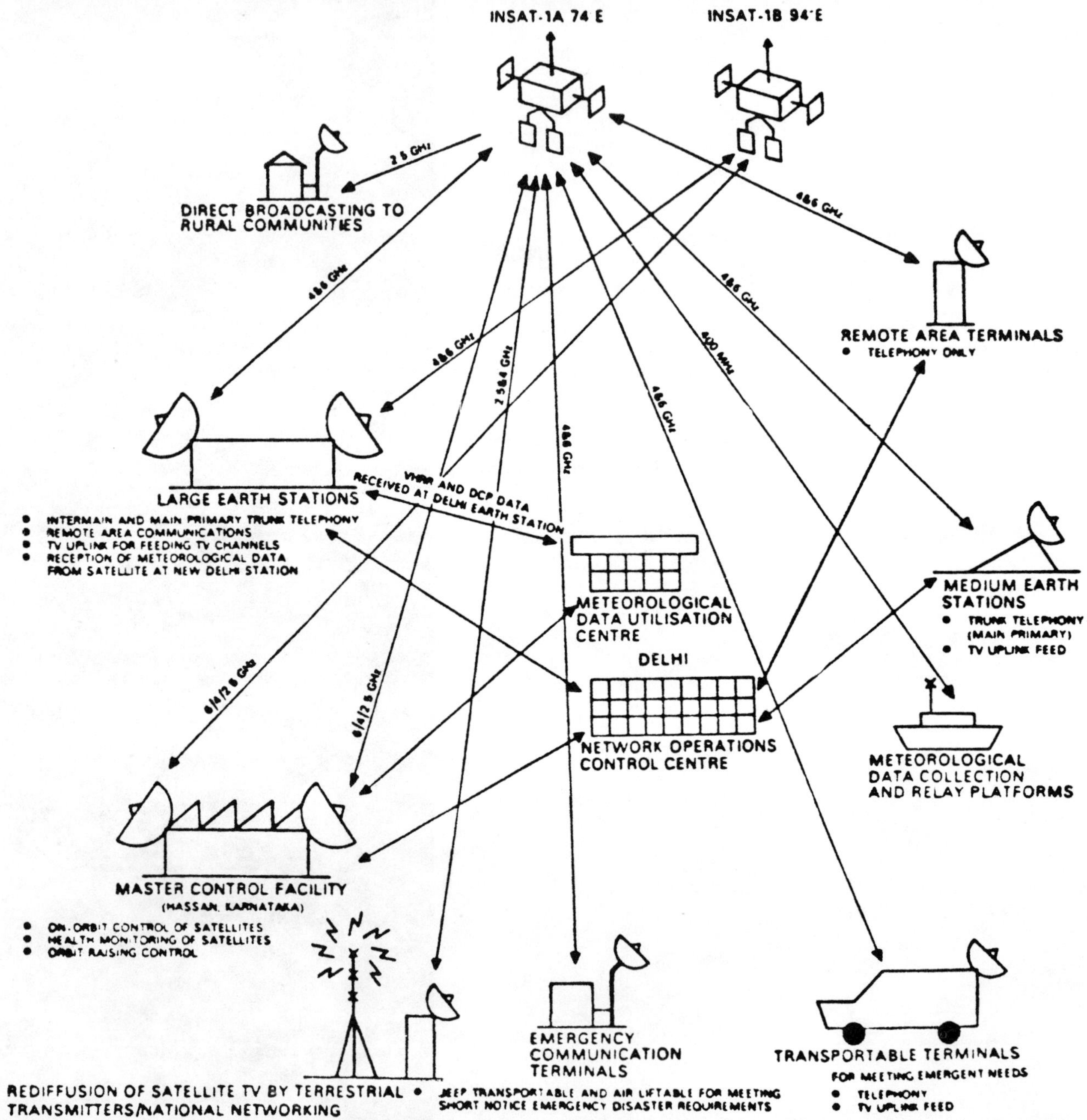

LEGEND

Large Earth Stations	5
Medium Earth Stations	12
Remote Area Terminals	12
Transportable Terminals	4
Emergency Communication	2

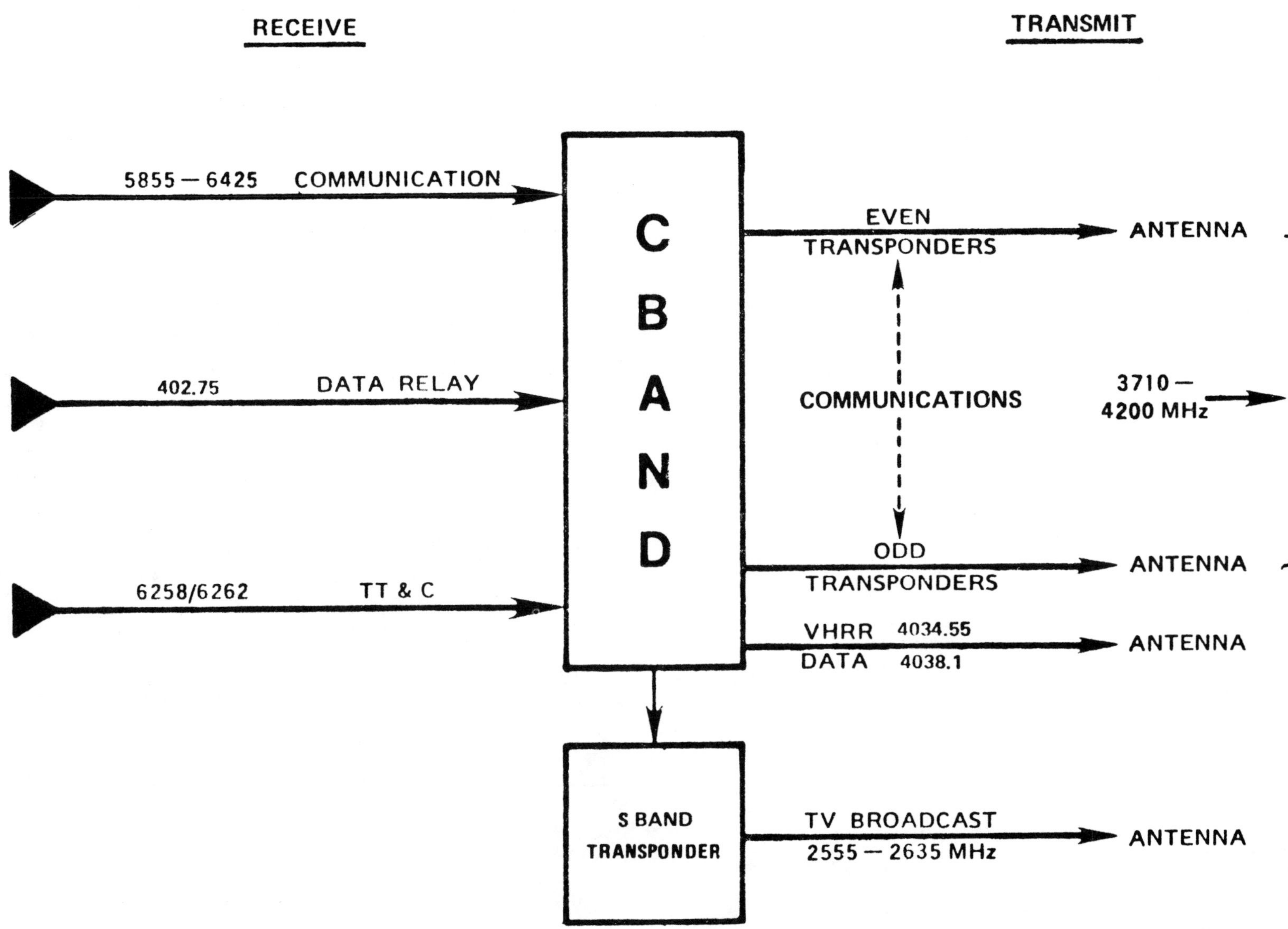

INSAT COMMUNICATION BLOCK DIAGRAM

Ramsawami Parthasarathy
IEEE Compendium

INSAT TECHNICAL PARAMETERS

General

Number of satellites	2
Design life	7 years
Launch vehicle	Delta 3910 & the space shuttle
Major contractor	FACC
Date of 1st launch	Early 1982
Location	74 degrees East, 94 degrees East

Physical description

Stabilization	Three axis
Antenna reflector	Deployable
Weight	
Launch	3350 kg
Initial orbit	1140 kg

Antenna configuration

Uplink	spot
Downlink	spot

Communications capability

Total usable bandwidth	490 MHz	
Frequency range	Television	Other
Uplink	5855 - 5935 MHz	5935 - 6425 MHz
Downlink	2555 - 2635 MHz	3710 - 4200 MHz
Repeaters	2	12
Bandwidth	36 MHz	36 MHz
RF Power	50 watts	5 watts
Available e.i.r.p. (spot)	42 dBW	32 dBW (minimum)
Receive G/T (minimum)	- 4.5 dB/K	
Total capacity		
Telephony	Over 8000 - 2 way circuits	
Television	Direct broadcast to augmented TV receivers	

Metereological capability

VHRR	4034.55 MHz
Data channel	From unattended data collection and transmission platforms, about 100 4038.1 MHz.

INSAT FREQUENCY PLAN

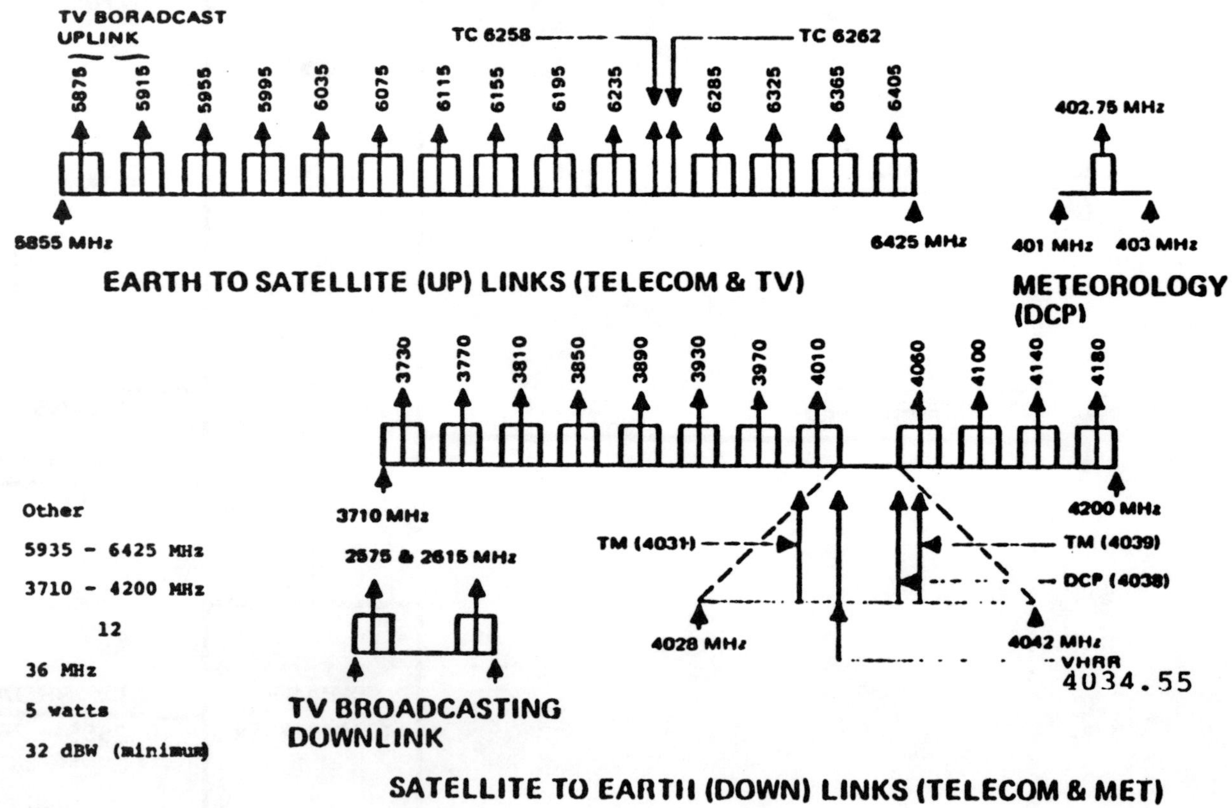

Ramsawami Parthasarathy
IEEE Compendium

PALAPA A
("National Unity")*

PHOTO COURTESY OF
HUGHES AIRCRAFT CORP.

Hans Dodel
IEEE Compendium

*or "Fruit of the Effort".

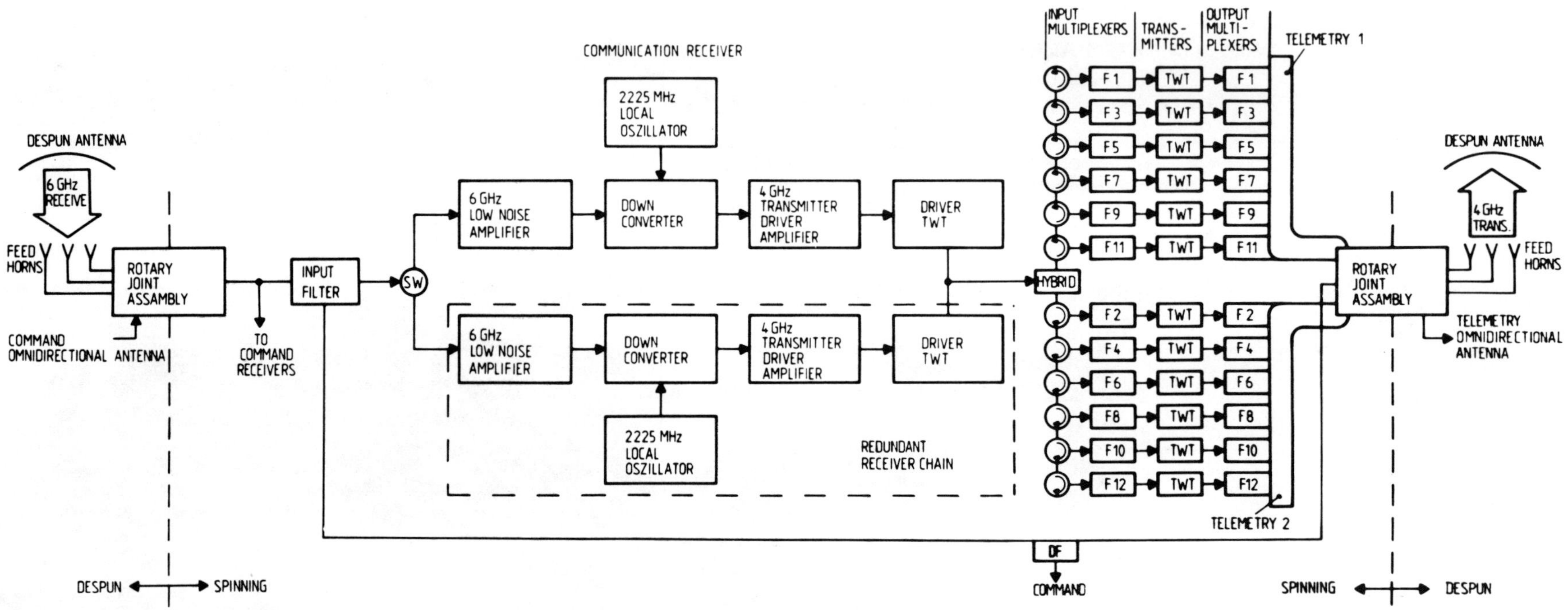

PALAPA A COMMUNICATION BLOCK DIAGRAM

Hans Dodel
IEEE Compendium

Satellite Position:

Primary Satellite 83°E

Secundary Satellite 77°E

Eclipse Capability:

10 out of 12 transponders

Transponder Configuration and Frequency plan:

Transponder No.	Transmit/Receive Frequency
1	3720 MHz / 5945 MHz
2	3760 MHz / 5985 MHz
3	3800 MHz / 6025 MHz
4	3840 MHz / 6065 MHz
5	3880 MHz / 6105 MHz
6	3920 MHz / 6145 MHz
7	3960 MHz / 6185 MHz
8	4000 MHz / 6225 MHz
9	4040 MHz / 6265 MHz
10	4080 MHz / 6305 MHz
11	4120 MHz / 6345 MHz
12	4160 MHz / 6385 MHz

Future outlook:
It is estimated that, by 1983, Palapa-1 and -2 will be fully loaded and Palapa-3 with 24 transponders should be launched.

PALAPA FREQUENCY PLAN

General	
No. of Satellites built	2 (P-1&P-2)
Design Life	7 years
Launch Vehicle	Delta 2914
Major Contractor	Hughes
Date of 1st Launch	July 1976
Physical Description	
Stabilization	Dual Spin (Despun Antenna Earth Oriented, Spinning Drum; 100 rpm)
Cylinder Diameter	190.5 cm (75 in.)
Cylinder Height	162.6 cm (64 in.)
Height W/antenna mast	242.9 cm (135 in.)
Weight	
Launch	572.0 (1260 lb)
Orbit	301.6 kg (665 lb)
Solar Power	300 WATTS
Antenna Configuration	One 60 in. offset feed Reflector optimized for Indonesia/ASEAN coverage for both receive and transmit.
Polarization	Uplink vertical pol.
	Downlink horizontal pol.

Communication Characteristics				
Total Usable Bandwidth				432 MHz
Assigned Frequency Range				
Up-Link (Receive)	5927	to		6418 MHz
Down-Link (Transmit)	3707	to		4193 MHz
Repeaters				
Number - on line				12
Number - redundant				12
Bandwidth				36 MHz
RF Power				5 Watts (7 dBW)
Minimum E.I.R.P. over Indonesia				30 dBW
Maximum E.I.R.P. in the center				34 dBW
Minimum G/T over Indonesia				-10 dB/K
Saturation Flux Density:				
DTWT mode				-80 dBW/m^2
MTA mode				-83 dBW/m^2
Communication Capability	2124 FDM/FM channels			
using 40 earth stations	266 SCPC-channels			
with 10 m diameter antennas	1 TV-channel			
Maximum Capacity	5760 two-way voice or			
	12 TV-channels			

PALAPA TECHNICAL PARAMETERS

Hans Dodel
IEEE Compendium

PALAPA B

PHOTO COURTESY OF
HUGHES AIRCRAFT CORP.

Hans Dodel
IEEE Compendium

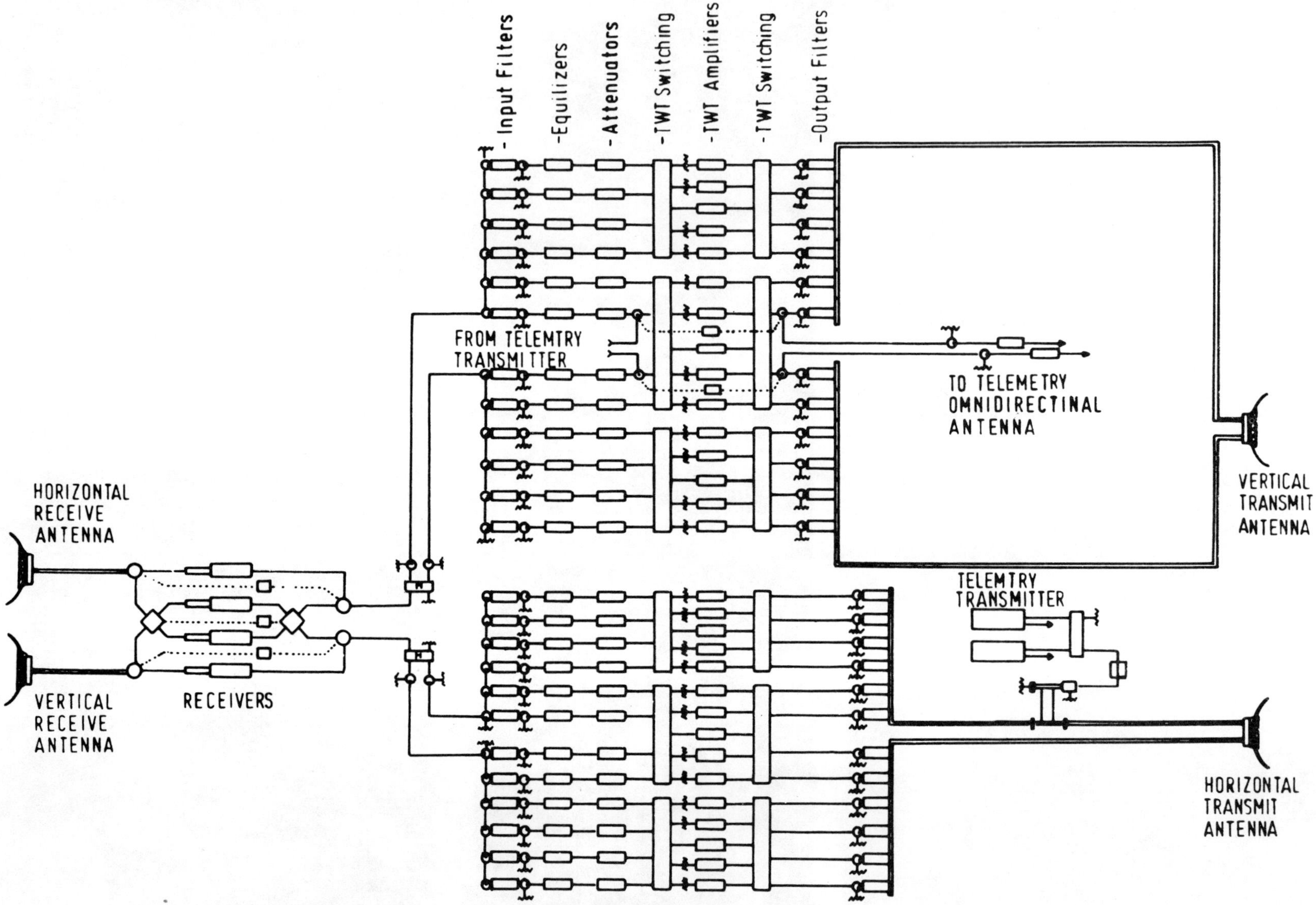

PALAPA B
COMMUNICATION BLOCK DIAGRAM

Hans Dodel
IEEE Compendium

Satellite Position:

Primary Satellite 108° E

Secundary Satellite 118° E

Eclipse Capability:

22 out of 24 transponders

Transponder Configuration and Frequency plan:

Transponder No.	Transmit/Receive Center Frequency
1,2	- 3720 MHz / 5945 MHz
3,4	- 3760 MHz / 5985 MHz
5,6	- 3800 MHz / 6025 MHz
7,8	- 3840 MHz / 6065 MHz
9,10	- 3880 MHz / 6105 MHz
11,12	- 3920 MHz / 6145 MHz
13,14	- 3960 MHz / 6185 MHz
15,16	- 4000 MHz / 6225 MHz
17,18	- 4040 MHz / 6265 MHz
19,20	- 4080 MHz / 6305 MHz
21,22	- 4120 MHz / 6345 MHz
23,24	- 4160 MHz / 6385 MHz

PALAPA B
FREQUENCY PLAN

General

No. of Satellites built	2
Design Life	8 years
Launch Vehicle	Delta 3910/3920 or alternatively Shuttle or ARIANE
Major Contractor	Hughes Aircraft Co.
Date of 1st Launch	1983

Physical Description

Satellite Type	HS 376
Stabilization	Dual Spin (Despun Antenna Earth Oriented, Spinning Drum.
Spinning Speed	as ANIK-D
Pointing Accuracy	0.05°
Ranging Accuracy	Better than 20 meters
Cylinder Diameter	216 cm
Cylinder Height at Launch	273 cm
Height W/antenna mast	683 cm
Weight	
Launch	1250 kg
Orbit (initinal)	625 kg
Solar Power at Equinox	1108 WATTS Begin of Life
Solar Power at Solstice	831 WATTS End of Life

Antenna Configuration

One 2.0 m deployable Reflector optimized for Indonesia/ASEAN coverage for both receive and transmit.

Polarization

Uplink and Downlink in vertical & horizontal pol. for complete frequency reuse through crosspol. Transmit/Receive Isolation in excess of 30 dB.

PALAPA B
TECHNICAL PARAMETERS

Communication Characteristics

Total Usable Bandwidth			432 MHz
Assigned Frequency Range			
Up-Link (Receive)	5927	to	6418 MHz
Down-Link (Transmit)	3707	to	4193 MHz
No. of Transponders			24
LNR redundancy			4 for 2
HPA redundancy			5 for 4
Bandwidth per Transp			36 MHz
RF Power per Transponder			10 Watts
Minimum E.I.R.P. over Indonesia			34 dBW
Minimum E.I.R.P. to ASEAN			32 dBW
Maximum E.I.R.P. in the center			36 dBW
Minimum G/T over Indonesia			- 5 dB/K
Minimum G/T to ASEAN			- 7 dB/K
Saturation Flux Density:			
DTWT mode			-80.0 dBW/m^2
MTA mode			-83.0 dBW/m^2
Maximum			-89.5 dBW/m^2

Communications Capability

1350	FDM/FM	channels (single Carrier)
700	FDM/FM	channels (multi Carrier)
1200	SCPC	channels (300 kHz spacing)
1600	SCPC	channels (22.5 kHz spacing)
2	TV	channels

Link Optimization effected thru input gain steps over a 9 dB range.

Future Outlook

Palapa - C is presently scheduled for a first Launch in 1988/89.

Hans Dodel
IEEE Compendium

PALAPA B TECHNICAL PARAMETERS (continued)

SATCOM

PHOTO COURTESY OF
RCA ASTRO-ELECTRONICS

Michael J. Robusto
IEEE Compendium

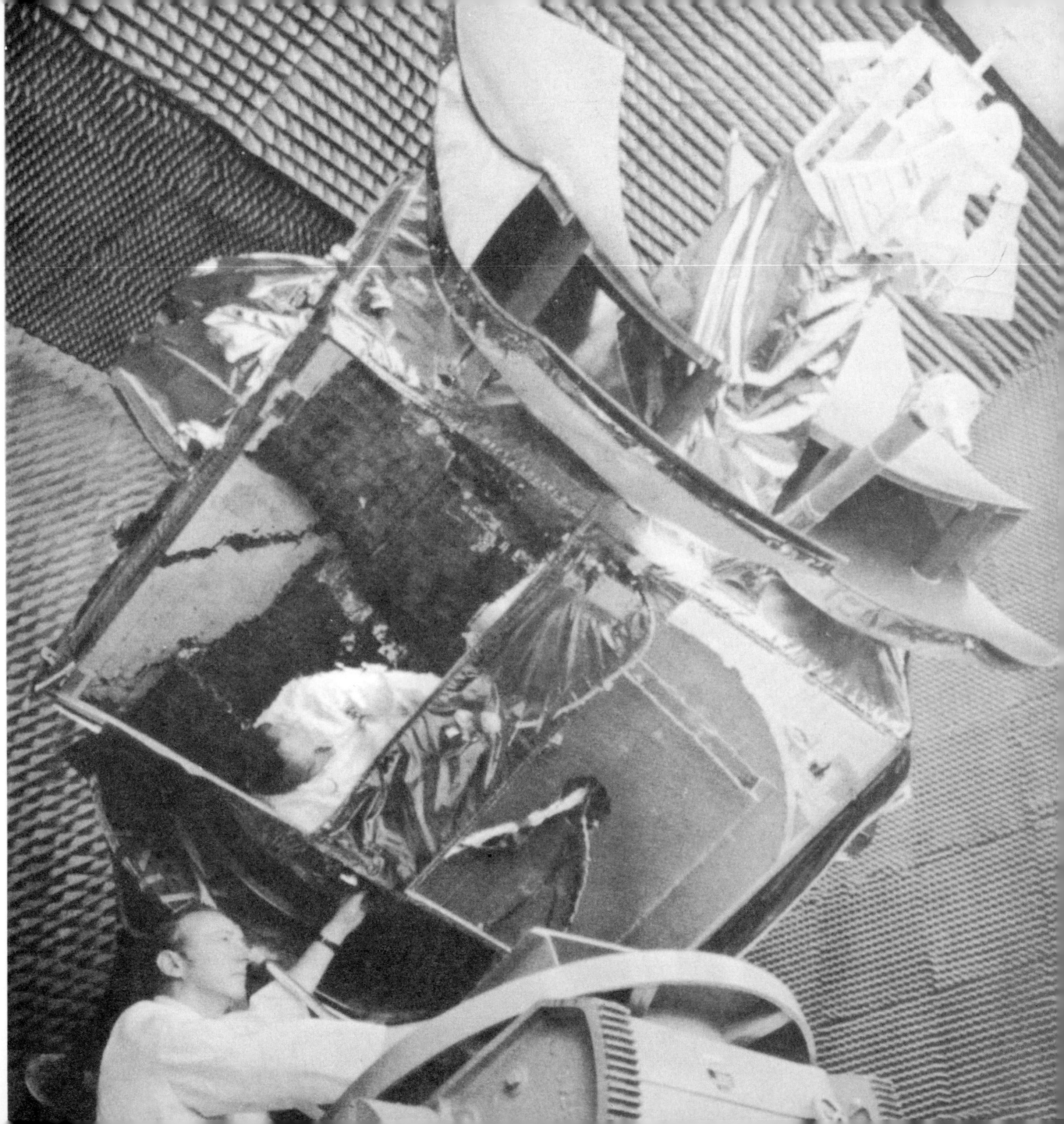

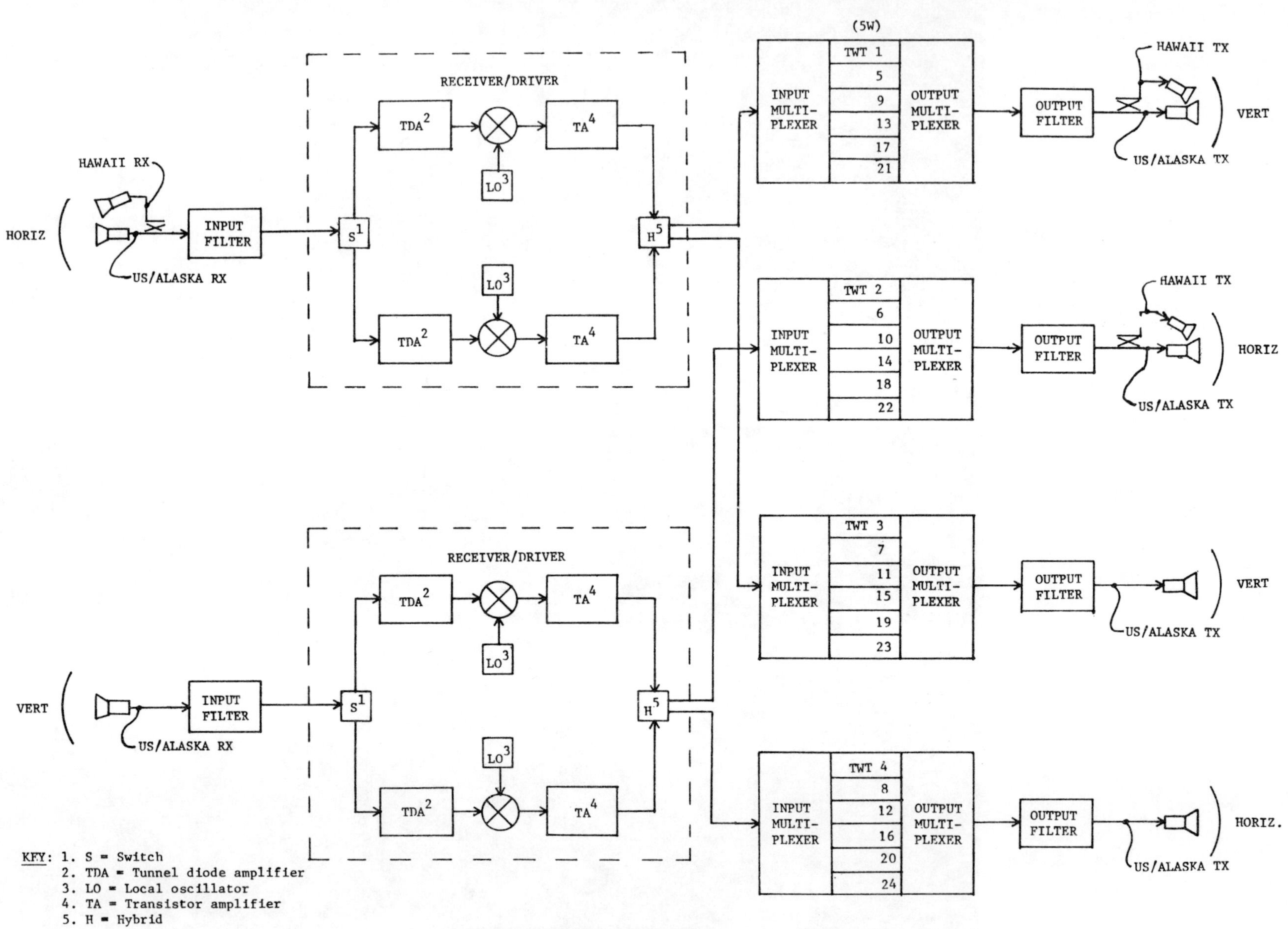

SATCOM COMMUNICATION BLOCK DIAGRAM

Michael J. Robusto
IEEE Compendium

SATCOM TECHNICAL PARAMETERS

General

No. satellites built	4
Design life	8 years
Orbit	Synchronous equatorial
Launch vehicle	McDonnell/Douglas 3 stage thrust augmented Thor/ Delta 3914
Apogee injection motor	Aerojet solid propellant
Major contractor	RCA Astro-Electronics Division
Date of first launch	12 December 1975

Physical Description

Main body	5.3x4.1x4.1ft(1.6 x1.2x1.2m)
Overall length (with solar panels deployed)	37 ft (11m)
Weight	
Launch	1991 lbs (903 Kg)
Orbit (initial)	1019 lbs (462 Kg)

Control System

Stabilization	Three-axis body stabilization with redundant momentum wheels
Reaction	Redundant hydrazine thrusters
Capability	8 yr station keeping life

Electrical Power

Solar array	
Power	770W max, 550W after 8 yrs
Size	71.5 sq. ft (6.6 sq.m) area of silicon solar cells
Batteries	3 Ni-Cad 12 amp-hr each providing 485W during eclipse operation

Command

Type	Ternary FSK
No. commands	248 commands

Telemetry

Type	PAM/FM/PM
No. channels	121 channels

Communications Antennas

Type	4 offset fed reflectors (2 for horizontal polarization, 2 for vertical polarization)
Configuration	Arranged in dual overlapping segments
Feeds	Six offset fed horns total
Polarization	Dual (horizontal and vertical)
Coverage	48 continental US states plus Alaska Hawaiian spot beam
Cross polarization isolation	
Receive	33 dB
Transmit	33 dB

Communications Capability

Total useable bandwidth	816 MHz (24 channels x 34 MHz/ channel)
Frequency range	
Uplink (receive)	5925-6425 MHz
Downlink (transmit)	3700-4200 MHz
Repeaters	
No. and Type	24 quasi-linear single conversion repeaters
Bandwidth	34 MHz each
Power amplifier	24 5W TWT's (1 for each channel)
Redundancy	
Receiver/Driver	2 for 1 redundancy per channel
Power amplifier	No redundancy
Receive G/T	-6 dB/K (48 states plus Alaska) -10 dB/K (Hawaii)
Available e.i.r.p.	32 dBW (48 states plus Alaska) 26 dBW (Hawaii)
Capacity	Each channel can handle one of the following: 1000 voice channels (FDM/FM) 500 voice channels (SCPC) 2 FM TV 60 Mb/s TDMA

Significant Features

Use of graphite-fiber epoxy composite for filters/multiplexors

Use of Thor/Delta 3914 launch vehicle which was developed specifically for this program

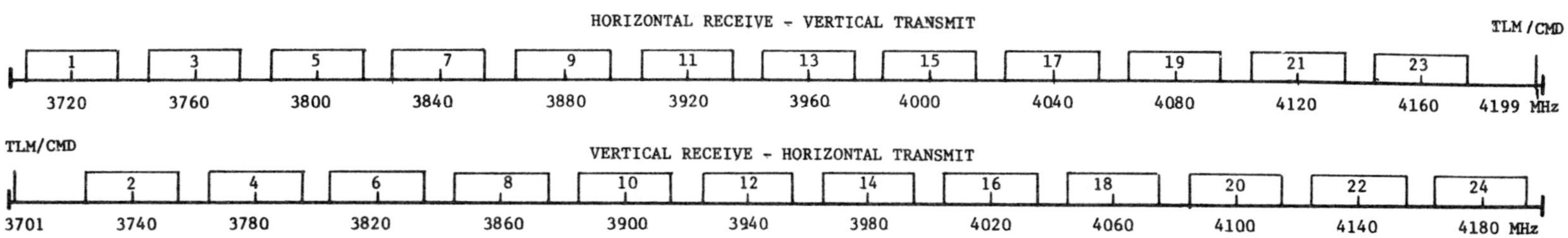

Notes:

1. Frequencies shown are channel center transmit frequencies. Receive frequencies are 2225 MHz higher.
2. Number above channel center frequency refers to transponder identity.
3. Each channel has 34 MHz of useable bandwidth.

SATCOM FREQUENCY PLAN

Michael J. Robusto
IEEE Compendium

SBS

PHOTO COURTESY OF
HUGHES AIRCRAFT CORP.

Laurence F. Gray
IEEE Compendium

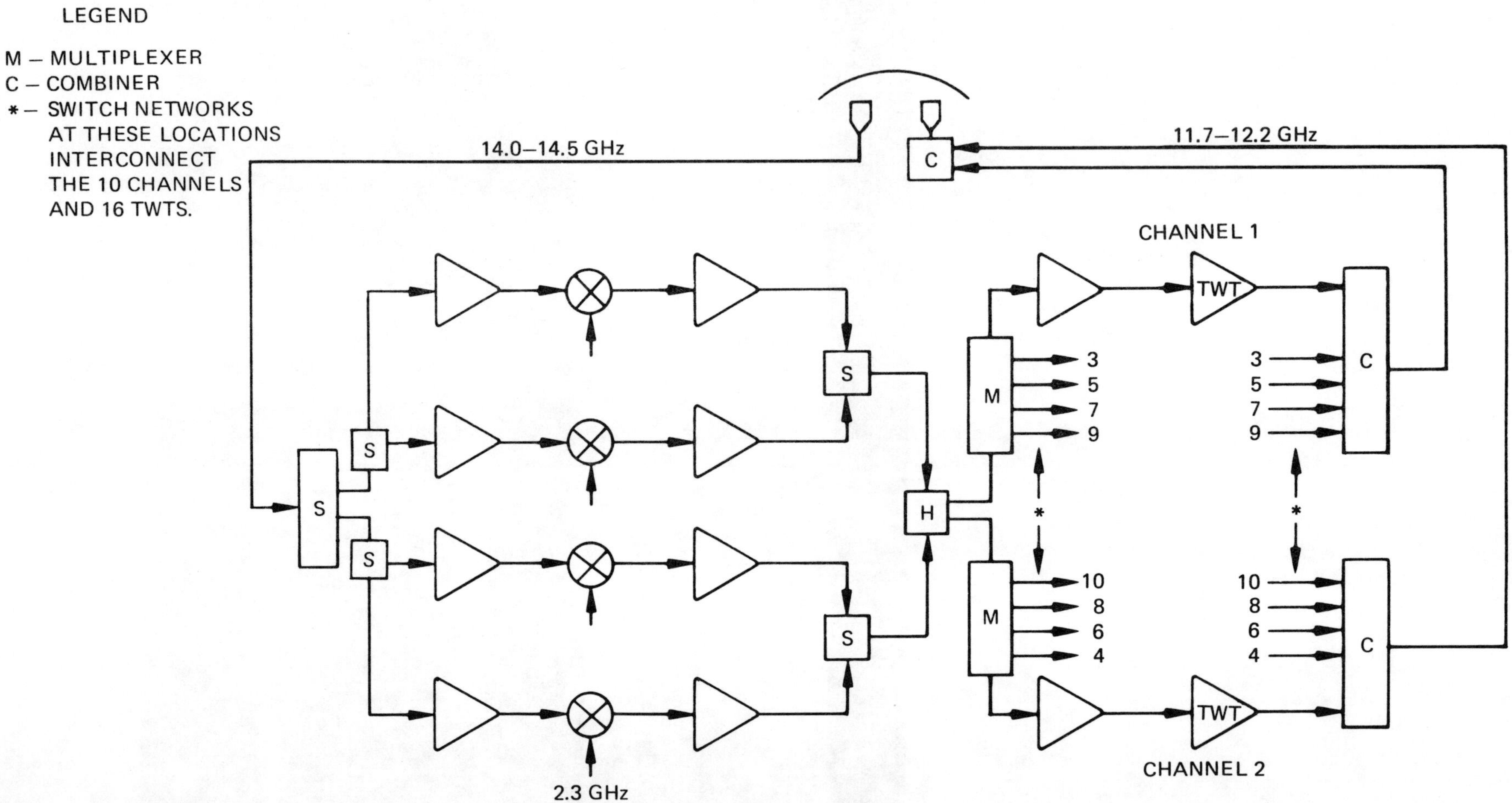

SBS COMMUNICATION BLOCK DIAGRAM

Laurence F. Gray
IEEE Compendium

CHANNEL NUMBER	RECEIVE/TRANSMIT FREQUENCIES (GHz)
1	14.025/11.725
2	14.074/11.774
3	14.123/11.823
4	14.172/11.872
5	14.221/11.921
6	14.270/11.970
7	14.319/12.019
8	14.368/12.068
9	14.417/12.117
10	14.466/12.166

SBS FREQUENCY PLAN

General

No. Satellites Built	3
Design Life	Shuttle 8 yrs or Delta 7 years
Launch Vehicle	Delta 3910/PAM or Shuttle
Major Contractor	Hughes
Date of 1st Launch	**15 Nov 1980**

Physical Description

Stabilization	Dual Spin (Despun Communication Platform Earth Oriented, Spinning Drum)
Cylinder Diameter	215.9 cm (85 in)
Cylinder Height-Lnch.	284.0 cm (112 in)
Height Deployed	660.4 cm (260 in)
Weight	
Launch (Shuttle)	4036 kg (8899 lb)
(Delta)	3348 kg (7381 lb)
Orbit (initial)	584 kg (1287 lb)

Antenna Configuration

Continental US with emphasis on eastern part.	Two 72-inch diameter paraboloids with 10 transmit and 15 receive horns

Polarization

Transmit	Linear (E-plane parallel to equatorial plane)
Receive	Orthogonal linear

Communication Capability

Total Usable Bandwidth	430 MHz
Assigned Frequency Range	
Up-Link (Receive)	14.0 to 14.5 GHz
Down-Link (Transmit)	11.7 to 12.2 GHz
Repeaters	
No. on line	10
No. redundant	6
Type	Telefunken TL-12026
Bandwidth	43 MHz
RF Power	20 watts (13 dBW)
Available e.i.r.p.	
Minimum	37.0 dBW
Over most of eastern third of CONUS	43.7 dBW
Receive G/T	
Minimum	-5.5 dB/K
Over most of eastern third of CONUS	+1.0 dB/K
Saturation Flux Density	
Minimum	$-80\ dBW/m^2$
Over most of eastern third of CONUS	$-87\ dBW/m^2$
Communication Capability	
TDMA 48 Mb/s	1300 equivalent voice circuits per transponder

SBS TECHNICAL PARAMETERS

Laurence F. Gray
IEEE Compendium

TELESAT (ANIK A)

PHOTO COURTESY OF TELESAT

Jahangir A. Tehrani
IEEE Compendium

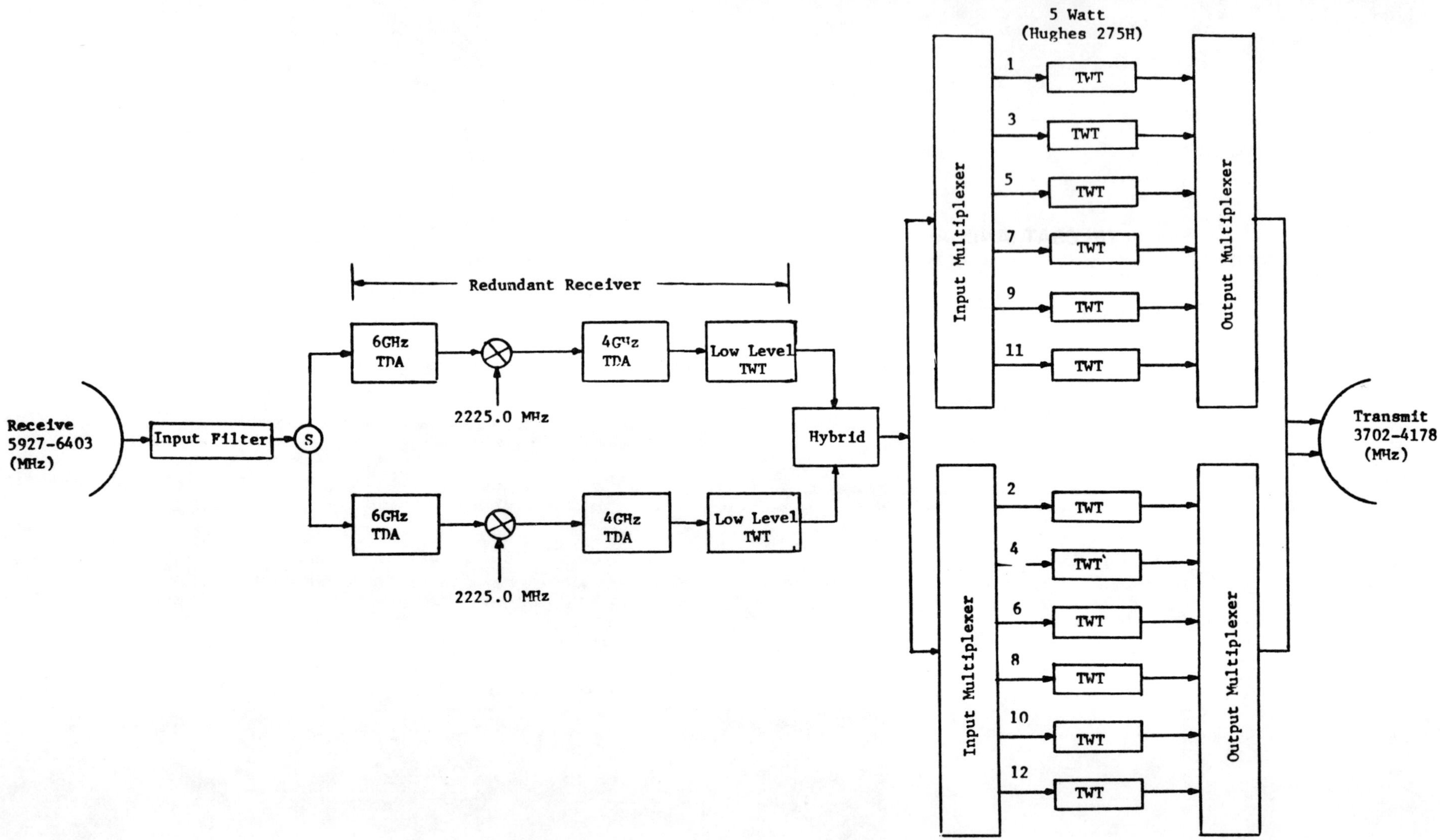

TELESAT (ANIK A) COMMUNICATION BLOCK DIAGRAM

Jahangir A. Tehrani
IEEE Compendium

Transp. No.	Receive/Transmit (MHz)
1.	5945/3720
2.	5985/3760
3.	6025/3800
4.	6065/3840
5.	6105/3880
6.	6145/3920
7.	6185/3960
8.	6225/4000
9.	6265/4040
10.	6305/4080
11.	6345/4120
12.	6385/4160

|←— 36 MHz —→|

Domestic Telephone, Telegraph
Data and Television

FREQUENCY PLAN

General

No. Satellites Built	3
Design Life	7 years
Launch Vehicle	Delta
Major Contractor	Hughes
Date of 1st Launch	November 9, 1972

Physical Description

Stabilization	Spin, Single Despun Antenna
Cylinder Diameter	190.5 cm (75 in)
Cyliner Height	170.2 cm (67 in)
Height w/antenna mast	353.1 cm (139 in)
Weight	
Launch	561.8 kg (1250 lbs)
Orbit (initial)	294.4 kg (655 lbs)

Antenna Configuration

Global	One 152.4 cm (60 in) diameter offset fed parabola, beam shaped to maximize gain over Canadian territory, 3° x 8° beamwidth, beam center tilted 7.85^o north of equatorial plane.

Polarization	linear

Communication Capability

Total Usable Bandwidth	432 MHz
Assigned Frequency Range	
Up-Link (Receive)	5927-6403 MHz
Down-Link (Transmit)	3702-4178 MHz
Repeaters	12 on line
Type	Hughes type 275H
Bandwidth	36 MHz(34 usable)
RF power	5 watts
Minimum e.i.r.p. per repeater	33 dBW
G/T (Receive)	>-7 dB/K
Saturated Flux Density	-80 dBW/m^2

Communication Capacity

Per Transponder	960 FDM/FM one way telephone or 1 TV or SCPC
Total Satellite Capacity	12 transponders available for various services such as FDM/FM, TV, or SCPC

TELESAT (ANIK A) TECHNICAL PARAMETERS

Jahangir A. Tehrani
IEEE Compendium

TELESAT (ANIK B)

PHOTO COURTESY OF TELESAT

Jahangir A. Tehrani
IEEE Compendium

Note: The 6/4 GHz section is the same as the Anik A Communication subsystem, except that TWTs are doubled in power and second TDA is replaced by transistor amplifier.

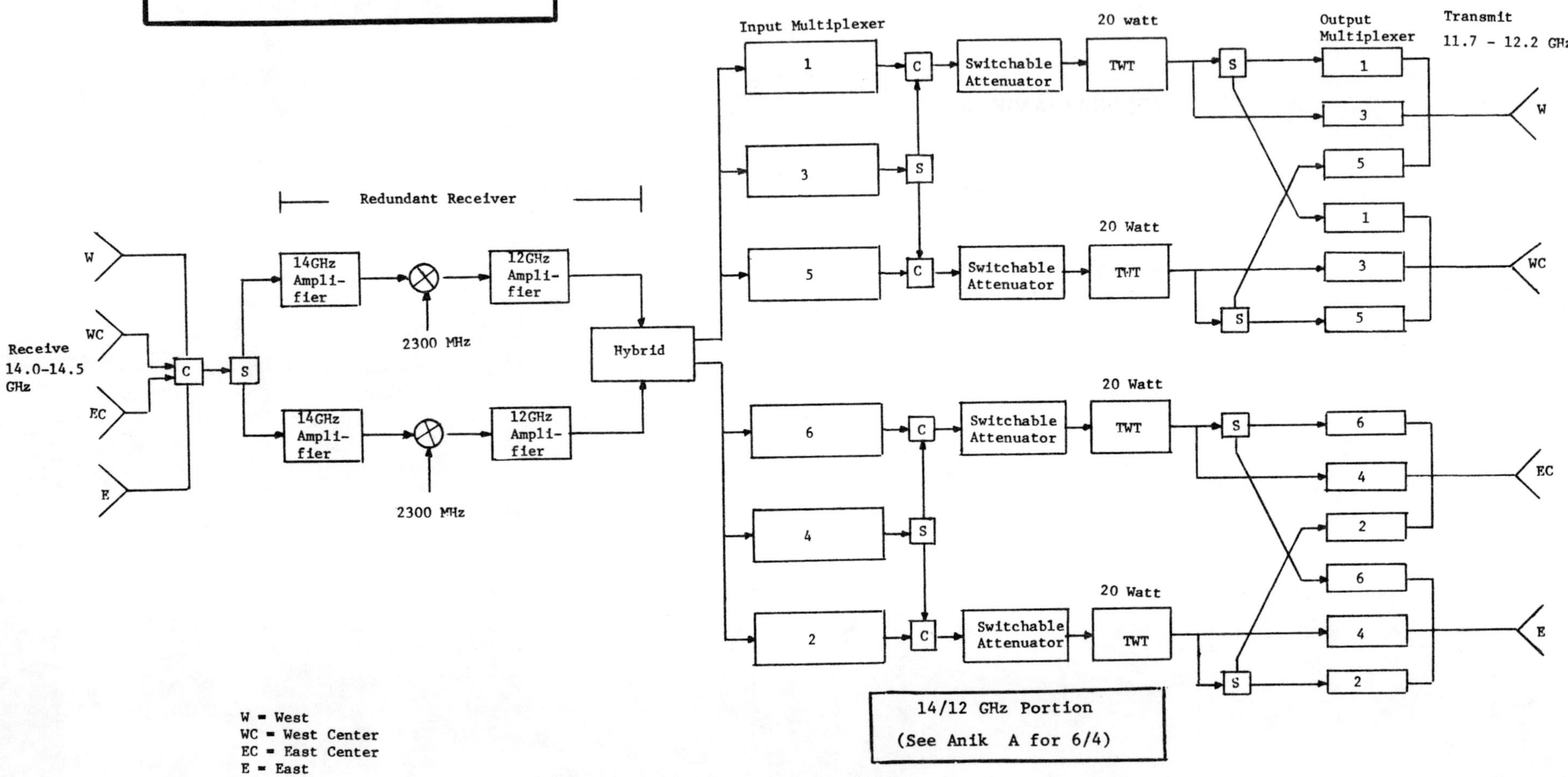

TELESAT (ANIK B) COMMUNICATION BLOCK DIAGRAM

Jahangir A. Tehrani
IEEE Compendium

Transp. No.	Receive/Transmit (MHz)
1.	5945/3720
2.	5985/3760
3.	6025/3800
.	.
.	.
.	.
.	.
.	.
12.	6385/4160
	← 36MHz →
1.	14040/11740
2.	14120/11820
.	.
.	.
6.	14440/12140
	← 72 MHz →

Domestic Telephone, Telegraph, Data and Television

FREQUENCY PLAN

General	
No. Satellites Built	1
Design Life	7 years (4/6), 4 years (12/14)
Launch Vehicle	Delta 3914
Major Contractor	RCA
Date of Launch	Dec 15, 1978
Physical Description	
Stabilization	3 axis
Height w/antenna mast	325 cm (128 in)
Weight	
Launch	898.9 kg (2000 lbs)
Orbit (initial)	456.7 kg (1016 lbs)
Antenna Configuration	
6/4 GHz	Offset fed parabola ~3° x 8° beam shaped to match Canadian land mass
12/14 GHz	Offset fed parabola, ~36 in x 48 in. One receive beam to match Canada, four 1.8^{o}x2^{o} transmit beams each covering one quarter of Canada.
Minimum Measured Gain Over Coverage Area (12/14GHz)	
Transmit	35.1 dB
Receive	29.4 dB
Polarization	
4/6 GHz	linear
12/14 GHz	linear

Communication Capability	
Total Usable Bandwidth	846 MHz
Assigned Frequency Range	
Up-Link (Receive) MHz	5927-6403 and 14004-14476
Down-Link (Transmit) MHz	3702-4178 and 11704-12176
Repeaters	
4/6 GHz	twelve 36 MHz bandwidth
12/14 GHz	six 72 MHz bandwidth
RF power	
4/6 GHz	One 10 watts TWT per repeater
12/14 GHz	four 20 watts TWT " "
Minimum e.i.r.p	
4/6 GHz	36 dBW
12/14 GHz	46.5 dBW
Receive G/T	
4/6 GHz	≥-6 dB/K
12/14 GHz	≥-1 dB/K
Single Carrier Saturation Flux Density	
4/6 GHz	-81 dBW/m^2
12/14 GHz	-86 dBW/m^2

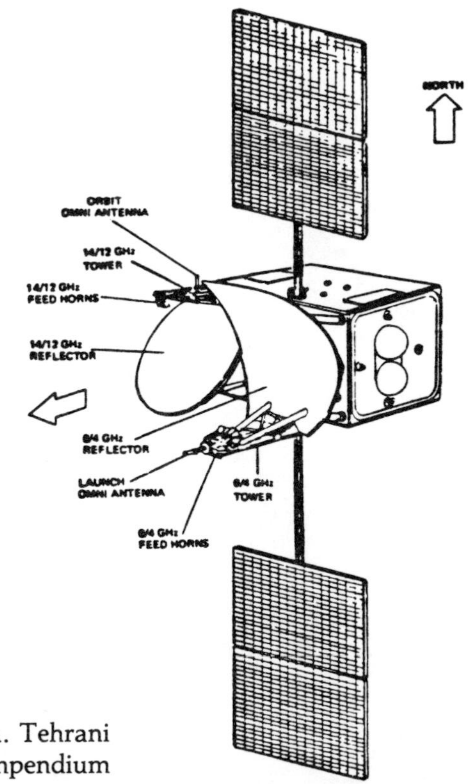

TELESAT (ANIK B) TECHNICAL PARAMETERS

Jahangir A. Tehrani
IEEE Compendium

TELESAT (ANIK C)

PHOTO COURTESY OF TELESAT

Jahangir A. Tehrani
IEEE Compendium

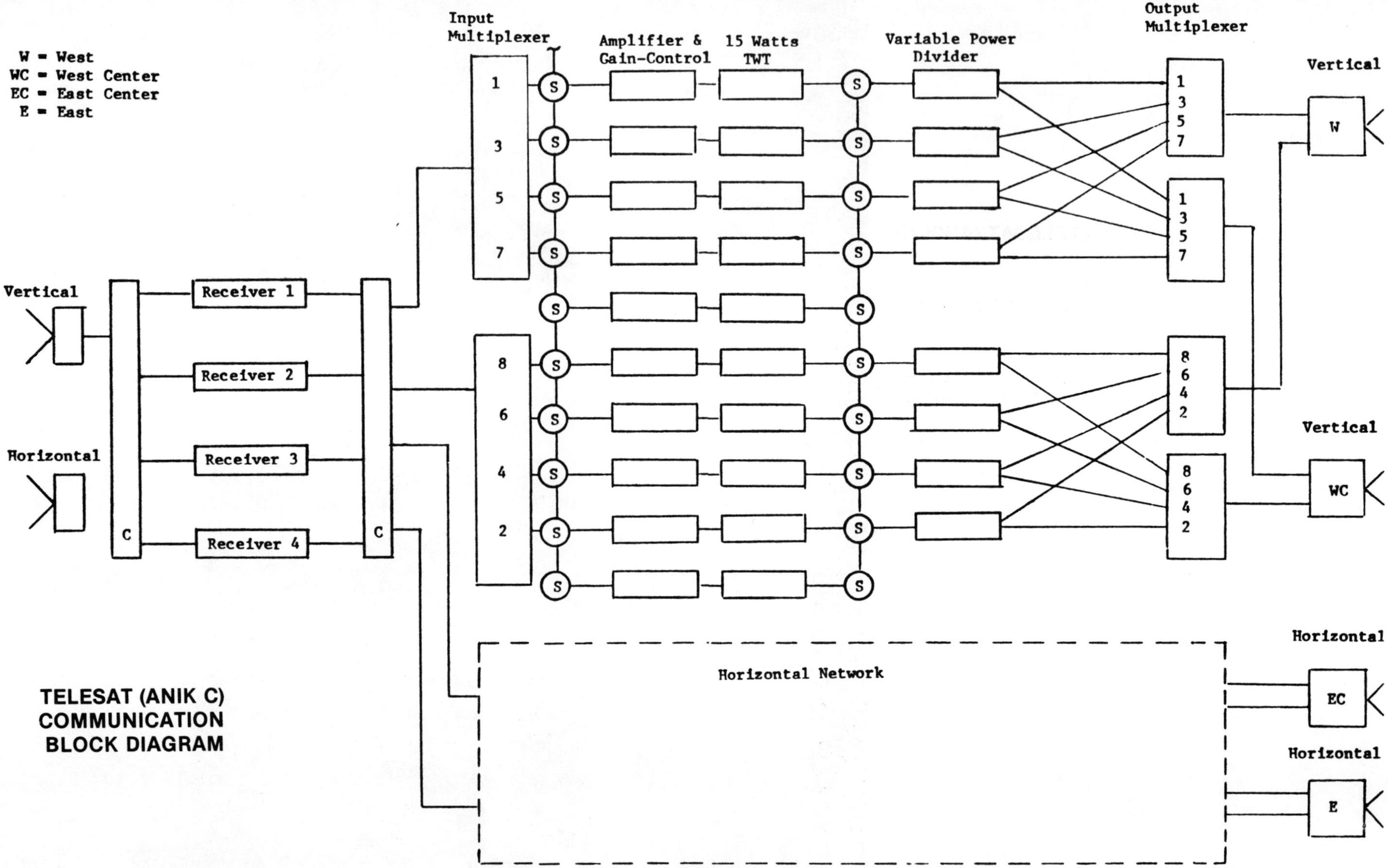

TELESAT (ANIK C) COMMUNICATION BLOCK DIAGRAM

Note: i) Three spare communication receivers are available to replace main receivers, by ground command, in the event of malfunction.

ii) The redundancy switching network connects sixteen RF channels to twenty TWT's with a minimum number of switches thus allowing up to four TWT failures without loss of an RF channel.

Jahangir A. Tehrani
IEEE Compendium

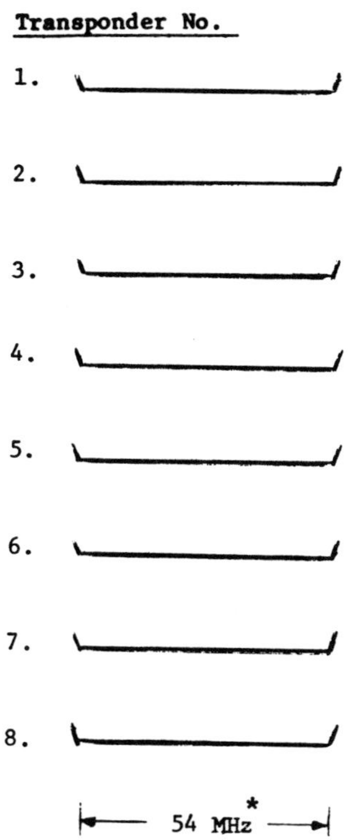

Domestic Telephone, Telegraph, data, and Television

FREQUENCY PLAN

General	
No. Satellites	3
Design Life	8 years
Launch Vehicle	Delta or Space Shuttle
Major Contractor	Hughes and Canadian Companies
Date of 1st launch	Planned for 1981
Physical Description	
Stabilization	Spin
Cylinder Diameter	215.9 cm (85 in)
Cylinder Height	642.6 cm (253 in)
Weight	
Launch	1078.7 kg (2400 lbs)
Orbit (initial)	543.86 kg (1210 lbs)
Antenna Configuration	
12/14 GHz	One 72 in diameter paraboloid, dual linear polarization in both receive and transmit, 2 Receive beam $1^\circ x8^\circ$, 4 contiguous transmit beams $\sim 1^\circ x2^\circ$ each, beams aimed to cover southern half of Canada
Polarization	linear

Communication Capability	
Total Usable Bandwidth	864 MHz
Assigned Frequency Range	
Up-Link (Receive)	14003-14497 MHz
Down-Link (Transmit)	11703-12197 MHz
Repeaters (Frequency Reuse)	16 (8V and 8H)
Bandwidth	54 MHz
RF power	15 watts
e.i.r.p. Per Repeater	48 dBW
Receive G/T	>3 dB/K
Single Carrier Saturation Flux Density	-80 dBW/m^2
Communication Capacity	
Per Repeater	91 Mbps (1344 Voice Channel) or 2 TV
Total communication Capacity	
Per 14 Repeaters	28 TV or 18816 one way message ckt

TELESAT (ANIK C) TECHNICAL PARAMETERS

*Center frequencies not available at the time of this writing.

Jahangir A. Tehrani
IEEE Compendium

WESTAR

PHOTO COURTESY OF
HUGHES AIRCRAFT CORP.

Richard G. Gould
IEEE Compendium

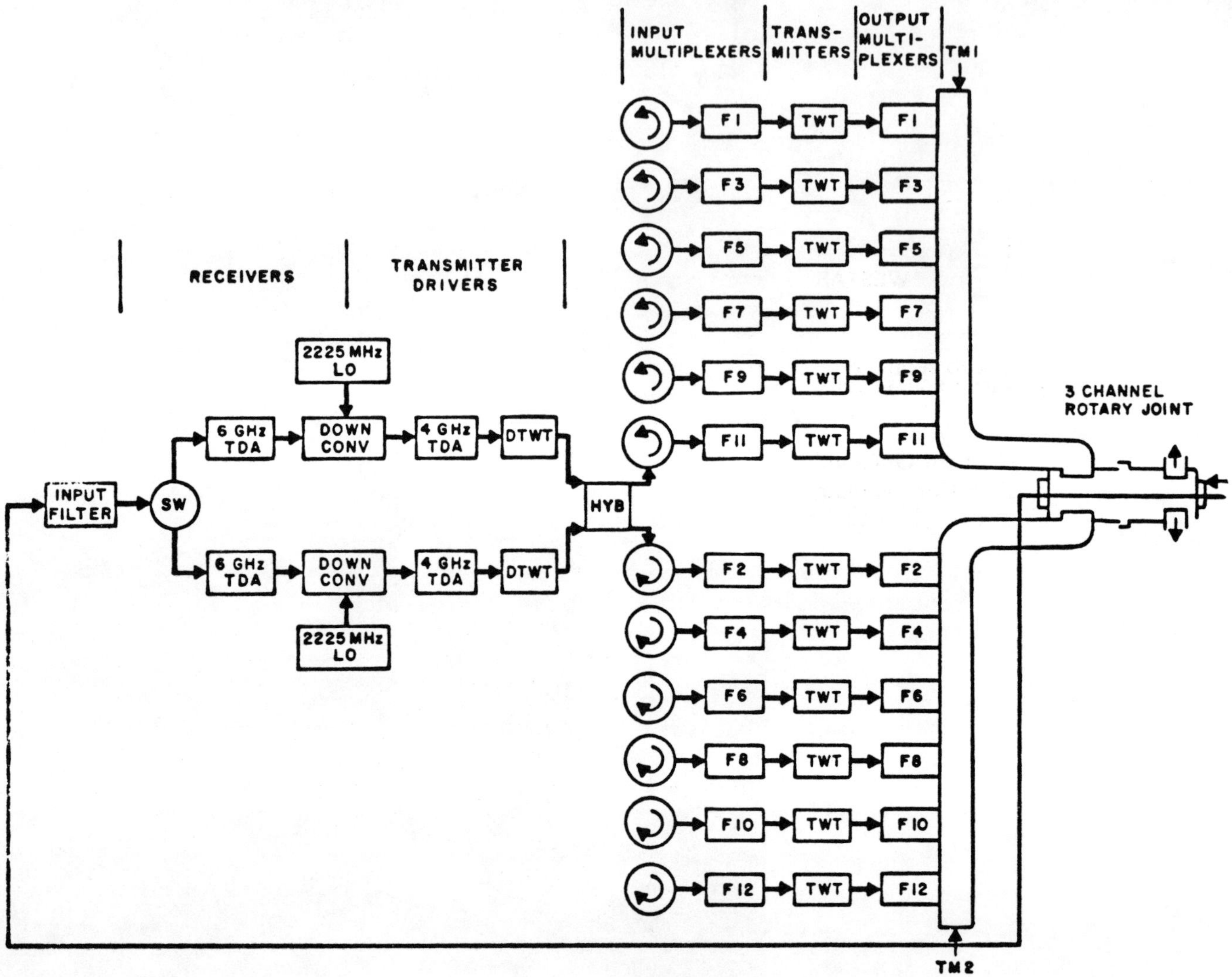

WESTAR COMMUNICATION BLOCK DIAGRAM

Richard G. Gould
IEEE Compendium

WESTAR TECHNICAL PARAMETERS

General

No. satellites built	3(all launched)
Design life	7 yrs (10 of 12 transponders operating during eclipse and at end of life)
Orbit	Synchronous, equatorial
Launch vehicle	McDonnell/Douglas 2914 Delta
Apogee injection motor	Solid propellant; 6028 fps
Major contractor	Hughes Aircraft Company
Date of first launch	April 13, 1974

Physical Description

Cylinder diameter	190 cm (75 in)
Cylinder height	153 cm (67 in)
Overall length	340 cm (139 in)
Weight	
Launch	574 kg (1262 lb)
Orbit (initial)	297 kg (653 lb)

Control System

Stabilization	Dual spin (single, despun earth-oriented antenna, spinning drum) to ± 0.1°
Configuration	2 independent N_2H_4 (hydrazine) systems
Capability	7 yr station-keeping life (± 0.1°)

Electrical Power

Solar array power	305W B.O.L. (260W E.O.L.)
No. and size	20,000 cells, 2x3 cm
Batteries	2 NiCd (each 28 cells)

Telemetry

Configuration	Dual-mode, toroidal beam, omni-directional, linearly-polarized 4 GHz used for launch and transfer orbit phase (comm. antenna used for telem. on-station) 22 functions; dual transmitters and encoders

Command

Configuration	Toroidal, omnidirectional, linearly-polarized 6 GHz used for launch and transfer orbit phase (comm. antenna used for cmd. on-station);
Decoders	Dual receivers and decoders
No. commands	32

Communication Antennas

Type	Mechanically-despun, 60 in. parabolic reflector w/multiple feeds
Beamwidth	6.8° x 3.5° to cover CONUS (lower e.i.r.p. over Alaska and Hawaii)
Polarization	
Receive	Linear, vertical
Transmit	Linear, horizontal

Communications Capability

Total usable bandwidth	480 MHz
Frequency range	
Uplink (receive)	5925-6425 MHz
Downlink (transmit)	3700-4200 MHz
Repeaters	
Type	12 independent, linear heterodyne single-conversion
Bandwidth	36 MHz each on 40 MHz centers (20 MHz between 4180-4200 MHz and 6405-6425 MHz for T&C)
Power amplifiers	12 5-watt TWT's (one for each ch.)
Saturation flux density	-80 dBW/m^2 for CONUS (-73 dBW/m^2 for Alaska and Hawaii)
Receive figure of merit (G/T)	-7.4 dB/K for CONUS (-14.4 dB/K for Alaska and Hawaii)
Available e.i.r.p.	33 dBW over CONUS (26.6 dBW over Alaska and Hawaii)
Capacity	18,000 one-way voice circuits or 12 TV channels, or a combination thereof, per spacecraft

Significant Features

Spacecraft similar to Palapa and Anik A

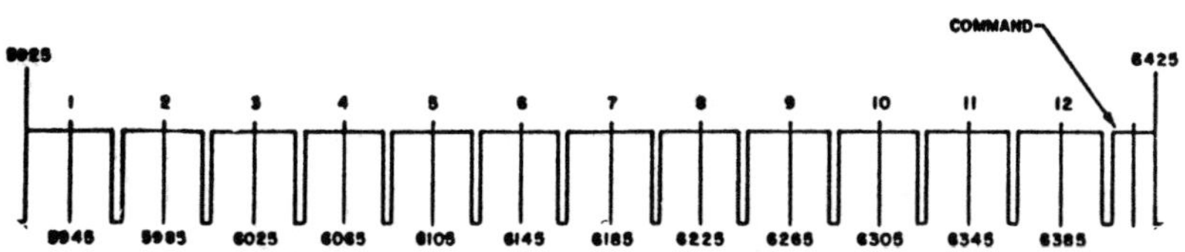

RECEIVE CHANNEL CENTER FREQUENCIES IN MHz

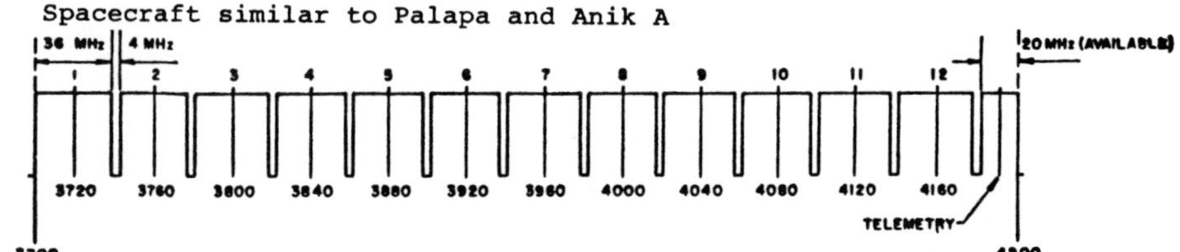

TRANSMIT CHANNEL CENTER FREQUENCIES IN MHz

WESTAR FREQUENCY PLAN

Richard G. Gould
IEEE Compendium

TDRSS/ADVANCED WESTAR

PHOTO COURTESY OF TRW

Richard G. Gould
IEEE Compendium

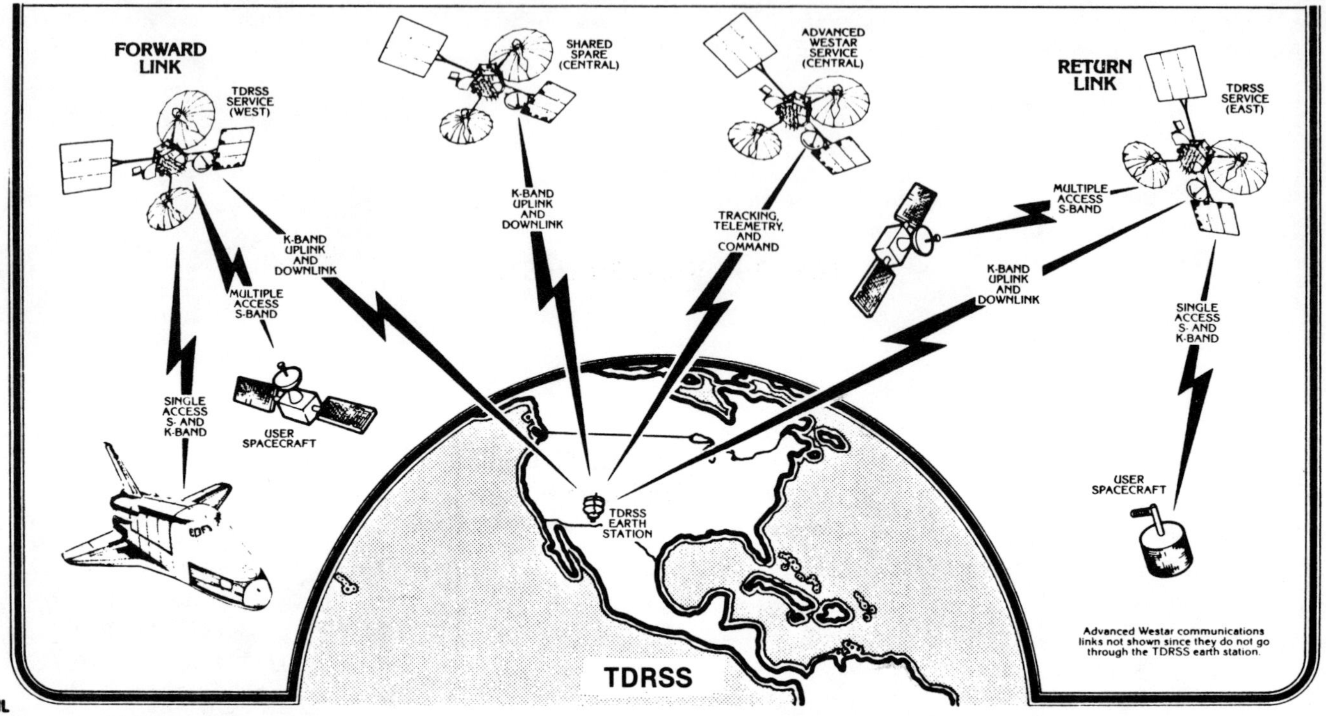

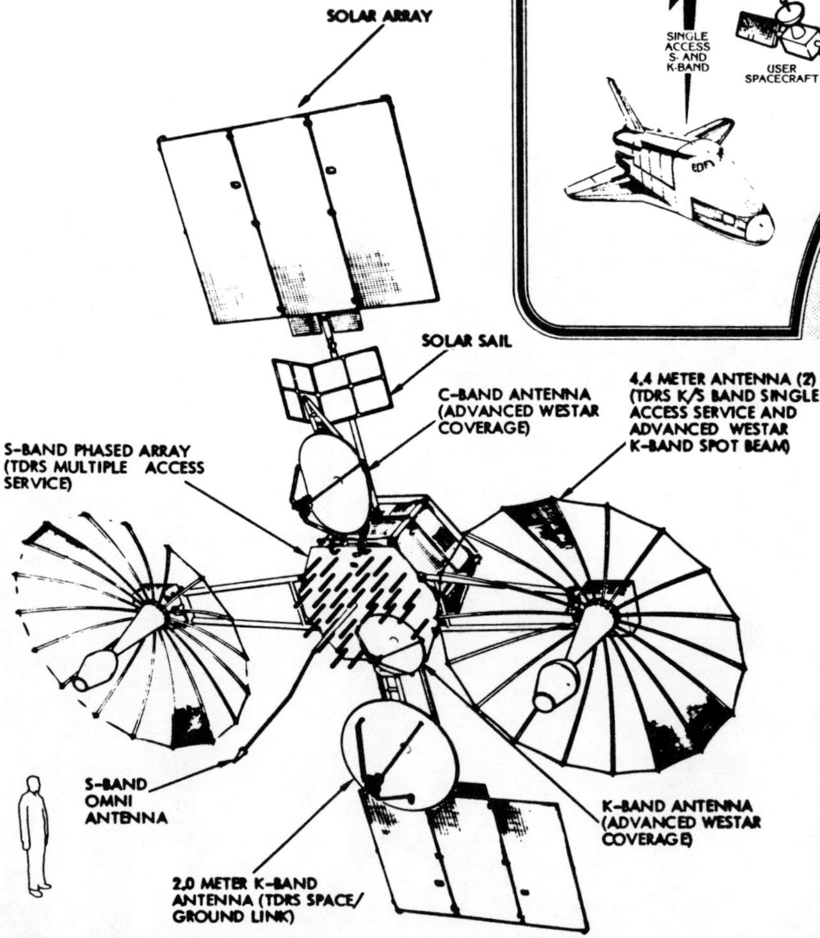

TDRSS/ADVANCED WESTAR

Richard G. Gould
IEEE Compendium

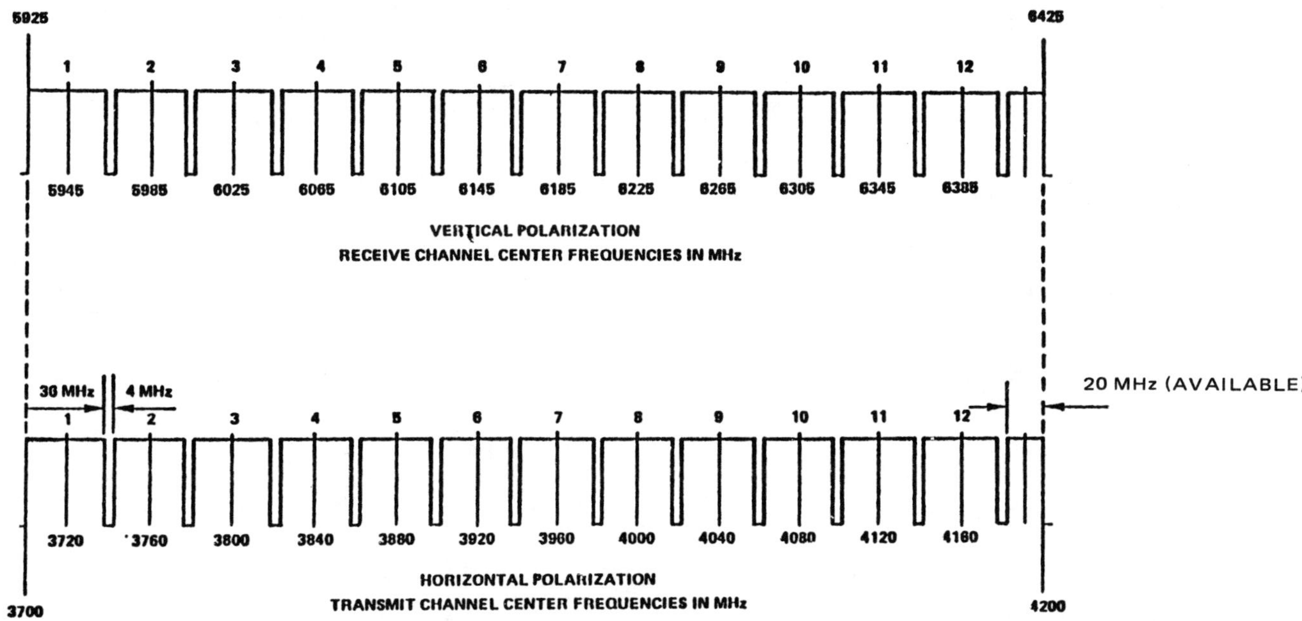

ADVANCED WESTAR
C-BAND FREQUENCY AND POLARIZATION PLAN

TDRSS/ADVANCED WESTAR FREQUENCY PLAN

Richard G. Gould
IEEE Compendium

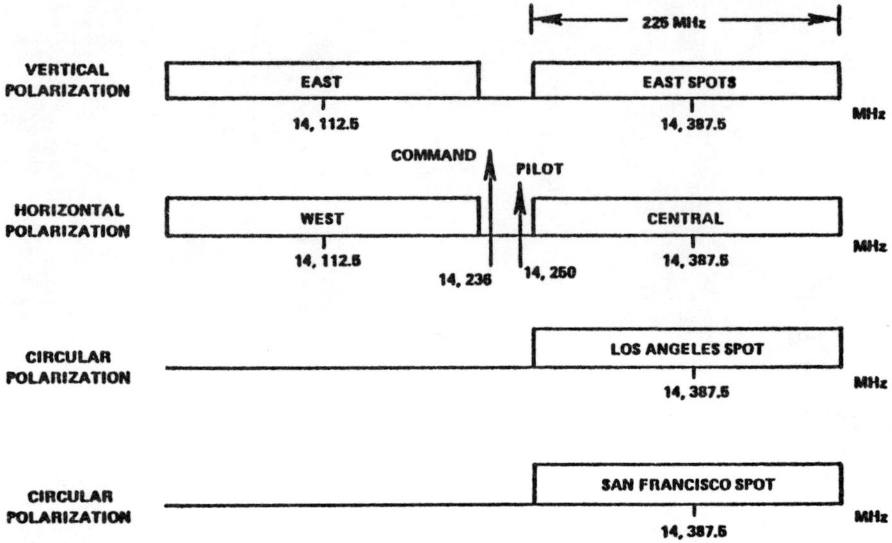

ADVANCED WESTAR
K-BAND FREQUENCY & POLARIZATION PLAN
UPLINK

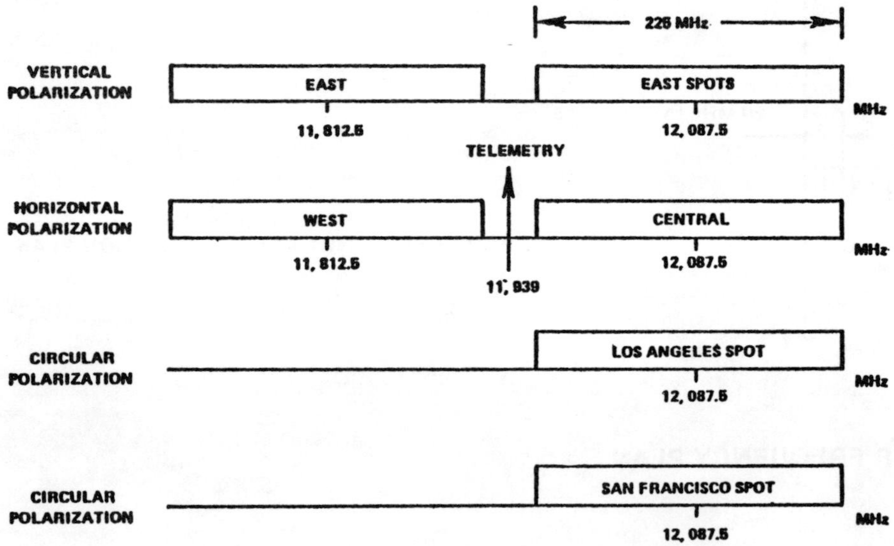

ADVANCED WESTAR
K-BAND FREQUENCY & POLARIZATION PLAN
DOWNLINK

TDRSS/ADVANCED WESTAR FREQUENCY PLAN (continued)

Richard G. Gould
IEEE Compendium

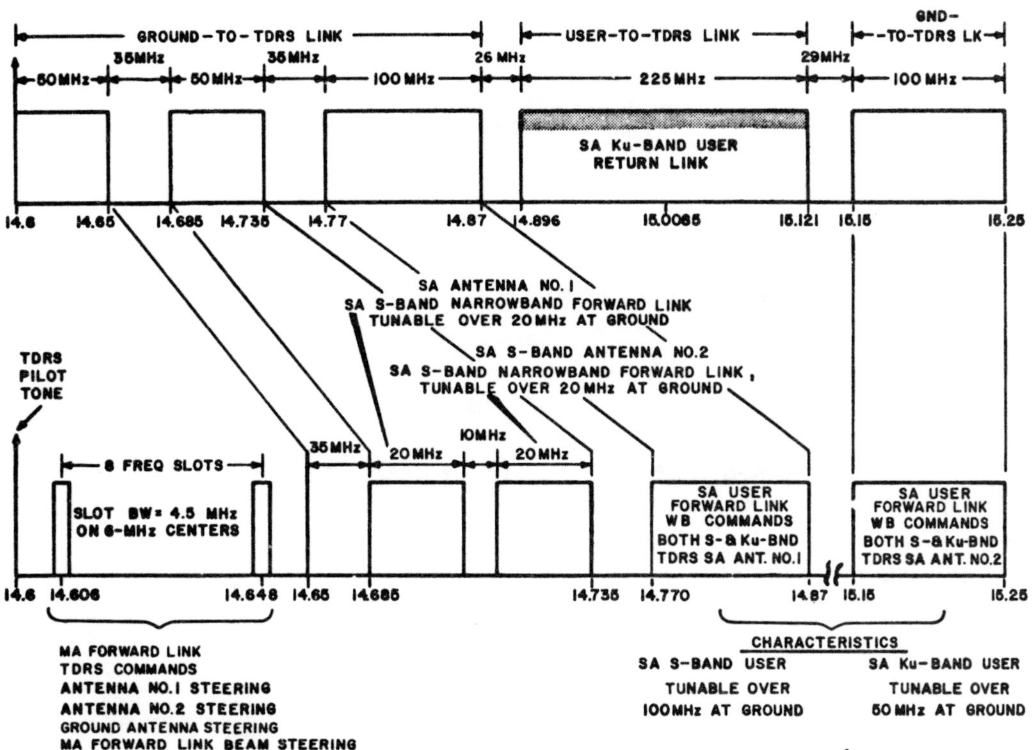

TDRS Ku-band Receive Frequency Plan

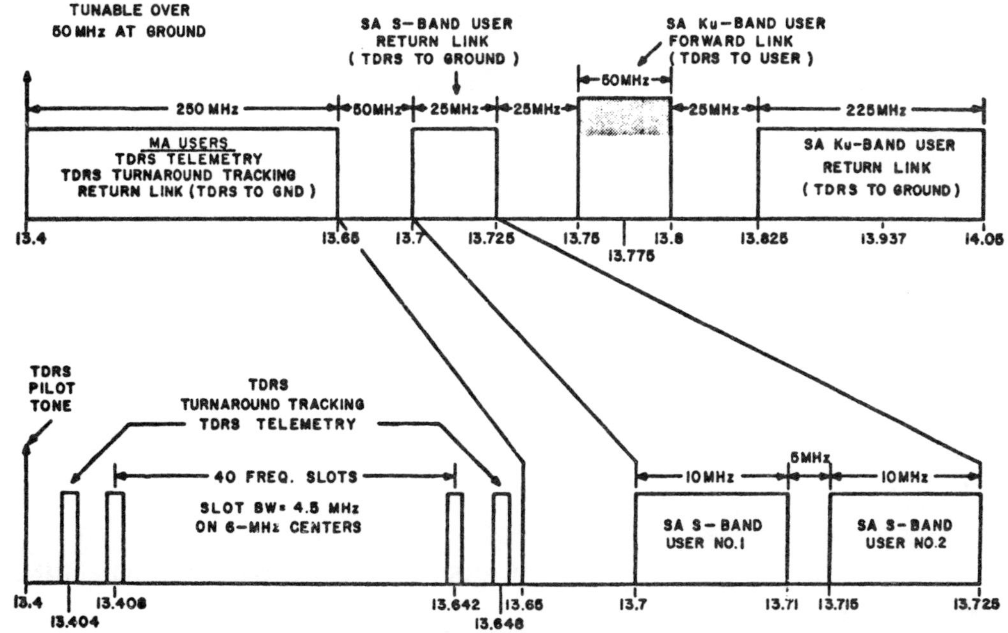

TDRS Ku-band Transmit Frequency Plan

TDRSS/ADVANCED WESTAR FREQUENCY PLAN (continued)

Richard G. Gould
IEEE Compendium

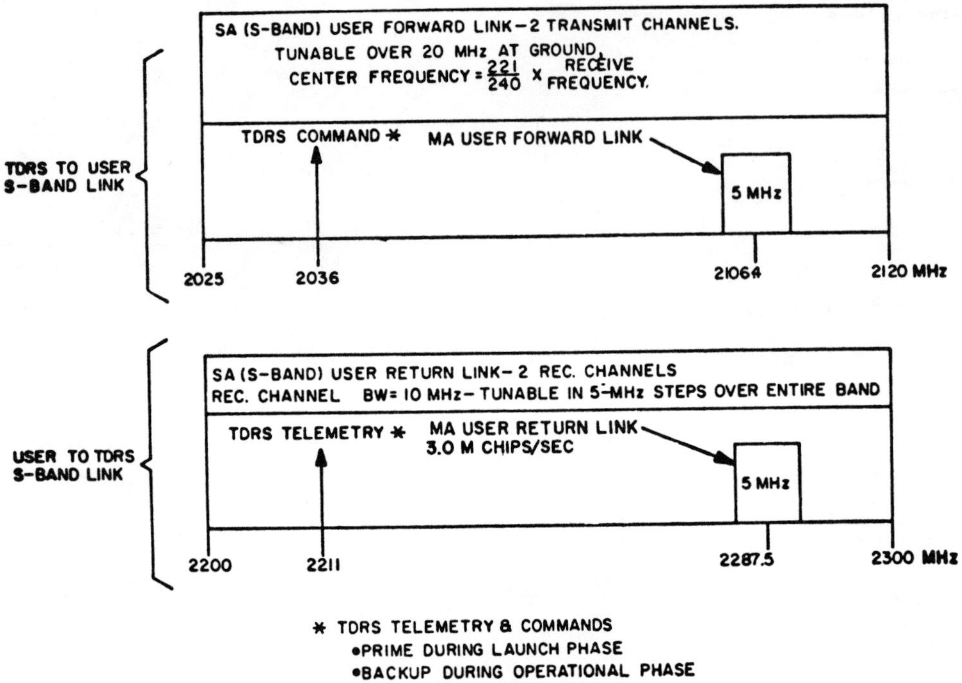

TDRS S-band Frequency Plan

TDRSS/ADVANCED WESTAR FREQUENCY PLAN (continued)

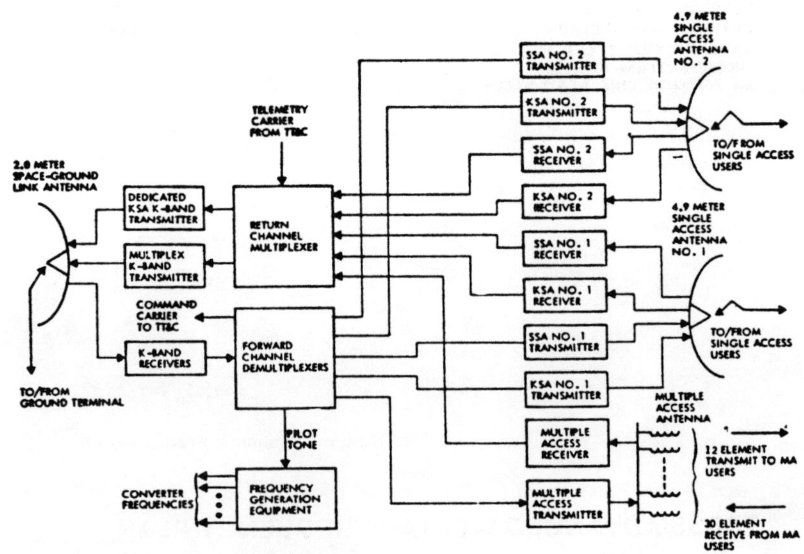

TDRSS/ADVANCED WESTAR COMMUNICATIONS BLOCK DIAGRAM

Richard G. Gould
IEEE Compendium

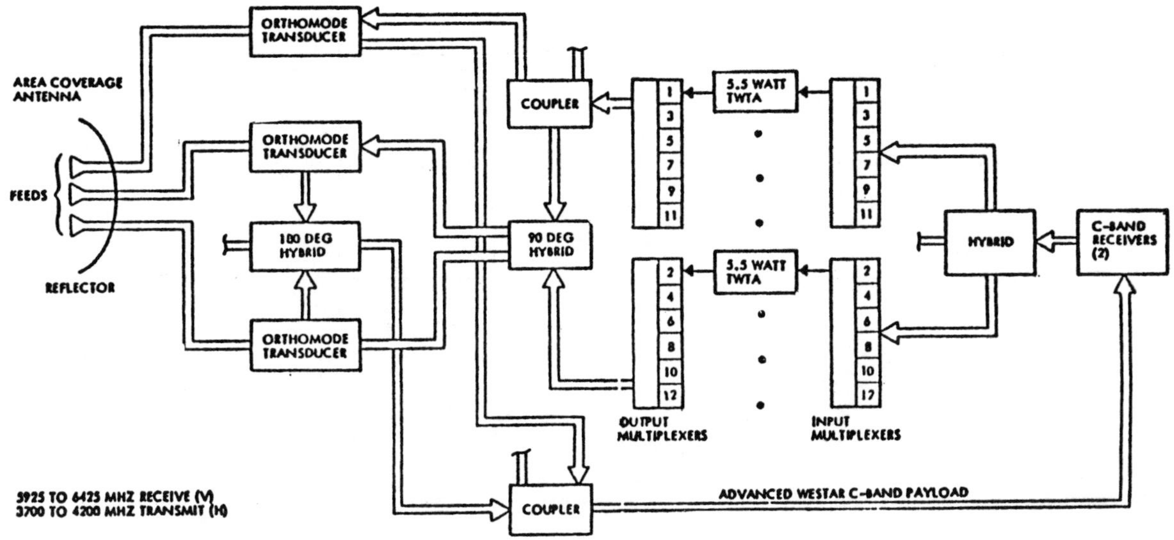

ADVANCED WESTAR C-BAND COMMUNICATIONS PAYLOAD

TDRSS/ADVANCED WESTAR COMMUNICATIONS BLOCK DIAGRAM (continued)

Richard G. Gould
IEEE Compendium

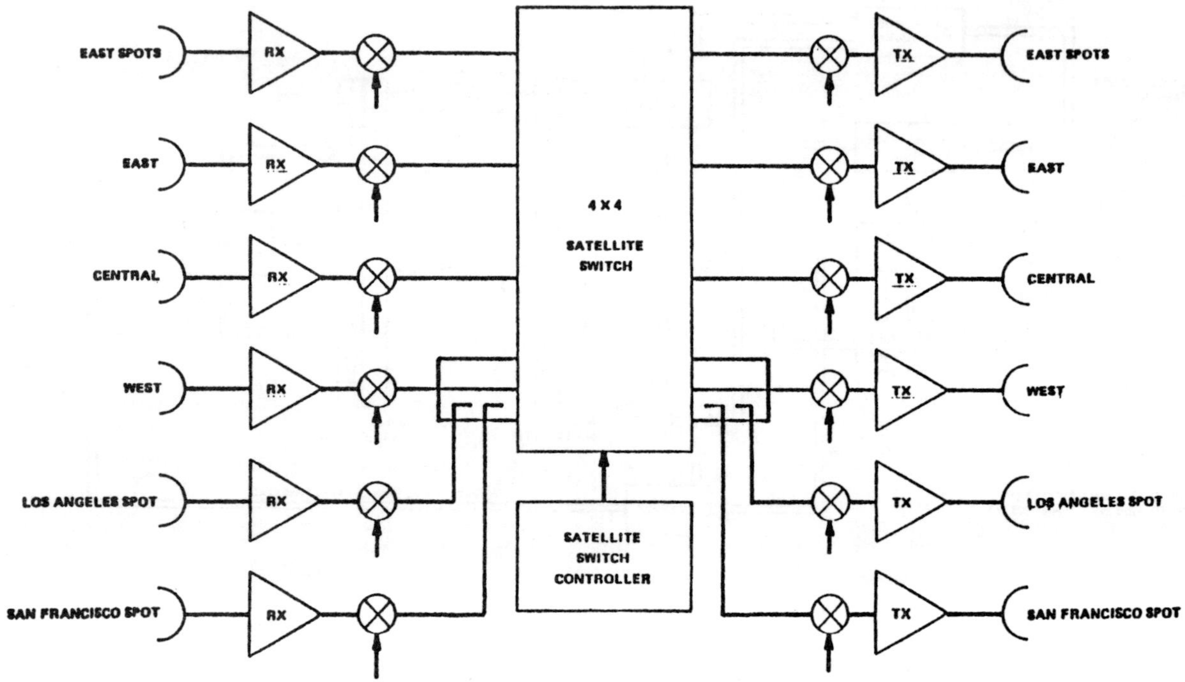

ADVANCED WESTAR K-BAND COMMUNICATIONS PAYLOAD

TDRSS/ADVANCED WESTAR COMMUNICATIONS BLOCK DIAGRAM (continued)

Richard G. Gould
IEEE Compendium

General

No. Satellites Built:	4(2 only TDRSS, 1 only AW, 1 in-orbit spare w/TDRSS and AW capability)
Design Life	10 yrs (full eclipse operation)
Orbit	Quasi-geostationary: inclined equatorial, 7° max. inclin.
Launch Vehicle	Space Shuttle (Titan 34D back-up)
Major Contractor	TRW, Inc.
Date of 1st Launch	1981(?)

Physical Description

Dimensions	Hexagonal prism 2.4m(8 ft)across, 4.6m(15 ft) high; 17.4m(57.2 ft) betw. tips of deployed solar arrays; 13m(42.6 ft) across large deployed antennas
Weight(in orbit)	2268 kg(5,000 lbs)
Stabilization	3-axis (±0.1° in pitch and roll, ±0.25° in yaw)

Electrical Power

Solar Array	Sun-tracking, 1700W E.O.L.
Batteries	NiCd

Telemetry

Configuration	2211 MHz (launch phase); omni antenna, circular polariz. 13.731GHz (in-orbit), TDRSS, 1 linear polarization 11.939 (in orbit), AW lin. pol. Data rates to 16,000 b/s

Command

Configuration	2035.9625 MHz (launch phase), conical spiral, omni ant. linear polarization 14.786 GHz (in orbit), TDRSS, linear polariz.(via SGL ant) 14.236 GHz (in orbit), AW lin. polariz. (via commun. ant.) PSK data modulation, 16 kHz PM sub-carrier modulation

	Advanced Westar (AW)	Tracking & Data Relay Satellite System (TDRSS)
Communications Antennas	C-Band: 1-1.2×1.6 meter(4×5 ft) reflector, linear polariz. w/3 beams, steerable by ground command for CONUS, Alaska & Hawaii. K-Band: 1-1.1m(3.6 ft) parabolic reflector for East & West Area beams; 1-2m (6.6 ft) parabolic reflector for Central Area and N.Y. & Miami spot beams (also used for TDRSS SGL) 2-4.9m (16 ft) parabolic reflectors for L.A. and San Francisco spot beams (also used for TDRSS SA) Polarizations: Vertical for East Area and for N.Y. and Miami spot beams; Horizontal for Central and Western Area: Circular for L.A. & S.F. spot beams.	S-Band: 30-element,phased array (MA) 2-4.9m (16 ft) parabolic reflectors (SA) (also used for AW and TDRSS K-Band), G=36.7 dB, Circ. Pol. K-Band: 2-4.9m (16 ft) parabolic reflectors (SA) (also used for AW and TDRSS S-Band), G=54.6 dB, Circ. Pol. 1-2.0m(6.6 ft) parabolic reflector SGL G=45.3 dB transmit, 46.0 dB receive; linear pol.
Communications Capability		
Total Usable Bandwidth	C-Band: 500 MHz K-Band: 500 MHz	S-Band: 100 MHz user to TDRS, 95 MHz TDRS to user K-Band: 325 MHz TDRS Transmit, 650 MHz TDRS Receive
Frequency Range	C-Band: Up-link: 5925-6425 MHz Down-link: 3700-4200 MHz (channelization identical to Westar) K-Band: Up-link: 14,002.5-14,227.5 GHz and 14,272.5-14.497.5 GHz Down-link: 11,702.5-11,927.5 GHz and 11,972.5-12,197.5 GHz	S-Band: MA; 2103.4-2109.4 MHz, Transmit SA; 2025-2120 MHz, Transmit K-Band: SA; 13.750-13.8 GHz,Transmit SGL:13.4-14.05 GHz, Transmit S-Band: MA; 2285-2290 MHz. Receive SA; 2200-2300 MHz, Receive K-Band: SA; 14.896-15121.1 GHz, Receive SGL;14.6-15.25 GHz, Receive
Repeaters		
Type	C-Band: 12-transmitters, 2 redundant receivers K-Band 4 transmitters, 2 spares w/30W TWT's 2 transmitters, 2 spares w 1.5W TWT's	S-Band: SA; 2 transmitters, 2 receivers MA; 1 transmitter, 1 receiver K-Band: SA; 2 transmitter, 2 receivers SGL; 1 dedicated transmitter, 1 multiplex transmitter, 1 rcvr.
Bandwidth	C-Band: 36 MHz on 40 MHz centers K-Band: 225 MHz	
Power Amplifiers	C-Band: 12-5watt TWT's K-Band: 4-30watt TWT's plus 2 spares 2-1.5watt TWT's plus 2 spares	C-Band: SA: 26 watt TWT's MA: 3.6W from 10 phased elements K-Band: SA: 1.5 W TWT's SGL: 2-18W TWT's
Saturation P.F.D.	C-Band: -78 dBW/m^2 K-Band: Not available	S-Band: Not available K-Band: Not available
Receiver Figure of Merit (G/T)	C-Band: -7 dB/K CONUS, -12.5 dB/K Alaska & Hawaii K-Band: -4 to 6.6 dB/K depending on beam and city served	S-Band: -15 dB/K per element at edge of coverage; -1 dB/K overall MA; 10.4 dB/K SA. K-Band: SA: 25.5 dB/K SGL: 10.0 dB/K
Available E.I.R.P.	C-Band: 33 dBW, CONUS, 28 dBW Alaska, 26 dBW Hawaii; K-Band: 42.5-50.3 dBW depending on beam and city served	S-Band: MA: 34 dBW SA: 46.6 dBW K-Band: SA: 50 dBW SGL:50 dBW
Capacity	C-Band: 16,800 one-way voice(or 62 Mb/s=481 T-1's) or 12 color TV channels, or a combination thereof, depending on the number of accesses. K-Band: 250 Mb/s per transponder(=1,000 Mbs total) or 33 one-way color TV channels: typical capacity for 100 stations is 10,000 one-way voice circuits.	S-Band: MA; 20 return links (50 kb/s each) 1 forward link (to 10 kb/s) SA: 2 duplex links (each to 300 kb/s forward, 12 kb/s return) K-Band: SA: a duplex links (each to 25 Mb/s forward, 300 Mb/s return Note: only 1 forward and 1 return link at one frequency at a time, except for users closer than 0.4°
Significant Features	K-Band modulation is quaternary PSK (QPSK) Access is SS/SDMA/TDMA	

NOTES: SA= Single Access
MA= Multiple Access
SGL= Satellite-Ground Link

TDRSS/ADVANCED WESTAR TECHNICAL PARAMETERS

Richard G. Gould
IEEE Compendium

SECTION V

MILITARY SATELLITES

- DSCS-II
- DSCS-III
- FLTSATCOM
- GALS
- GORIZONT*
- IDCSP
- LEASAT
- LES 1 & 2
- LES 3
- LES 4
- LES 5
- LES 6
- LES 7
- LES 8 & 9
- NATO II**
- NATO III
- SKYNET I
- SKYNET II
- TACSAT

*See STATSIONAR Series in International/Domestic Section.
**NATO I = IDCSP

DSCS-II

PHOTO COURTESY OF TRW

William H. Curry, Jr.
IEEE Compendium

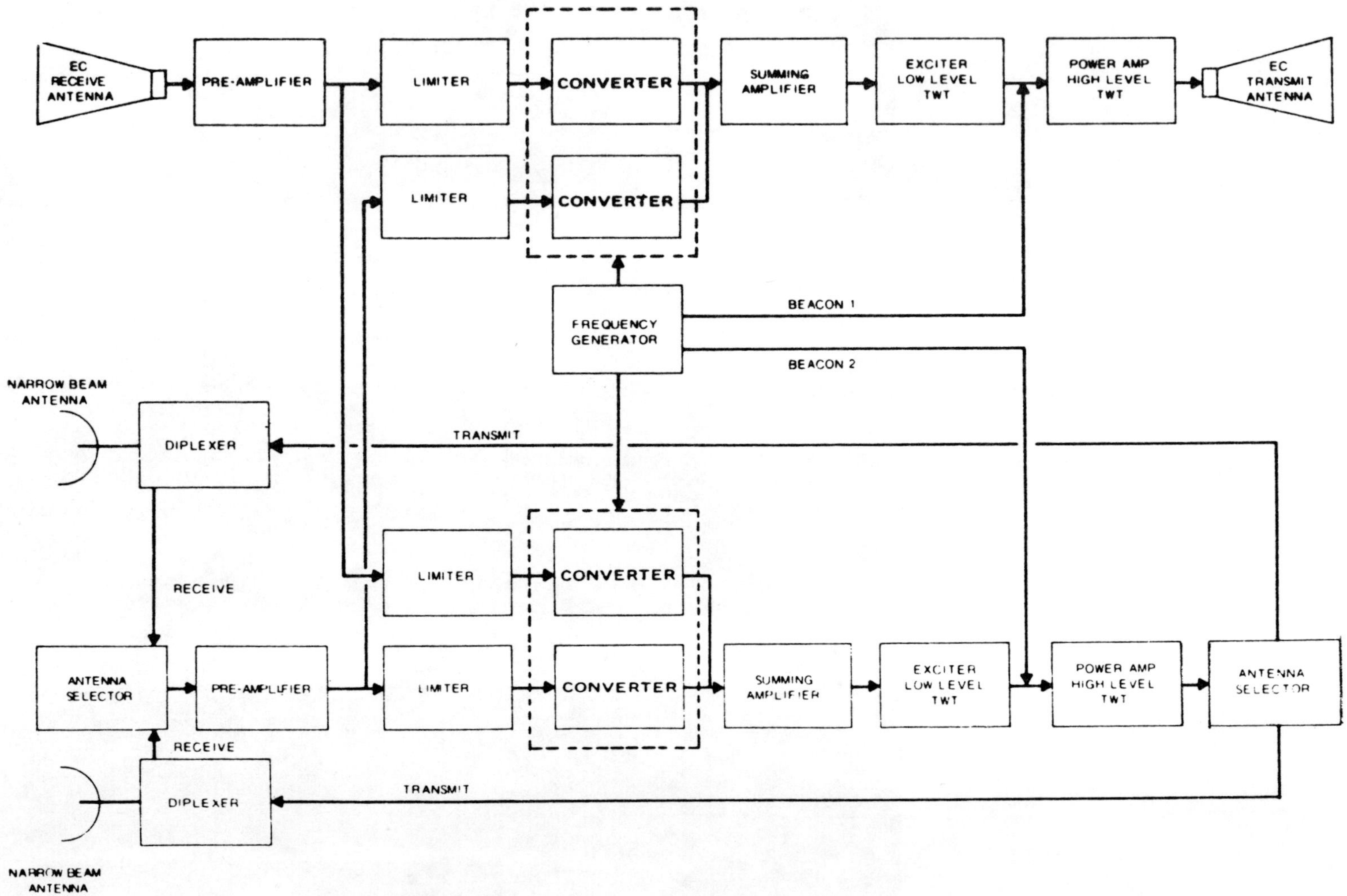

DSCS-II COMMUNICATION BLOCK DIAGRAM

William H. Curry, Jr.
IEEE Compendium

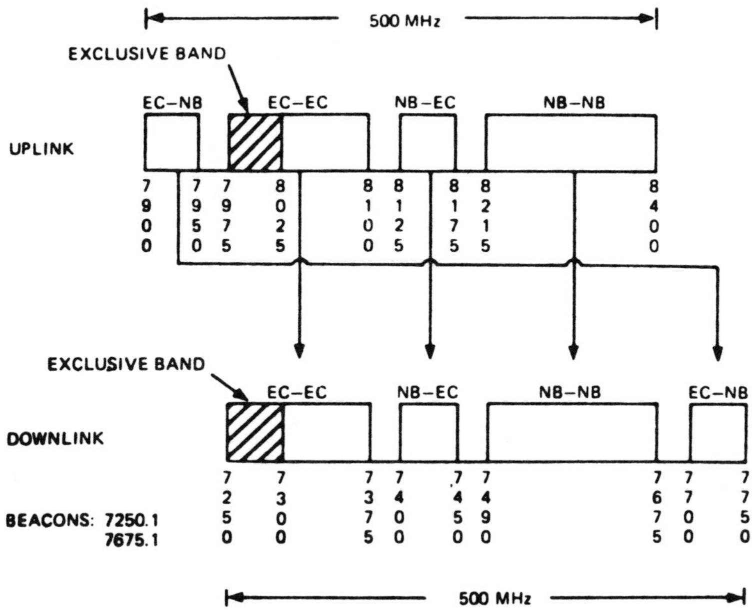

EC - Earth Coverage
NB - Narrow (Spot) Beam

DSCS-II FREQUENCY PLAN

General

No. built - **12**

Design life - 5 years.

Launch vehicle - TITAN IIIC
(2 spacecraft per launch)

Major contractor - TRW

Date first launch - Nov 71

Physical description

Stabilization - Spinning drum with earth-oriented antenna platform.

Cylinder diameter - 274.3cm (108 in).

Cylinder height - 396.2cm (156 in).

Weight (launch and in-orbit) - approximately 545 kg (1,200 lbs).

Antenna configuration

Global - 1-transmit
1-receive

Spot - 1-west, 1-east transmit/receive

Polarization - circular

Communications capability

Usable bandwidth - 390 MHz

Frequency range -
up-link (receive) 7900 - 8400 MHz
down-link (transmit) 7250 - 7750 MHz

Repeaters

No. on line - two.

No. redundant - two.

Type - TWT

Bandwidth - 200 MHz

RF power - 20W (13 dBW)

e.i.r.p.
global beam - 28 dBW
1 narrow beam on - 43 dBW
2 narrow beams on - 40 dBW

Receivers

Type - Tunnel diode

Noise figure - 7 dB

Communications capacity

Approximately 100 Mbs digital data
or
equivalent of 1300 two-way voice circuits.

DSCS-II TECHNICAL CHARACTERISTICS

William H. Curry, Jr.
IEEE Compendium

DSCS-III

PHOTO COURTESY OF
GENERAL ELECTRIC CO.

Richard W. Duesing
IEEE Compendium

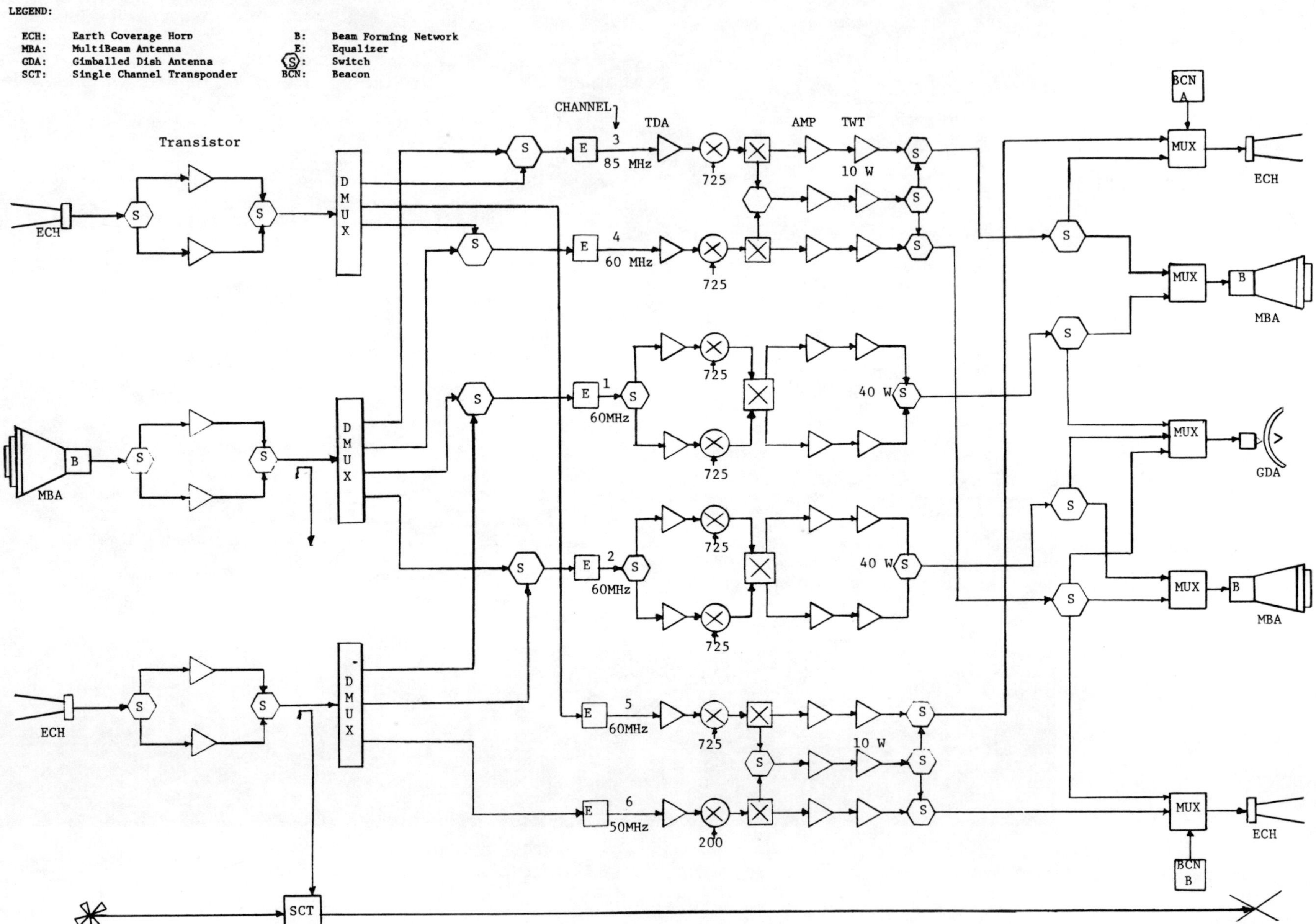

DSCS-III COMMUNICATIONS BLOCK DIAGRAM

Richard W. Duesing
IEEE Compendium

General

No. Satellites to be Built	2 Development Flight Models 10-12 Production Models
Design Life	10 years
Launch Vehicle	Titan IIIC, Titan 34D, or Shuttle/IUS
Major Contractor	General Electric
Scheduled Date of 1st Launch	1981

Physical Description

Stabilization	3-Axis Stabilized, 0.08° accuracy in pitch & roll, 0.8° in yaw, 0.2° antenna pointing accuracy
Length	207.01 cm (81.5 in)
Width	193.04 cm (76 in)
Depth	195.58 cm (77 in)
Each Solar Array Length (with Yoke)	484.63 cm (190.8 in)
Weight	
Dry	719.2 kg (1600 lbs)
Propulsion System Capacity	269.7 kb (600 lbs)

Antenna Configuration*

Transmit	2 MBAs : 28 in in aperture 19 beams, selective coverage 1 GDA : Reflector, 33 in diameter 2 ECH : earth coverage horns 1 SCT:crossed bow tie (UHF)
Receive	1 MBA : 45 in aperture, 61 beams, selective coverage and anti-jam protection 2 ECH : earth coverage horns 1 SCT : crossed dipole (UHF)

*MBA:Multibeam Antenna
GDA:Gimballed Dish Antenna
ECH:Earth Coverage Horn antenna
SCT:Single Channel Transponder

Polarization	All circular

Communication Capability

Total Usable Bandwidth	375 MHz	
Assigned Frequency Range		
Up-Link (Receive)	7900-8400 MHz	
Down-Link (Transmit)	7250-7750 MHz	
UHF Receive	300 to 400 MHz	
UHF Transmit	225 to 260 MHz	
Repeaters		
No. on Line	6	
No. Redundant	4	
Bandwidth	4:60 MHz, 1:85 MHz 1:50 MHz	
RF Power	4:40 Watts, 6:10 Watts	
Noise Figure	4.0 dB	
Available e.i.r.p. (nominal)	Earth Coverage	Narrow Coverage
CH1,2(MBA)	29.0 dBW	40.0 dBW
CH3,4(MBA)	23.0 dBW	34.0 dBW
CH3-6(ECH)	25.0 dBW	
CH1,2(GDA)	44.0 dBW	
CH4 (GDA)	37.5 dBW	
Receive G/T (Beam Edge)		
CH1-4(MBA)	-16 dB/K	-1 dB/K
CH3-6(ECH)	-15 dB/K	
Gain Control Range	39 dB	
Gain		
MBA	15.5 dB	27 dB
Horn	17.4 dB	
Narrow Coverage Capability		
MBA	1.0° Diameter Circle on Earth	
GDA	3.0° Diameter Circle on Earth	
Communication Capability	Ground/ship/airborne, protected service to WWMCCS, with large or small FDMA or TDMA terminals.	

DSCS-III TECHNICAL PARAMETERS

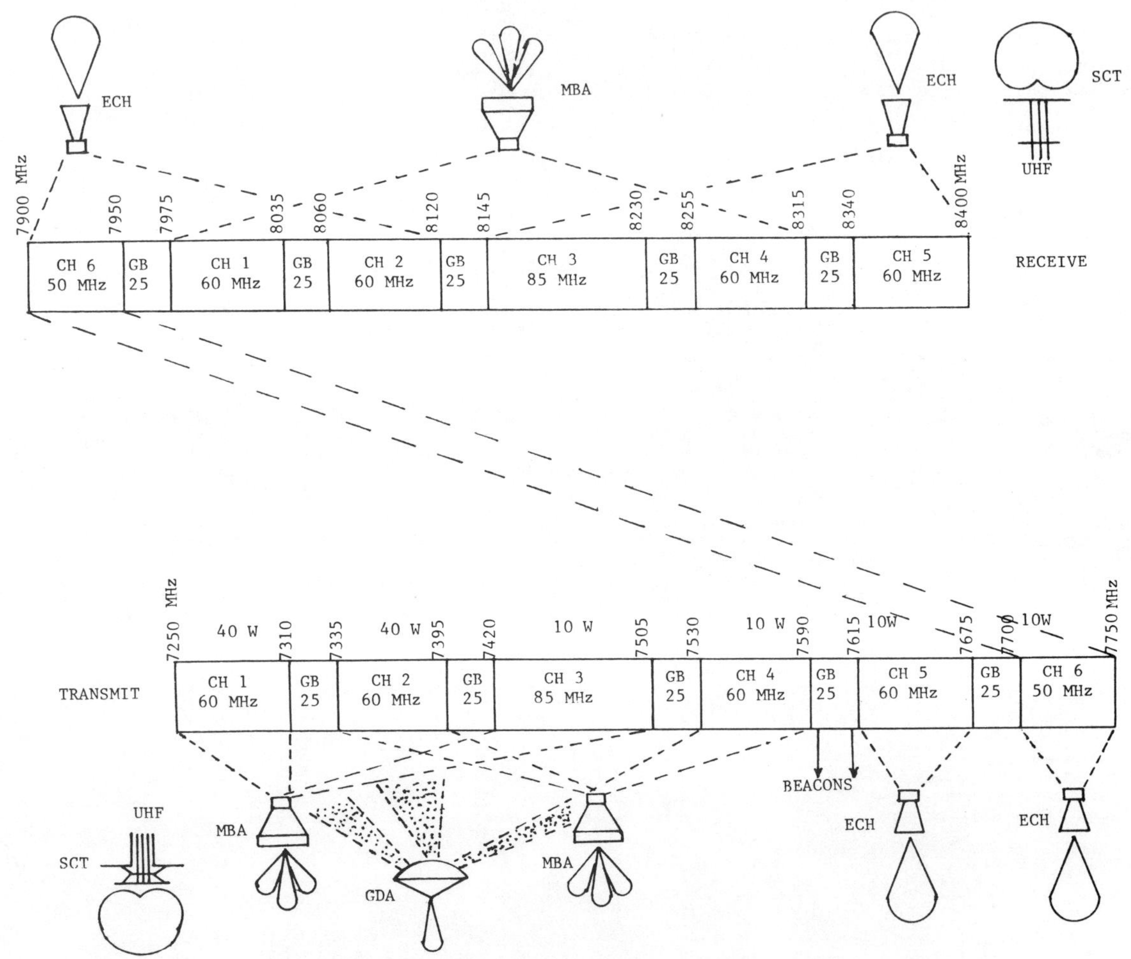

DSCS-III FREQUENCY PLAN

Richard W. Duesing
IEEE Compendium

FLTSATCOM

PHOTO COURTESY OF TRW

William H. Curry Jr.
IEEE Compendium

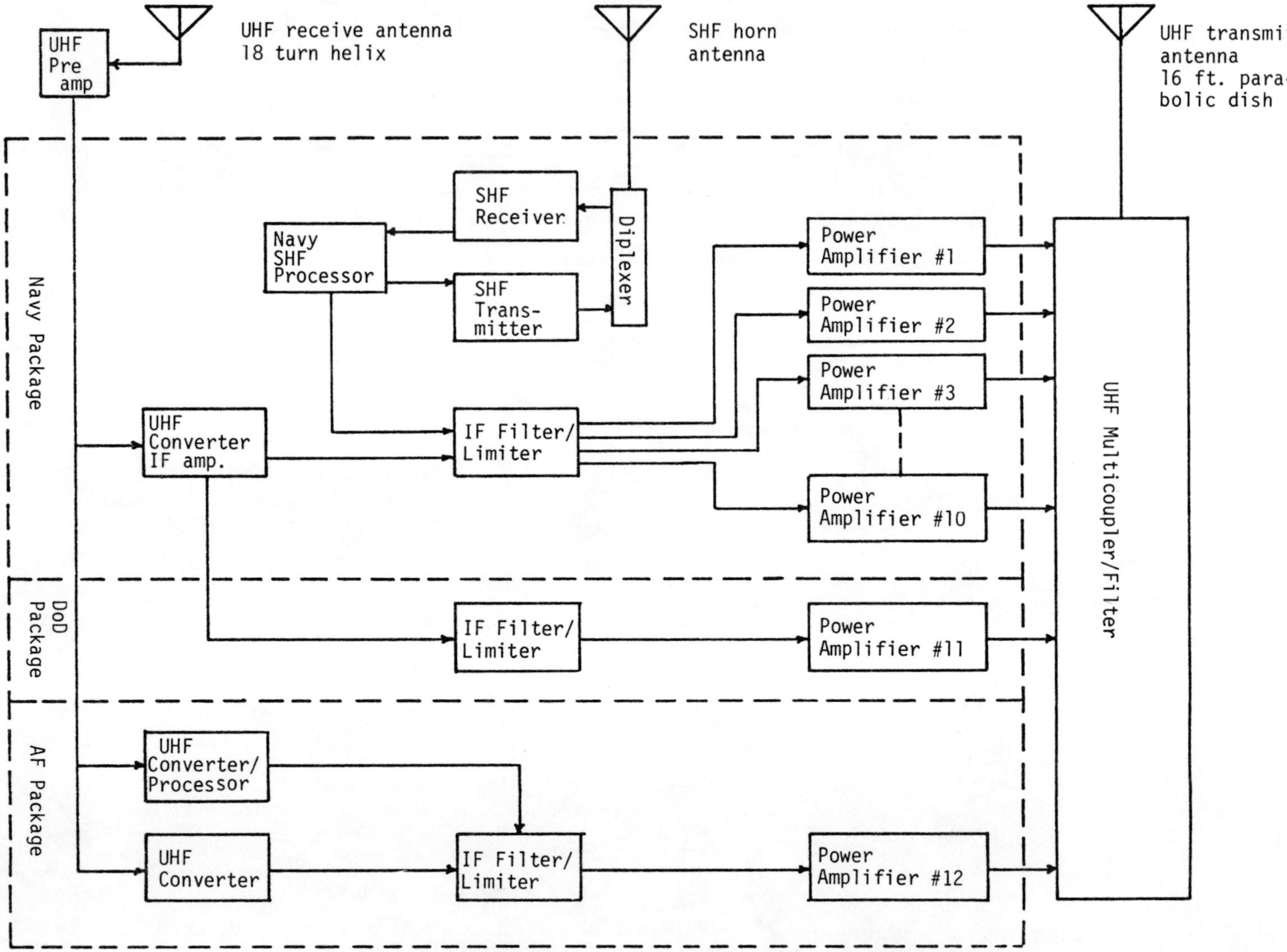

FLTSATCOM COMMUNICATION BLOCK DIAGRAM

William H. Curry, Jr.
IEEE Compendium

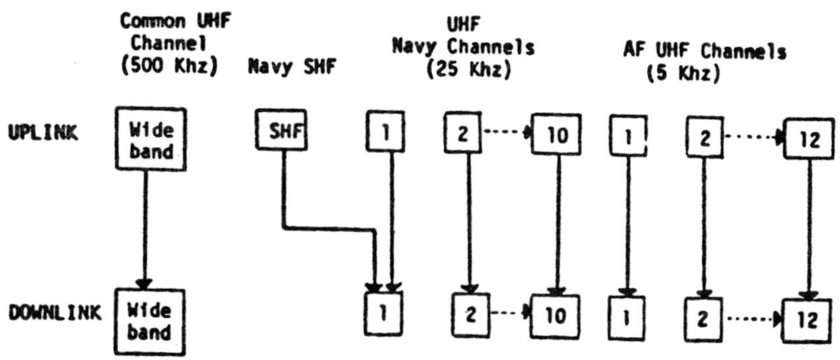

Note: FLEETSAT operates in 225-400 MHz military UHF band. Frequency assignments vary for each spacecraft to minimize interference.

FLTSATCOM FREQUENCY PLAN

General

No. spacecraft 4

Design life - 5 years.

Launch vehicle - Atlas Centaur

Major contractor - TRW

Date of first launch - **Feb 1978**

Physical description

Stabilization - 3-axis Body fixed reaction wheels

Body dimensions - Hexagon

Panel Width - 122cm (48 in)

Panel Height - 163cm (64 in)

Weight
Launch - 1818 kg (4000 lbs)
On-Orbit - 909 kg (2000 lbs)

Antenna configuration

Global

UHF transmit - 488cm (192 in) parabola

UHF receive - 18 turn helical

Communications capability

Total usable bandwidth - 800 kHz

Repeaters-12

Type - Solid state

Capacity - Varies over approximately 40 dB dynamic range depending on earth terminal mix and operating modes.

FLTSATCOM TECHNICAL CHARACTERISTICS

GALS 1 THROUGH 4 SERIES*

("Tack")

*To the author's knowledge, a picture of this satellite has not been released.

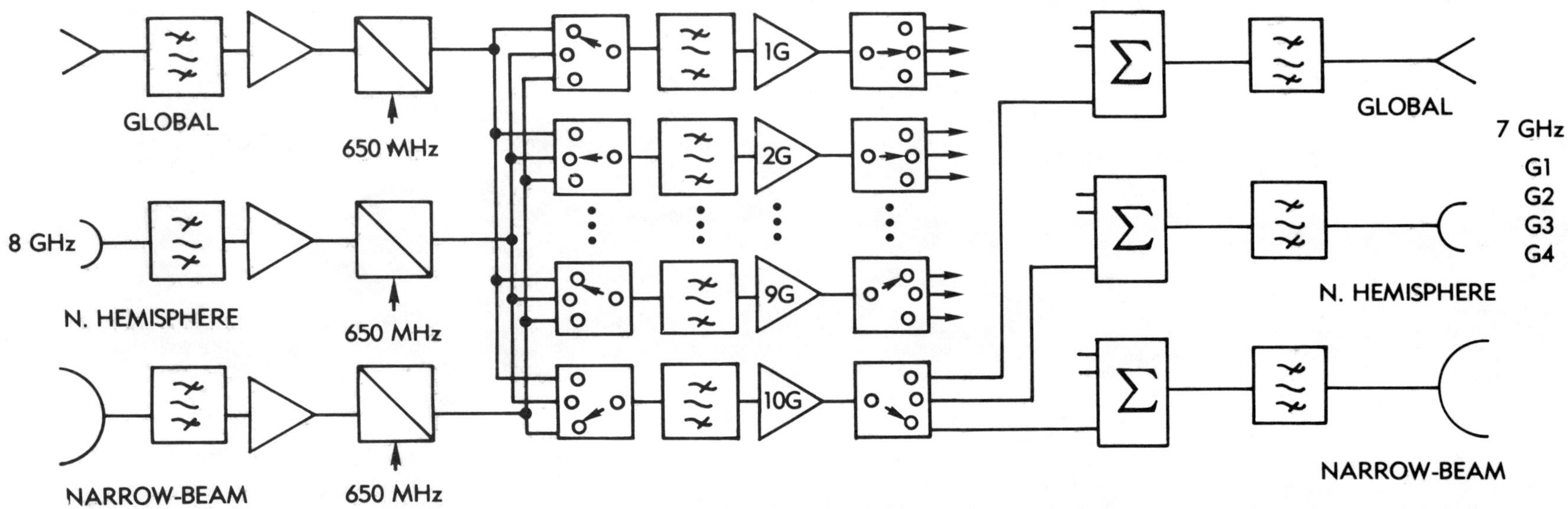

GALS 1 THROUGH 4 COMMUNICATION BLOCK DIAGRAM

Walter L. Morgan
IEEE Compendium

GALS 1 THROUGH 4, SPACE SEGMENT

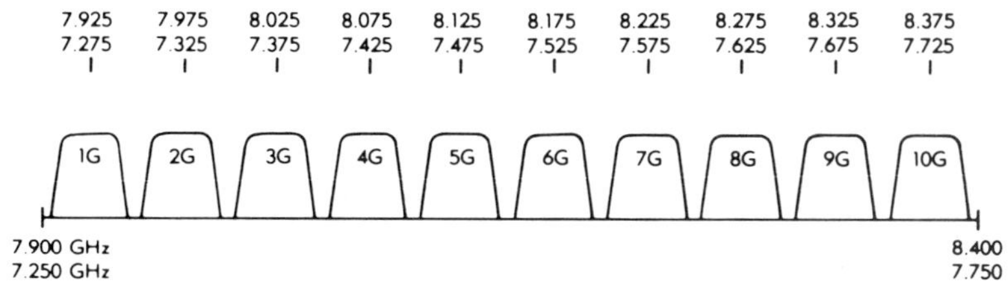

General (to 3/1/80)

The GALS system, which consists of four networks, each with one geostationary satellite and a complex of earth stations, is designed to provide government services.

No. Launched to Date	0 *
Longitude Locations	
GALS 1	335° ± 0.5°E
GALS 2	45° ± 0.5°E
GALS 3	85° ± 0.5°E
GALS 4	190° ± 0.5°E
Inclination Control	±0.5°

Antenna Configuration

Global	19 dBi
Northern Hemisphere	23 dBi
Narrow Beam	35 dBi

Coverage Provided

Antenna Beam	Global	Northern Hemisphere	Narrow Beam
GALS 1	Mexico-Iran	Chicago-Moscow	Mid-North Atlantic
GALS 2	Europe-Hong Kong	Europe-Hong Kong	P.R. China
GALS 3	East Europe-Australia	East Europe-Japan	P.R. China
GALS 4	East China-Mexico	East USSR-Denver	Mid-North Atlantic

Communications Capability

Assigned Frequency Range	
Up-Link (receive)	7.90-8.40 GHz
Down-Link (transmit)	7.25-7.75 GHz
No. of On-Line Repeaters	10
Spectral Power Density Radiated (maximum)	-40 dBW/Hz
Receive G/T (nominal for each beam)	
Beam Edge	-17, -14, +2.77 dB/K
Beam Center	-14, -11, +0.23 dB/K
System Noise Temperature (minimum)	2000, 2500, 3000K

GALS 1 THROUGH 4 FREQUENCY PLAN AND TECHNICAL PARAMETERS

*Gorizont contains elements of GALS equipment at 8/7 GHz.

IDCSP

PHOTO COURTESY OF
FORD AEROSPACE AND
COMMUNICATIONS CORP.

Khodadad Betaharon
IEEE Compendium

IDCSP

PHOTO COURTESY OF
FORD AEROSPACE AND
COMMUNICATIONS CORP.

Khodadad Betaharon
IEEE Compendium

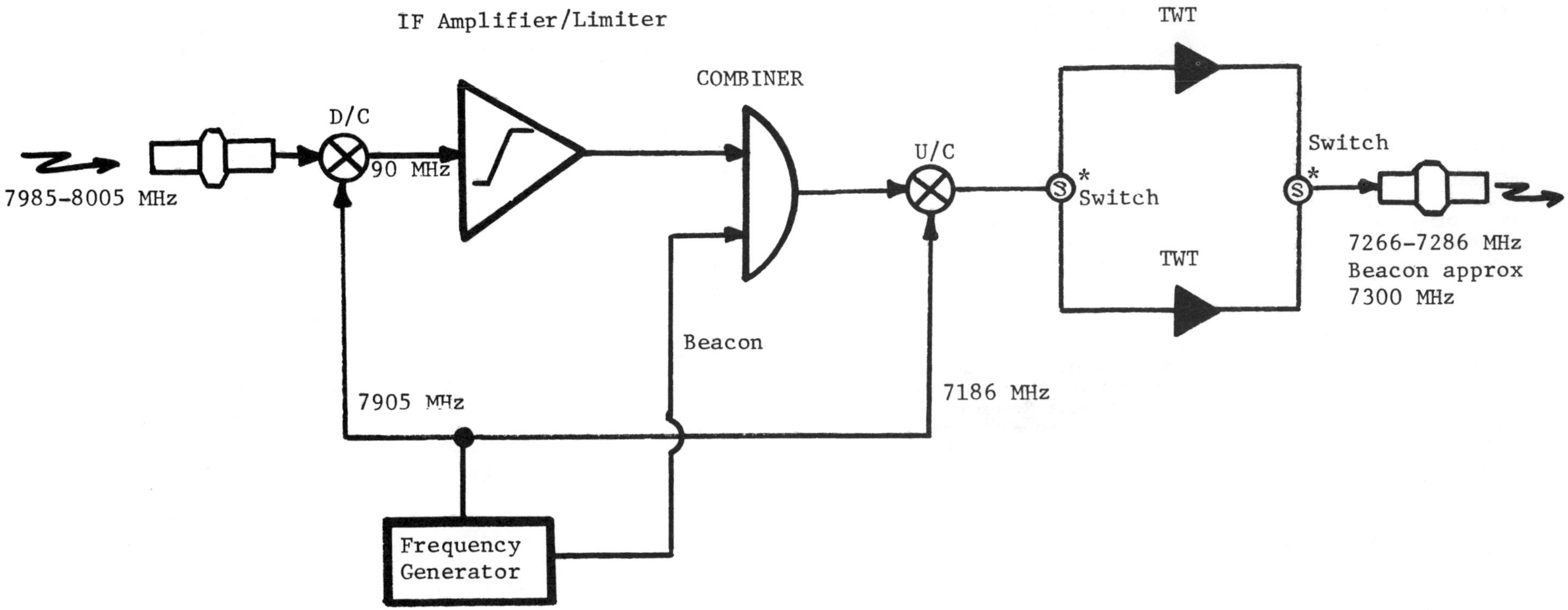

Note:

1. *Switch-over to redundant TWT occurs automatically in case of TWT failure. Switch can only occur once.
2. An automatic power shut-off circuit activates 6 years after launch.
3. Telemetry is carried on a separate antenna at approx 400 MHz.

INITIAL DEFENSE COMMUNICATIONS SATELLITE PROGRAM (IDCSP) COMMUNICATIONS BLOCK DIAGRAM

General

No. Satellites Built	38 (8 per launch)
Design Life	3 years (goal)*
	1.5 years (required)
Launch Vehicle	Titan IIIC
Major Contractor	Philco-Ford (DOD/DCA/ SAMSO)
Date of 1st Launch	16 June 1966

Physical Description

Stabilization	Spin Stabilized (Despun Biconal Horns)
Cylinder Diameter	91.4 cm (36 in)
Cylinder Height	81.3 cm (32 in)
Weight	45 kg (100 lbs)
Batteries	None

Antenna Configurations

Earth Coverage	One Transmit (LHCP) and One Receive (RHCP) both are: Biconal Hours- Toroidal 360^ox28^o with 3-5 dBi gain

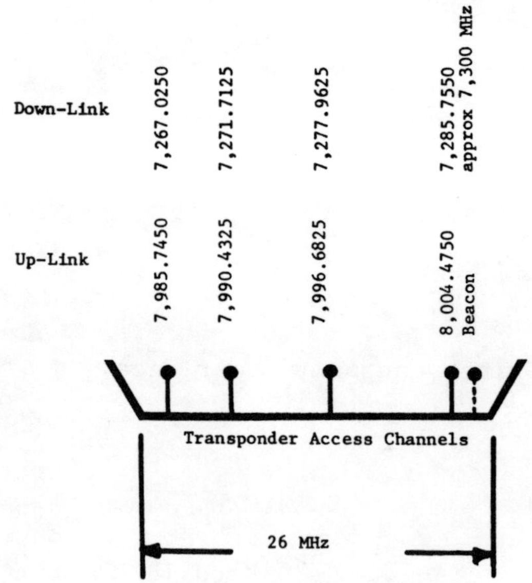

IDCSP FREQUENCY PLAN

Communication Capability

Total Usable Bandwidth	26 MHz
Assigned Frequency Range	
Up-Link (Receive)	7985-8005 MHz
Down-Link (Transmit)	7266-7286 MHz
Repeaters	
No. On Line	One
No. Redundant	One
TWT Power	3 watt
e.i.r.p.	7 dBW (min.)
Receiver/System Noise Figure	10 dB

Communication Capability

FDM/FM, Pseudonoise (PN) DPSK and MFSK	Up to 5 Commercial Quality Voice or Up to 11 Tactical Quality Voice or 1550 teletype (for improved link terminal configuration)
PN for antijam	

*Satellites drift randomly (depending on exact altitude) from West to East at approximately 30° per day. A single satellite stays within view of a ground station for about 4.5 days.

IDCSP TECHNICAL PARAMETERS

Khodadad Betaharon
IEEE Compendium

LEASAT

PHOTO COURTESY OF
HUGHES AIRCRAFT CORP.

Gerald L. Dutcher
IEEE Compendium

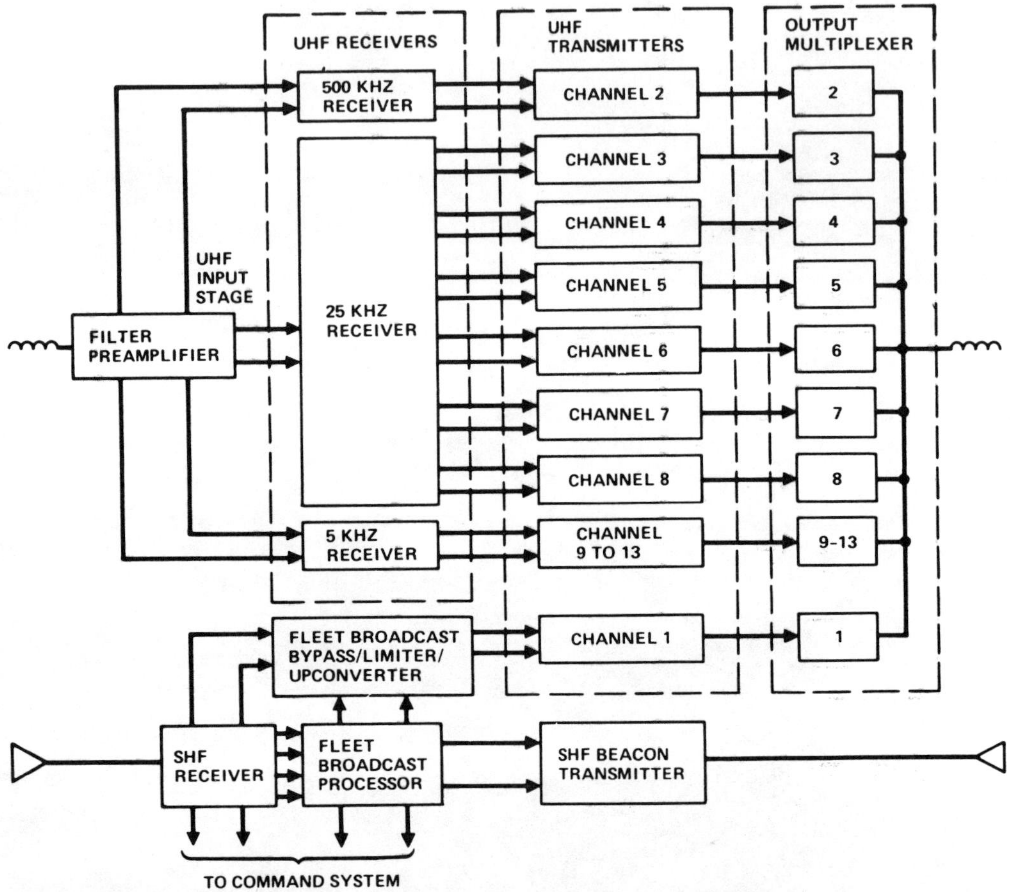

LEASAT COMMUNICATION BLOCK DIAGRAM

Gerald L. Dutcher
IEEE Compendium

General	
No. of Satellites	4 in orbit, 1 ground spare
Design Life	10 years
Launch Vehicle	STS
Major Contractor	Hughes
Planned Date of 1st Launch	1982
Physical Description	
Stabilization	Dual Spin (Despun Antenna Earth Oriented, Spinning Drum)
Cylinder Diameter	422 cm (166 in)
Cylinder Height	290 cm (114 in)
Height w/Antenna Mast	616 cm (243 in)
Weight	
At Separation From STS	6898 kg (15175 lbs)
On Orbit (Dry)	1234 kg (2715 lbs)
Antenna Configuration (Global)	
UHF	1 Transmit Helix 1 Receive Helix
SHF	1 Transmit Horn 1 Receive Horn
Polarization	
UHF	RHC
SHF	
Receive	RHC
Transmit	LHC

Communication Capability	
Frequency Range	
Up-Link (Receive)	
UHF	293-318 MHz
SHF	8000 MHz
Down-Link (Transmit)	
UHF	243-266 MHz
SHF	7300 MHz
Repeaters	
Relay	
No. of Channels	6
Bandwidth	25 kHz
e.i.r.p.	26 dBW
G/T	-18 dB/K
Wideband	
No. of Channels	1
Bandwidth	500 kHz
e.i.r.p.	28 dBW
G/T	-18 dB/K
Narrowband	
No. of Channels	5
Bandwidth	5 kHz
e.i.r.p.	16.5 dBW
G/T	-18 dB/K
Fleet Broadcast	
No. of Channels	1
Bandwidth	Onboard Processing
e.i.r.p.	26 dBW
G/T	-20 dB/K
Redundancy	Two for one on all electronics units

LEASAT TECHNICAL PARAMETERS

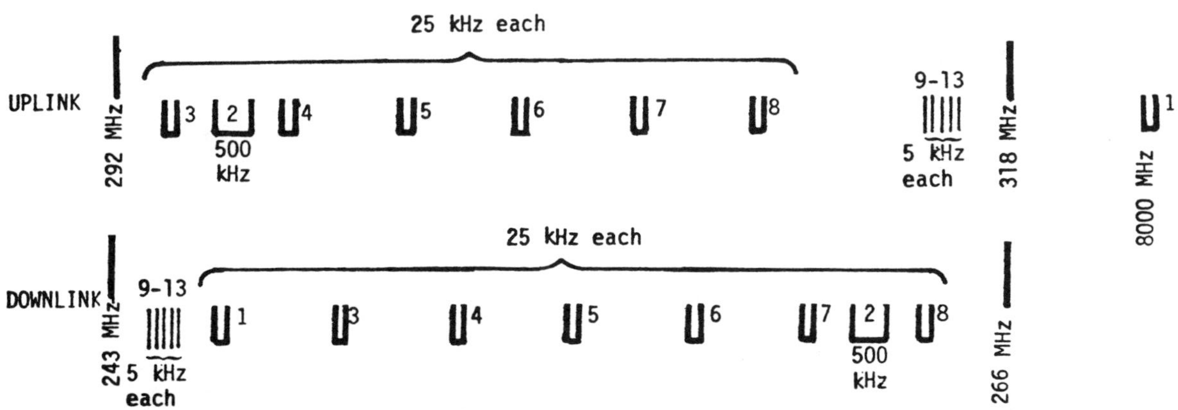

LEASAT FREQUENCY PLAN

Gerald L. Dutcher
IEEE Compendium

LES 1 or 2 (IDENTICAL)

PHOTO COURTESY OF
LINCOLN LABORATORY

William W. Ward
IEEE Compendium

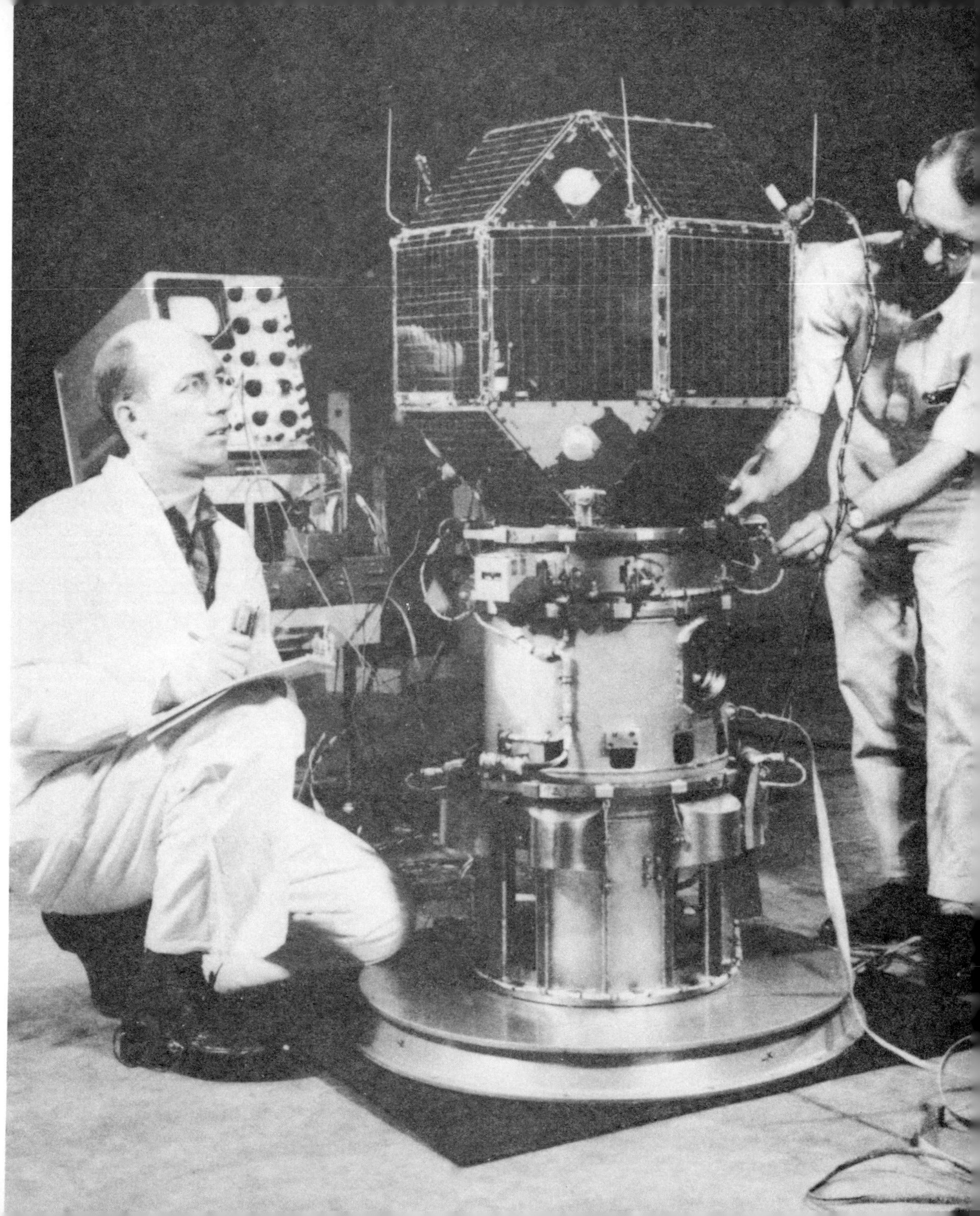

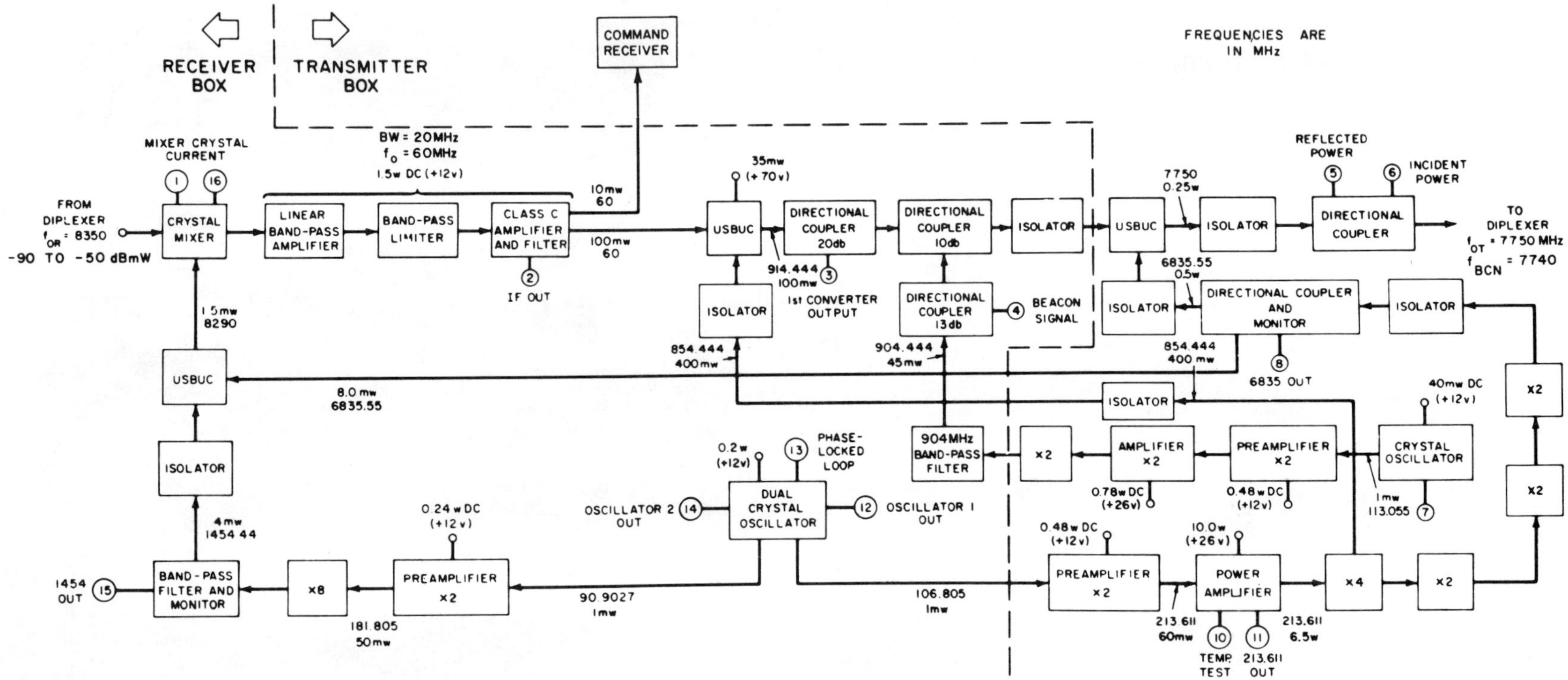

LES 1, 2, 4
COMMUNICATION BLOCK DIAGRAM

William W. Ward
IEEE Compendium

General

Number built	2
Design life	2 years (timer)
Launch vehicle	Titan III-A
Major contractor	MIT Lincoln Labratory
Dates of launch	11 February 1965 (LES-1) 6 May 1965 (LES-2)
Orbits	Circular, 2800-km (1500-nm) altitude, 32° inclination (LES-1) Elliptical, 2800-km (1500-nm) perigee, 15,000-km (8000-nm) apogee, 32° inclination (LES-2)
SDC object number	1002 (LES-1) 1360 (LES-2)
COSPAR designation	1965-8C (LES-1) 1965-34B (LES-2)

Physical Description

Stabilization	Spinning at 180 rpm (around axis of AKM) LES-1's AKM was not fired LES-2's AKM was fired Autonomous magnetic stabilization system
Shape	26-sided polyhedron 18 square faces carrying solar cells 8 triangular faces carrying sensors and antennas
Size	About 60 cm (2 ft) between opposite faces
Weight	37 kg (82 lb)

Antenna Configuration

SHF reception and transmission	8 lens-horn antennas on the 8 triangular faces Selection of beam (octant) by autonomous control system
UHF transmission	4 short stubs on the 4 triangular faces away from the AKM (toroidal antenna pattern)

Polarization

Receive communications (SHF)	RHCP
Transmit communications (SHF)	LHCP
Transmit telemetry (UHF)	Linear

Communication Capability

Total usable bandwidth	20 MHz
Assigned center frequency	
Uplink (receive)	8350 MHz
Downlink (transmit)	7750 MHz
Repeaters	1
Bandwidth	20 MHz
Beacon signal	CW tone at 7740 MHz, 17 dB below downlink
RF power (BOL)	-9 dBW (120 mW) total (downlink, beacon, and intermods)
Telemetry signal	237.000 MHz biphase-modulated at 100 bit/sec
Available EIRP (maximum)	-7 dBW (200 mW) total (downlink, beacon, and intermods)
Receive G/T (maximum)	-37 dB/K

Power

Solar cells on polyhedron faces, 25 W maximum (BOL)
No eclipse power

Status

On-board timer (set for 21 January 1967) failed to turn LES-1 OFF. LES-1 is believed to have ceased transmission in June 1967.

On-board timer (set for 2 May 1967) worked, turning LES-2 OFF on schedule.

Reference

"Lincoln Experimental Satellite Program (LES-1, -2, -3, -4)", H. Sherman, D. C. MacLellan, R. M. Lerner, P. Waldron, Journal of Spacecraft & Rockets, 4(1967)1448-1452

LES 1 or 2
TECHNICAL PARAMETERS

LES 3

PHOTO COURTESY OF
LINCOLN LABORATORY

William W. Ward
IEEE Compendium

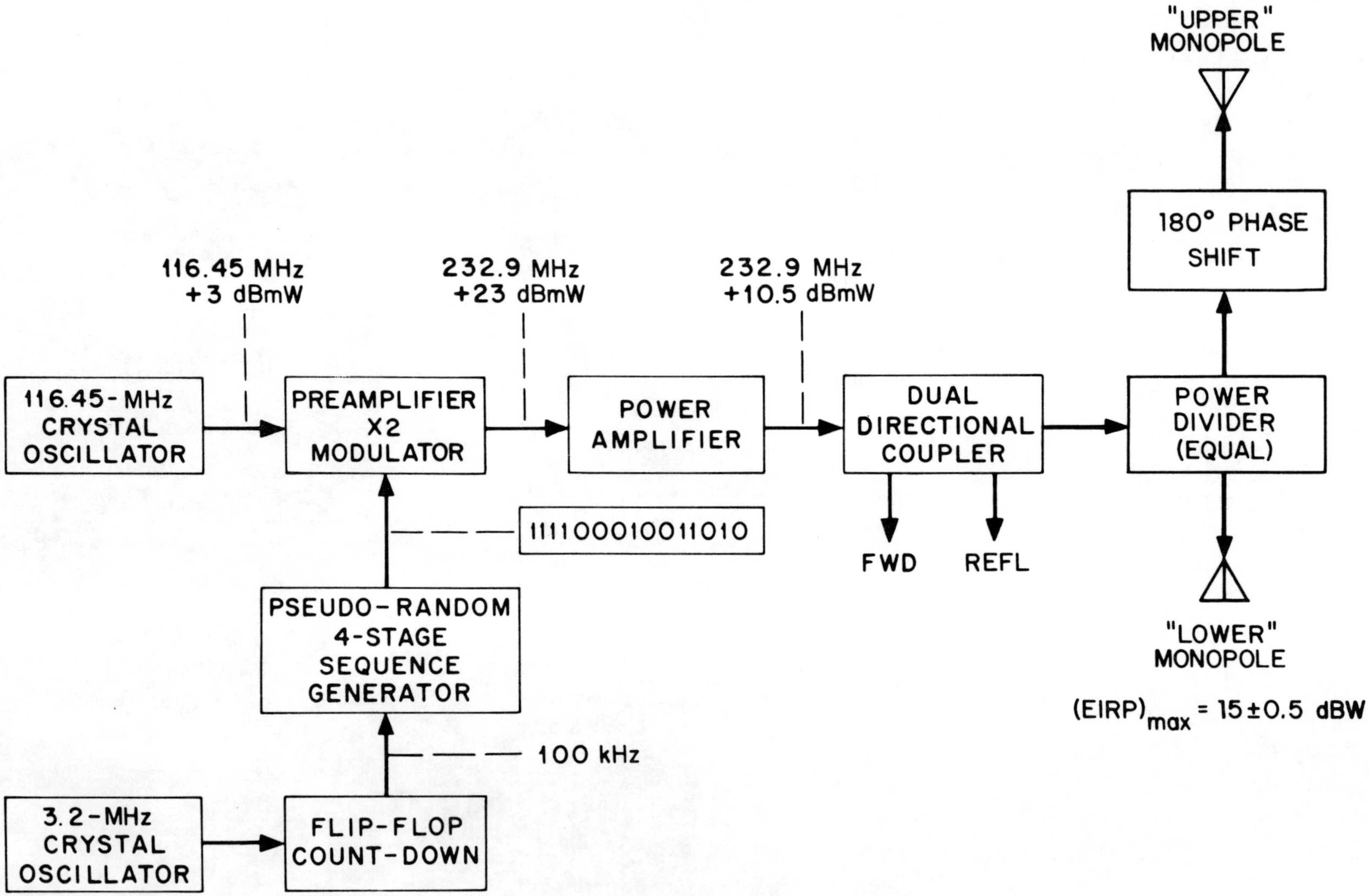

LES-3 BEACON
COMMUNICATION BLOCK DIAGRAM

William W. Ward
IEEE Compendium

General

Number built	1
Design life	1 year (timer)
Launch vehicle	Titan III-C
Major contractor	MIT Lincoln Laboratory
Date of launch	21 December 1965
Intended orbit	Circular, subsynchronous, equatorial
Actual orbit	Booster's transfer ellipse
Initially	195-km (105-nm) perigee 33,700-km (18,200-nm) apogee 26° inclination
SDC object number	1941
COSPAR designation	1965-108D

Physical Description

Stabilization	Spinning at 137 rpm (axis was intended to be perpendicular to orbit plane)
Shape	26-sided polyhedron 18 square faces carrying solar cells 8 blank triangular faces
Size	About 60 cm (2 ft) between opposite faces (exclusive of antennas) About 120 cm (4 ft) antenna tip to antenna tip
Weight	15.5 kg (34.1 lb)

Antenna Configuration

Quarter-wave monopoles (stubs) extending from opposite faces, fed equally and in phase opposition

Monopoles lie along the satellite's spin axis

Toroidal antenna pattern

Polarization

Transmit only	Linear

Communication Capability

Signal generator (beacon) for studies of multipath propagation

Carrier at 232.9 MHz biphase-modulated by 15-bit maximal-length shift-register sequence at 100 kbit/sec

Available EIRP (maximum) 15 dBW (30 W)

No telemetry

No commands

Beacon turned ON automatically at orbital insertion

Beacon intended to be turned OFF by timer in November, 1966 (after about 1 year in orbit). Timer failed to work.

Power

Solar cells on polyhedron faces, 25 W maximum (BOL)

No eclipse power

Status

Believed to have burned up due to reduction in perigee of eccentric orbit

Reference

"Lincoln Experimental Satellite Program (LES-1, -2, -3, -4)", H. Sherman, D. C. MacLellan, R. M. Lerner, P. Waldron, Journal of Spacecraft & Rockets, 4(1967)1448-1452

LES 3 TECHNICAL PARAMETERS

LES 4

PHOTO COURTESY OF
LINCOLN LABORATORY

William W. Ward
IEEE Compendium

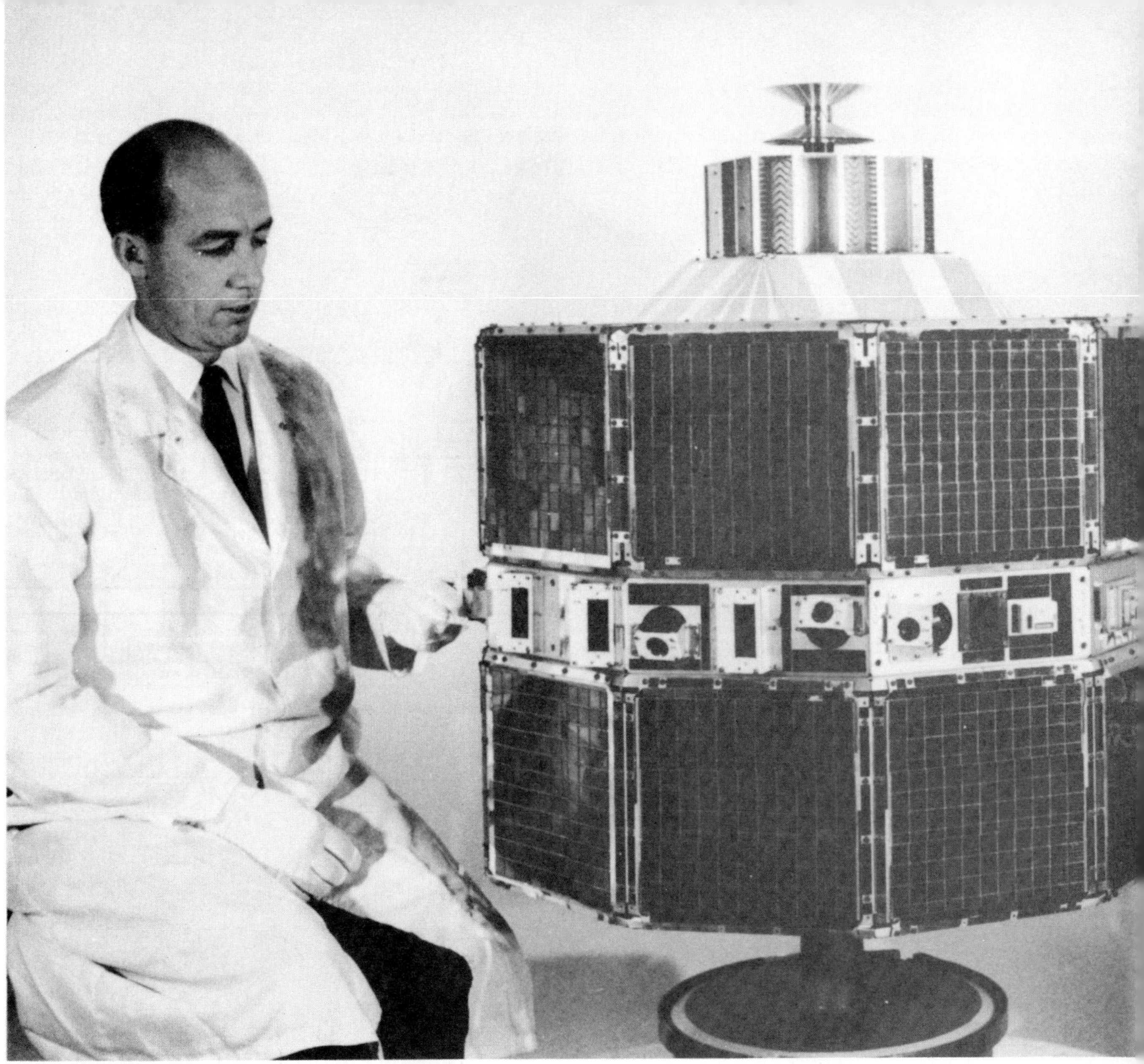

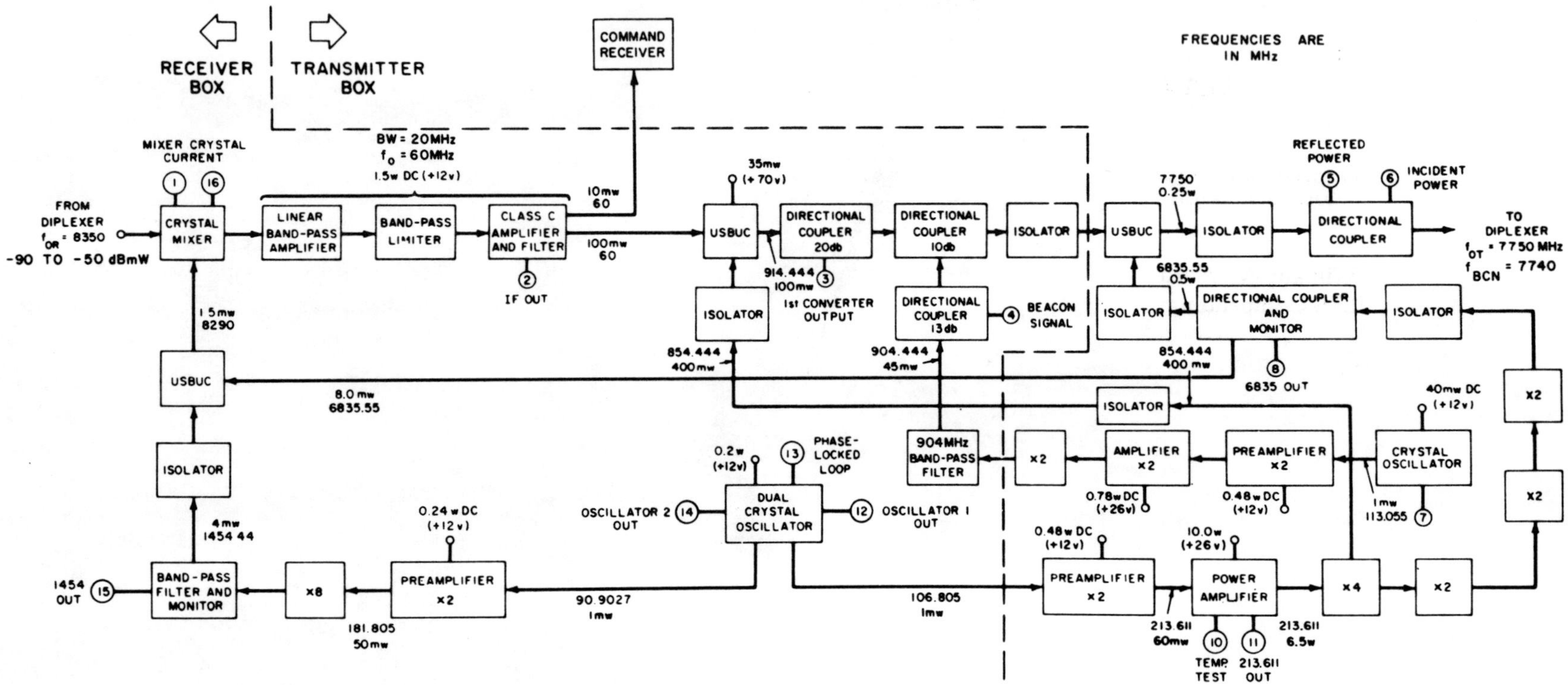

LES 1, 2, 4
COMMUNICATION BLOCK DIAGRAM

William W. Ward
IEEE Compendium

General

Number built	1
Design life	3 years (timer)
Launch vehicle	Titan III-C
Major contractor	MIT Lincoln Laboratory
Date of launch	21 December 1965
Intended orbit	Circular, subsynchronous, equatorial
Actual orbit	Booster's transfer ellipse
Initially	195-km (105-nm) perigee 33,700-km (18,200-nm) apogee 26° inclination
SDC object number	1870
COSPAR designation	1965-108 B

Physical Description

Stabilization	Spinning at 11 rpm (axis intended to be perpendicular to orbit plane) Autonomous magnetic stabilization system (for intended orbit)
Shape	10-sided cylinder (body) carrying solar cells and sensors X-band/UHF antenna system atop body
Size	
Body	About 63 cm (25 in.) high About 79 cm (31 in.) in diameter
Antenna assembly	About 25 cm (10 in.) square by 20 cm (8 in.) high
Overall height	About 83 cm (33 in.)
Weight	53 kg (116.5 lb)

Antenna Configuration

SHF reception	Biconical horn at top of antenna assembly (toroidal pattern)
SHF transmission	8 horns looking out around bottom of antenna assembly Electronic despin of beam by autonomous control system
UHF transmission	Center-fed dipole consisting of (a) antenna for SHF reception and (b) antenna for SHF transmission plus balance of spacecraft (toroidal pattern)

Polarization

Receive communications (SHF)	RHCP
Transmit communications (SHF)	LHCP
Transmit telemetry (UHF)	Linear

Communication Capability

Total usable bandwidth	20 MHz
Assigned center frequency	
Uplink (receive)	8350 MHz
Downlink (transmit)	7750 MHz
Repeaters	1
Bandwidth	20 MHz
Beacon signal	CW tone at 7740 MHz, 17 dB below downlink
RF power (BOL)	-7 dBW (210 mW) total (downlink, beacon, and intermods)
Telemetry signal	237.000 MHz biphase-modulated at 100 bit/sec
Available EIRP (maximum)	+3 dBW (2 W) total (downlink, beacon, and intermods)
Receive G/T (maximum)	-29 dB/K

Power

Solar cells on cylinder faces, 36 W intended minimum (BOL)
No eclipse power

Status

Believed to have burned up due to reduction in perigee of eccentric orbit

Reference

"Lincoln Experimental Satellite Program (LES-1, -2, -3, -4)", H. Sherman, D. C. MacLellan, R. M. Lerner, P. Waldron, Journal of Spacecraft & Rockets, 4(1967)1448-1452

LES 4
TECHNICAL PARAMETERS

William W. Ward
IEEE Compendium

LES 5

PHOTO COURTESY OF
LINCOLN LABORATORY

William W. Ward
IEEE Compendium

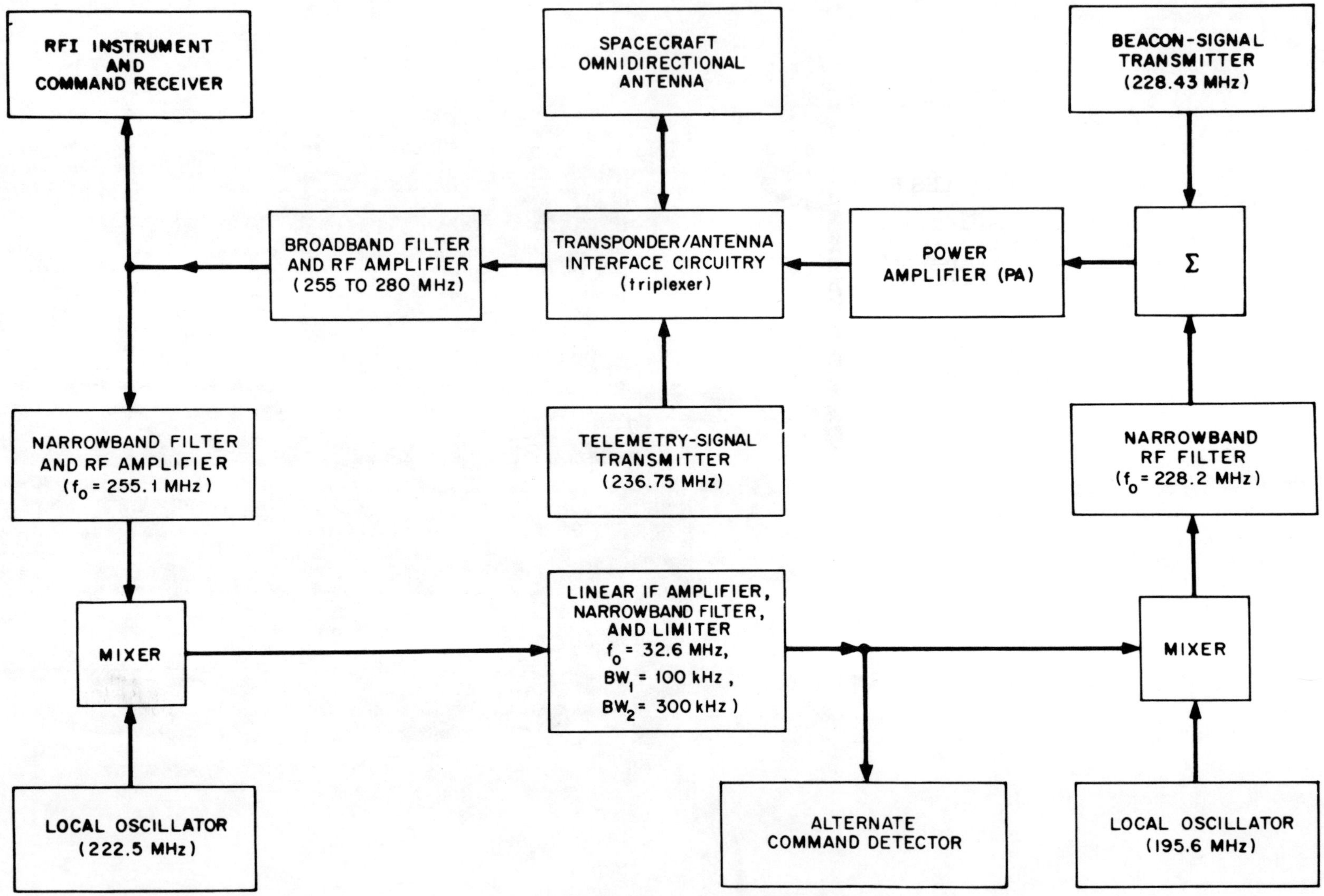

LES 5
COMMUNICATION BLOCK DIAGRAM

William W. Ward
IEEE Compendium

General

Number built	1
Design life	5 years (timer)
Launch vehicle	Titan III-C
Major contractor	MIT Lincoln Laboratory
Date of launch	1 July 1967
Orbit	Circular, subsynchronous (33°/day drift) 7° inclination (initially)
SDC object number	2866
COSPAR designation	1967-66E

Physical Description

Stabilization	Spinning at 10 rpm (axis perpendicular to orbit plane) Autonomous magnetic erection of spin axis
Cylinder diameter	122 cm (48 in.) excluding antennas
Cylinder height	163 cm (64 in.)
Weight	94 kg (207.5 lb)

Antenna Configuration

8 pairs of colinear half-wave dipoles deployed parallel to the cylinder in orbit
8 cavity-backed axial slots (each slot in 2 parts, above and below the view band)
Toroidal pattern

Polarization

Receive communications	RHCP
Transmit communications	RHCP
Transmit beacon	RHCP
Transmit telemetry	RHCP

Power

Solar cells on cylinder	136 W maximum (BOL)
No eclipse power	

Communication Capability

Total usable bandwidth	300 kHz or 100 kHz
Assigned center frequency	
Uplink (receive)	255.1 MHz in both modes
Downlink (transmit)	228.2 MHz in both modes
Repeaters	1
Bandwidth	300 kHz or 100 kHz
Beacon signal	228.43 MHz biphase-modulated at 800 bit/sec, 11 dB below downlink
RF power (BOL)	15.5 dBW (35.5 W) total (downlink, beacon, and intermods)
Telemetry signal	236.750 MHz biphase-modulated at 100 bit/sec (separate transmitter)
Available EIRP (nominal) on satellite equator (BOL)	
Downlink	16.3 dBW (43 W)
Beacon	5.4 dBW (3.5 W)
Telemetry	-3.3 dBW (0.47 W)
Receive G/T (nominal) on satellite equator (BOL)	-26 dB/K
Spectrum-scanning capability	RFI-measurement instrument scans from 283 to 253 MHz in 256 120-kHz steps over 11 minutes

Status

On-board timer (set for April 1972) failed to turn LES-5 OFF.
G/T degrades by 17 dB around aphelion, recovers around perihelion.
Communication/beacon transmitter failed in May 1970.
Telemetry transmitter is ON.

References

"Lincoln Experimental Satellites 5 and 6", D. C. MacLellan, H. A. MacDonald, P. Waldron, H. Sherman, AIAA Paper 70-494, Progress in Astronautics and Aeronautics, 26(1971)375-398

"Lincoln Experimental Satellite-5 (LES-5) Transponder Performance in Orbit", W. W. Ward, B. E. Nichols, Lincoln Laboratory TN 1968-18, 1 November 1968

LES 5
TECHNICAL PARAMETERS

LES 6

PHOTO COURTESY OF
LINCOLN LABORATORY

William W. Ward
IEEE Compendium

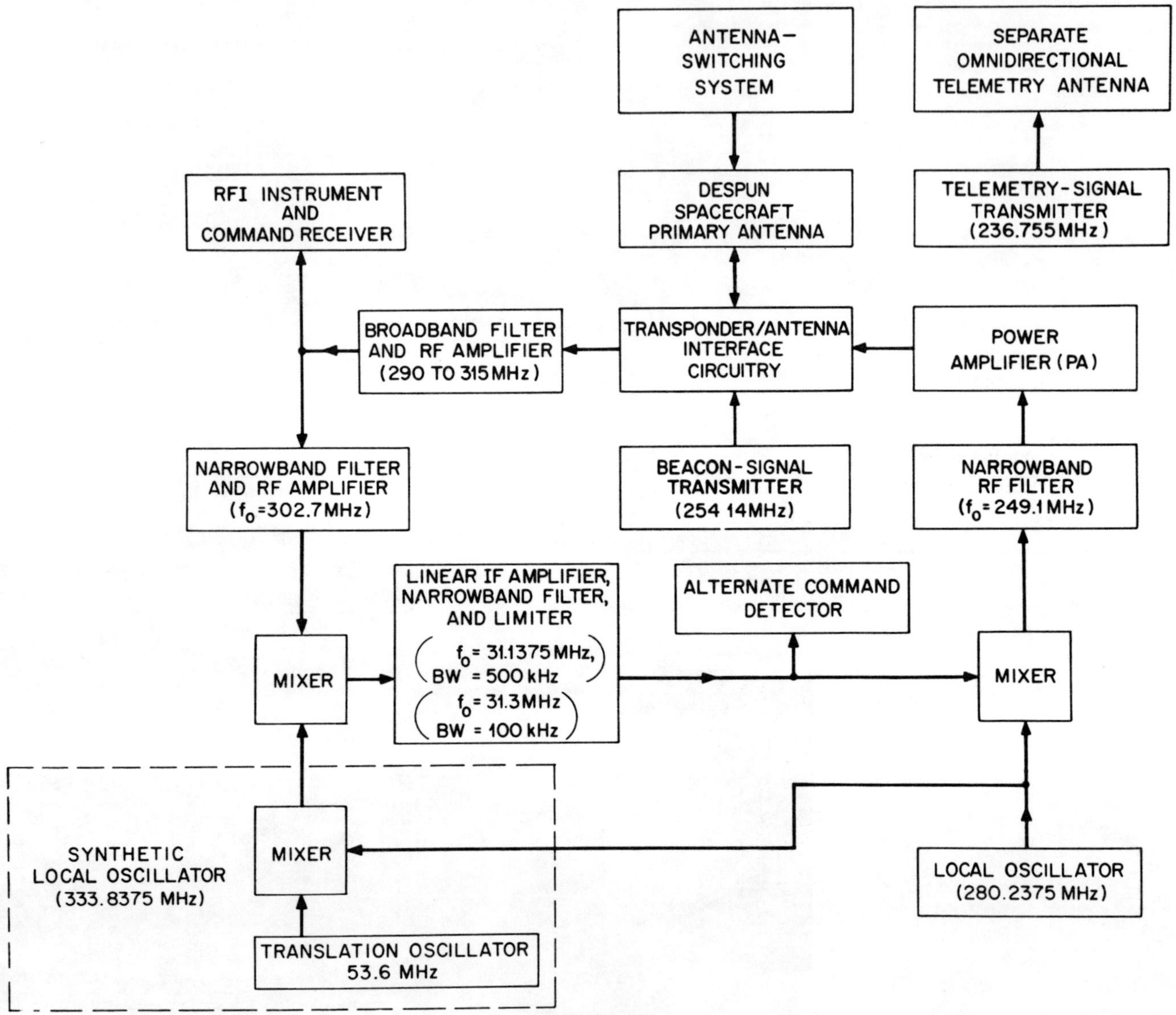

LES 6
COMMUNICATION BLOCK DIAGRAM

William W. Ward
IEEE Compendium

LES 6 TECHNICAL PARAMETERS

General

Number built	1
Design life	Indefinitely long
Launch vehicle	Titan III-C
Major contractor	MIT Lincoln Laboratory
Date of launch	26 September 1968
Orbit	Circular, synchronous, 3° inclination (initially)
SDC object number	3431
COSPAR designation	1968-81D

Physical Description

Stabilization	Spinning at 8 rpm (axis perpendicular to orbit plane) Erection of spin axis by commanded gas thrusting Autonomous magnetic-stabilization system available
Cylinder diameter	122 cm (48 in.) excluding communications antennas
Cylinder height	163 cm (64 in.) excluding telemetry antenna
Weight	169.4 kg (372.7 lb) including 5.1 kg (11.3 lb) of ammonia propulsion fuel

Antenna Configuration

Communications & beacon	8 pairs of colinear half-wave dipoles rigidly mounted parallel to the cylinder 8 cavity-backed axial slots (each slot in 2 parts, above and below the view band) Pairs of slots and pairs of dipoles are excited to form a directive beam in 1 of 16 positions Electronic de-spin of beam by autonomous control system
Telemetry	Quarter-wave monopole (along spin axis) above ground plane covering one end of cylinder (toroidal pattern)

Polarization

Receive communications	RHCP
Transmit communications	RHCP
Transmit beacon	RHCP
Transmit telemetry	RHCP

Power

Solar cells on cylinder	220 W maximum (BOL)
Limited eclipse power	Keep-alive batteries

Communication Capability

Total usable bandwidth	500 kHz or 100 kHz
Assigned center frequency	
Uplink (receive)	302.7 MHz in 500-kHz mode 302.5375 MHz in 100-kHz mode
Downlink (transmit)	249.1 MHz in 500-kHz mode 248.9375 MHz in 100-kHz mode
Repeaters	1
Bandwidth	500 kHz or 100 kHz
Beacon Signal	254.14 MHz biphase-modulated at 800 bit/sec (separate transmitter), 14 dB below downlink
Telemetry Signal	236.755 MHz biphase-modulated at 100 bit/sec (separate transmitter)
RF power (BOL)	
Downlink	20 dBW (100 W)
Beacon	6 dBW (4 W)
Telemetry	0 dBW (1 W)
Available EIRP (nominal) on satellite equator (BOL)	
Downlink (beam nose)	29.5 dBW (890 W)
Beacon (beam nose)	15.5 dBW (36 W)
Telemetry (toroidal)	0 dBW (1 W)
Receive G/T (nominal) on satellite equator, beam nose (BOL)	-20 dB/K
Spectrum-scanning capability	RFI-measurement instrument scans from 316 to 286 MHz in 256 120-kHz steps over 11 minutes

Status

LES-6 was put in reserve status upon launch of the first GAPFILLER (March 1976). Test of the transponder in 1978, a decade after launch, showed it to be still operable and useful.
Beacon transmitter is ON.
Telemetry transmitter is ON.

Reference

"Lincoln Experimental Satellites 5 and 6", D. C. MacLellan, H. A. MacDonald, P. Waldron, H. Sherman, AIAA Paper 70-494, Progress in Astronautics and Aeronautics, 26(1971)375-398

LES 7

PHOTO COURTESY OF
LINCOLN LABORATORY

William W. Ward
IEEE Compendium

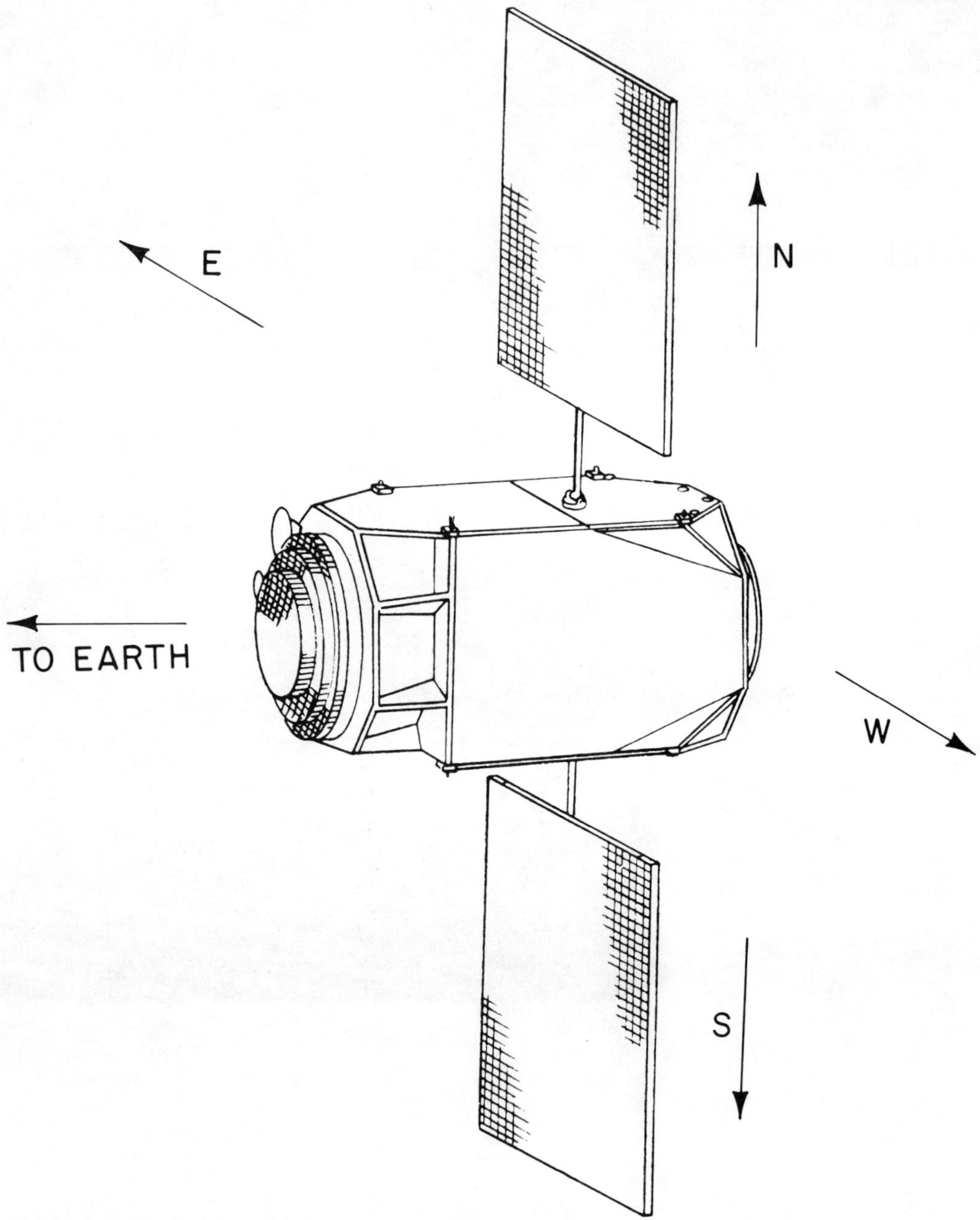

DESIGN STUDY FOR LES-7

William W. Ward
IEEE Compendium

General

Number built	0
Design life	5 to 10 years
Launch vehicle	Titan III-C (planned)
Major contractor	MIT Lincoln Laboratory
Orbit	Circular, synchronous, 2.5° inclination (planned)

Physical Description

Stabilization	3-axis control toward Earth Autonomous attitude-control system Single gimballed momentum wheel Pulsed plasma thrusters for secondary propulsion
Body dimensions	1.1 m (43 in.) x 1.1 m (43 in.) x 1.5 m (60 in.)
Solar-cell panels (each of two)	1.6 m (63.4 in.) x1.7 m (60 in.)
Overall height (panel tip to panel tip)	6 m (236.3 in.)
Weight (initial)	320 to 410 kg (700 to 900 lb)

Antenna Configuration

Multiple-beam antenna (MBA) for SHF reception	0.76 m (30 in.) diameter waveguide lens, f/d = 1, 19-port feedhorn cluster
Horn for SHF transmission	Earth coverage
Cassegrain paraboloid for EHF reception	0.46 m (18 in.) diameter
Aperture for S-band transmission	Unspecified

Polarization

Receive SHF communications	RHCP
Transmit SHF commuications	LHCP
Transmit SHF beacon	LHCP
Receive EHF	Unspecified
Transmit S-band telemetry	Unspecified

Communication Capability

Total usable bandwidth	60 MHz
Assigned frequency range	
Uplink (receive)	7975-8035 MHz
Downlink (transmit)	7250-7310 MHz
Repeaters	1
Bandwidth	60 MHz
Beacon signal	7320 MHz biphase-modulated at 800 bit/sec, 10 dB below downlink
RF power (BOL)	13.8 dBW (24 W) total (downlink, beacon, and intermods)
Telemetry signal	2295 MHz biphase-modulated at 100 or 10,000 bit/sec
EHF receive frequency	14.4 GHz
Available EIRP (nominal maximums)	
Downlink	30.5 dBW (1120 W)
Beacon	20.5 dBW (112 W)
Receive G/T (maximum, full aperture of MBA)	-1.5 dB/K
Control of receive pattern	By ground command

Power

Solar-cell arrays oriented toward Sun

Available DC power (BOL)	500 W (full Sun)
Limited eclipse power	Keep-alive batteries

Status

The LES-7 program ended early in 1970, it being ahead of its time. The MBA technology development was completed, and the results were reported. They ultimately supplied the basis for incorporating MBAs in the DSCS-III satellite.

Reference

"A Variable-Coverage Satellite Antenna System", A. R. Dion, L. J. Ricardi, Proc. IEEE, 59(1971)252-262

LES 7
TECHNICAL PARAMETERS

LES 8

PHOTO COURTESY OF
LINCOLN LABORATORY

William W. Ward
IEEE Compendium

LES 8 and 9

PHOTO COURTESY OF
LINCOLN LABORATORY

William W. Ward
IEEE Compendium

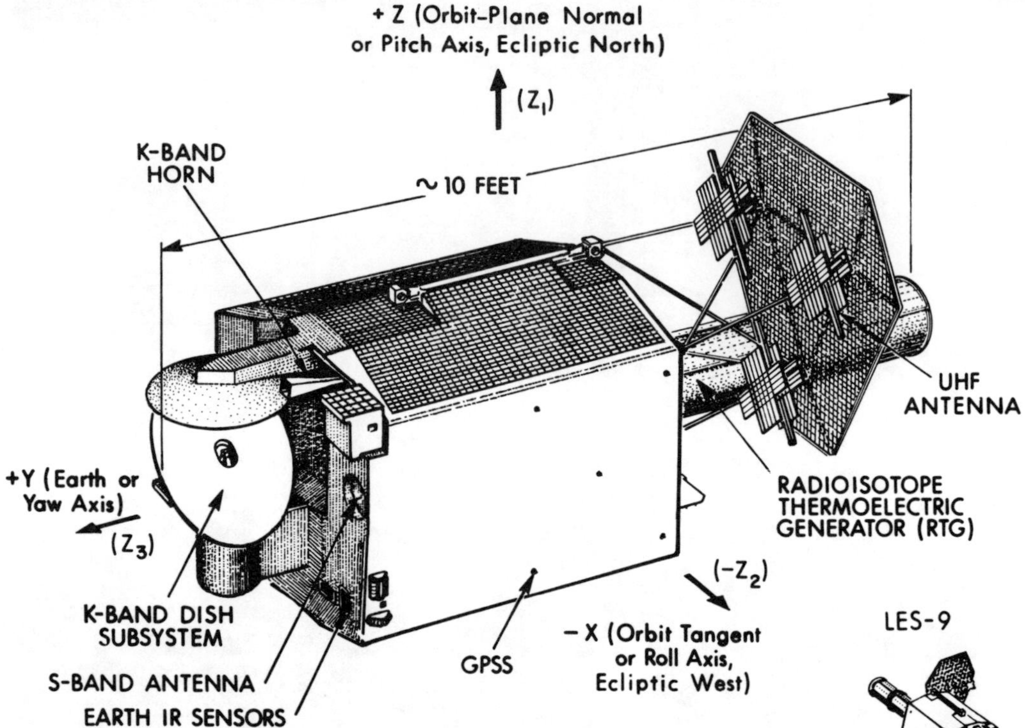

ILLUSTRATED ARRANGEMENT LES-9

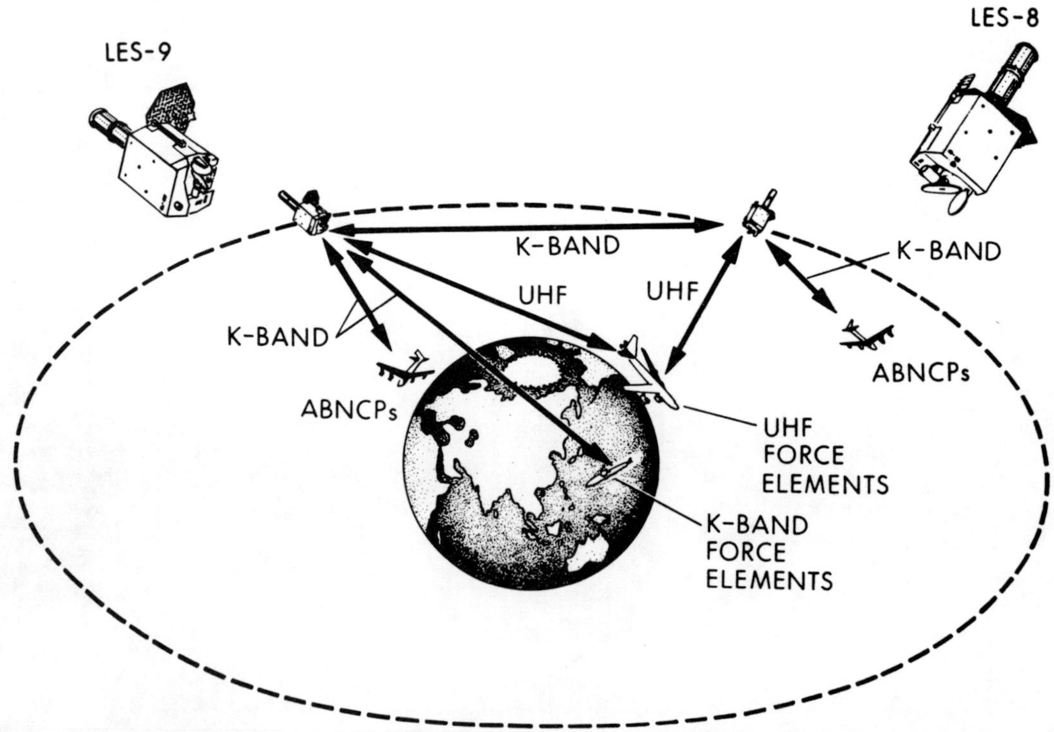

LES 8 and LES 9
COMMUNICATIONS CONFIGURATION

William W. Ward
IEEE Compendium

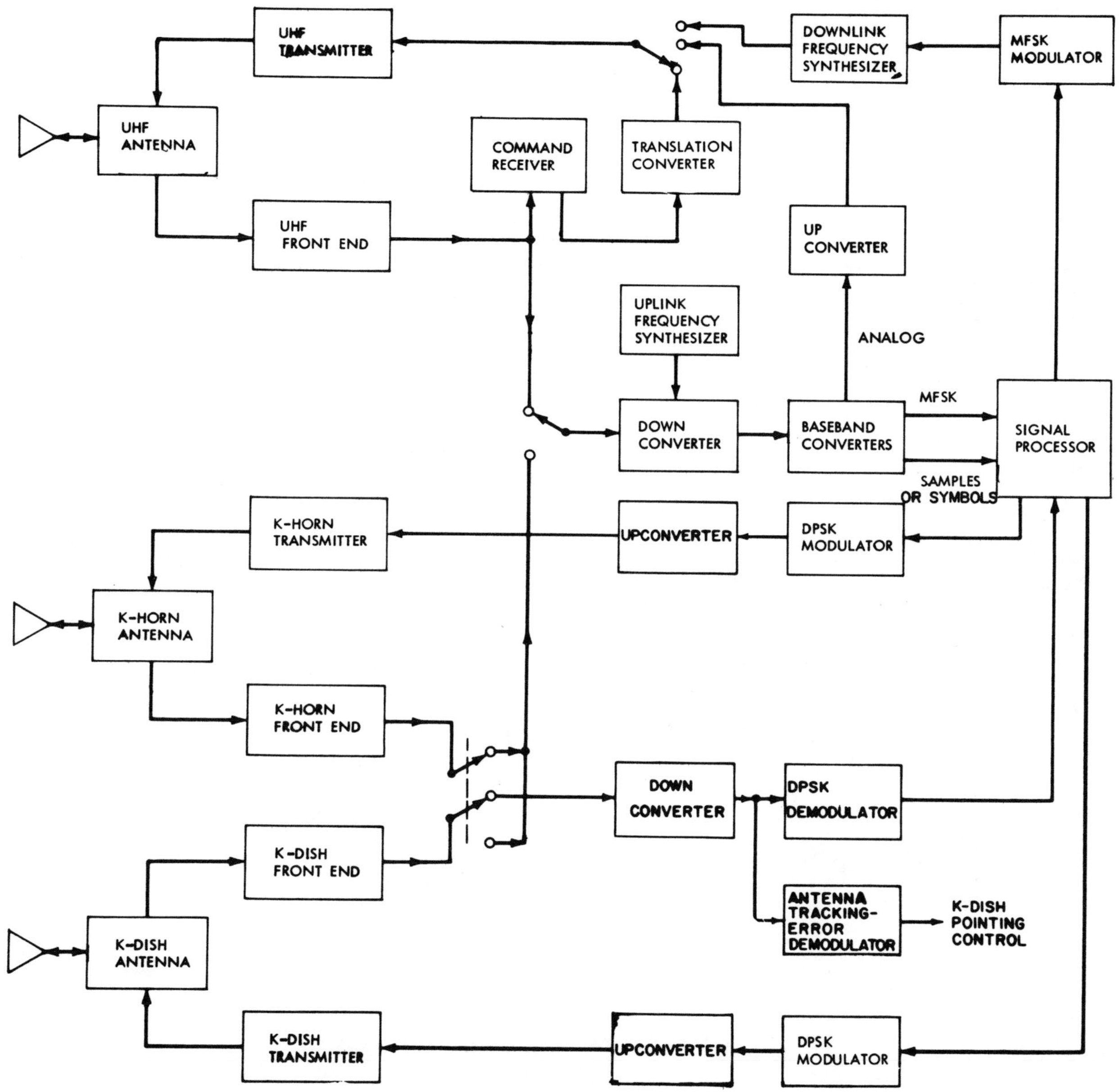

LES 8 and 9
COMMUNICATION BLOCK DIAGRAM

William W. Ward
IEEE Compendium

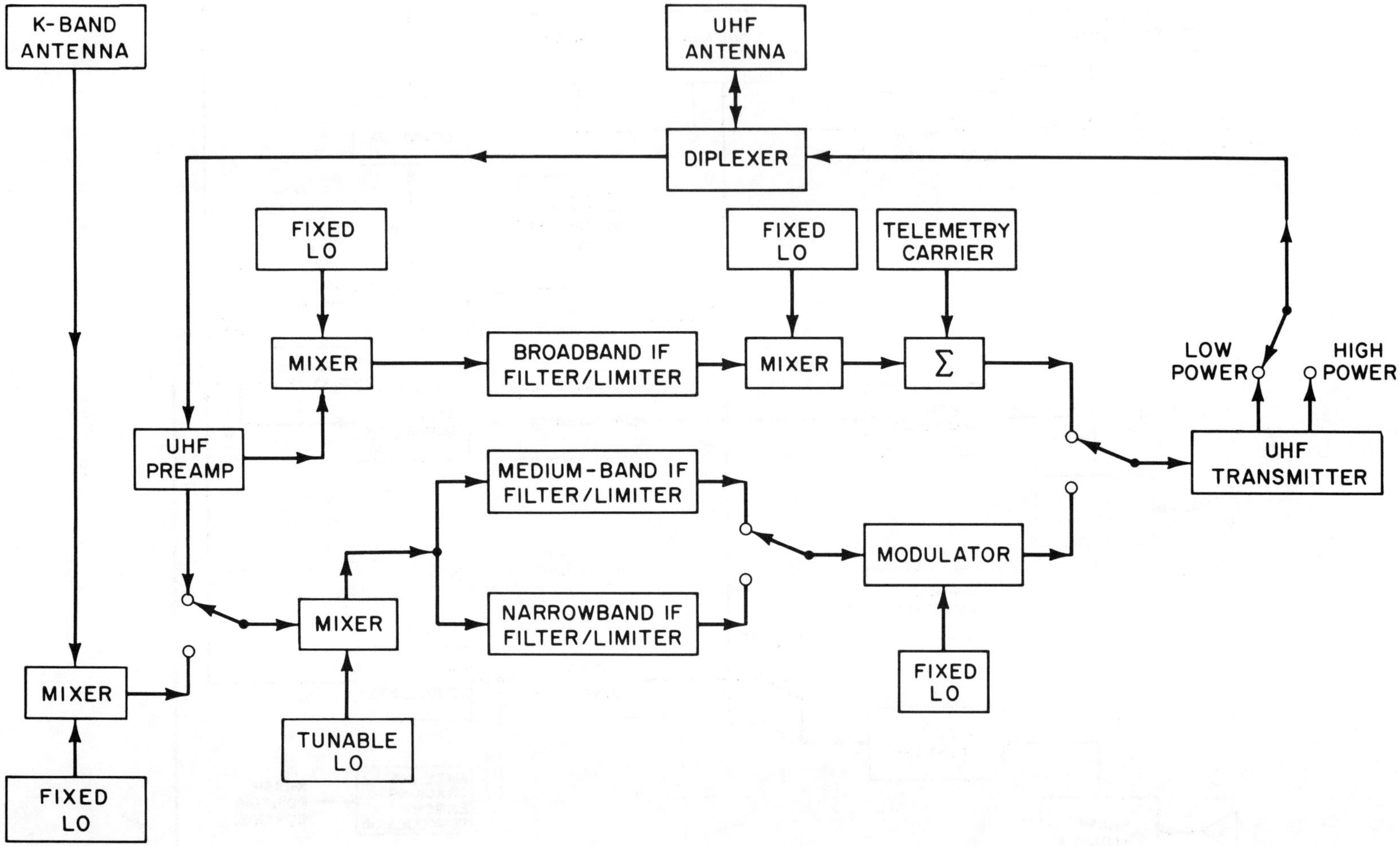

LES 8 and 9
ANALOG-TRANSPONDER MODES
UHF-TO-UHF
K-BAND-TO-UHF

William W. Ward
IEEE Compendium

LES 8 and 9
TECHNICAL PARAMETERS

General

Number built	2
Design life	Indefinitely long
Launch vehicle	Titan III-C
Major contractor	MIT Lincoln Labratory
Date of launch	14 March 1976
Orbit	Circular, synchronous, 25° inclination (near ecliptic plane)
SDC object numbers	8746 (LES-8) 8747 (LES-9)
COSPAR designations	1976-23A (LES-8) 1976-23B (LES-9)

Physical Description

Stabilization	3-axis control toward Earth Yaw axis toward geocenter Pitch axis perpendicular to orbit plane Roll axis tangent to orbit Autonomous attitude-control system Single gimballed momentum wheel Cold gas secondary propulsion
Approximate dimensions	
Yaw axis	3 m (10 ft)
Pitch axis	1.8 m (6 ft) } Includes UHF antenna array
Roll axis	2 m (6.8 ft) } Includes UHF antenna array
Approximate weight	450 kg (1000 lb) each, including 34 kg (75 lb) of ammonia propulsion fuel

Antenna Configuration

UHF communications	3 crossed dipoles mounted above ground plane, cantilevered out from spacecraft body
K-band communications	Horn antenna rigidly mounted on spacecraft body, squinted off yaw axis for best uplink/downlink coverage from orbital station Dish antenna (paraboloid and movable flat plate) for crosslink or uplink/downlink service
S-band telemetry	Cup-dipole on Earth-facing end of spacecraft Helix on Earth-opposite end of spacecraft (backup telemetry during drastic misorientation of spacecraft)

Polarization

UHF transmit	RHCP
UHF receive	RHCP
K-band transmit	LHCP (LES-8) RHCP (LES-9)
K-band receive	RHCP (LES-8) LHCP (LES-9)
S-band transmit	RHCP

Power

2 multi-hundred-watt radioisotope thermoelectric generators (MHW RTGs) per spacecraft (∿ 300 W at BOM)
Full operation during daily ∿ 70-min-long eclipses

William W. Ward
IEEE Compendium

Communications Capability

Primary operating modes	Spectrum-spreading (frequency hopping) for interference immunity
	On-board signal-processing functions
	Demodulation and remodulation
	Multiplexing and demultiplexing
	Data-stream options for retransmission
Test operating modes	
UHF uplink (receive)	302.8 MHz (LES-8) "HELP" frequencies
	303.6 MHz (LES-9) "HELP" frequencies
UHF downlink (transmit)	249.2 MHz (LES-8) "HELP" frequencies
	249.6 MHz (LES-9) "HELP" frequencies
K-band uplink (receive)	36.8332 GHz (LES-8) "HELP" frequencies
	38.0476 GHz (LES-9) "HELP" frequencies
K-band downlink (transmit)	37.98 GHz (LES-8 horn antenna)
	38.04 GHz (LES-8 dish antenna)
	36.90 GHz (LES-9 horn antenna)
	36.84 GHz (LES-9 dish antenna)
K-band crosslinks (transmit and receive)	38.04 GHz (LES-8 to LES-9)
	36.84 GHz (LES-9 to LES-8)
Repeaters	1 (3 selectable options)
Wide band (UHF-to-UHF)	500 kHz bandwidth
Uplink center frequency	302.7 MHz (LES-8); 303.4 MHz (LES-9) — LES-8 frequencies same as LES-6 and the 3 GAPFILLERs; LES-9 frequencies same as TACSAT
Downlink center frequency	249.1 MHz (LES-8); 249.6 MHz (LES-9)
Medium band (UHF-to-UHF, K-band-to-UHF)	∿ 25 kHz bandwidth
Uplink center frequency	Selectable by command
Downlink center frequency	248.86 MHz (LES-8)
	249.36 MHz (LES-9)
Narrow band (UHF-to-UHF, K-band-to-UHF)	∿ 5 kHz bandwidth
Uplink center frequency	Selectable by command
Downlink center frequency	248.86 MHz (LES-8)
	249.36 MHz (LES-9)
Telemetry signal (transmit)	2.24 GHz (LES-8)
	2.25 GHz (LES-9)
	Biphase-modulated at 100 or 10,000 bit/sec
	Wide-band-repeater downlink signal includes carrier signal at 248.86 MHz (LES-8) or 249.36 MHz (LES-9) biphase-modulated by telemetry data at 100 or 10,000 bit/sec, 11 dB below downlink

RF power (BOL)	
UHF downlink	
Low-power mode	8.6 dBW (7.3 W) (LES-8)
	9.2 dBW (8.4 W) (LES-9)
High-power mode	15 dBW (33 W) (LES-8)
	15 dBW (31 W) (LES-9)
K-band horn downlink	-3.4 dBW (0.5 W) (LES-8)
	-3.2 dBW (0.5 W) (LES-9)
K-band dish downlink or crosslink	-3.9 dBW (0.4 W) (LES-8)
	-4.0 dBW (0.4 W) (LES-9)
Available EIRP (BOL, maximum)	
UHF downlink	
Low-power mode	18.2 dBW (66 W) (LES-8)
	18.3 dBW (85 W) (LES-9)
High-power mode	24.7 dBW (295 W) (LES-8)
	25.0 dBW (315 W) (LES-9)
K-band horn downlink	21.8 dBW (150 W) (LES-8)
	21.5 dBW (140 W) (LES-9)
K-band dish downlink or crosslink	39.4 dBW (7.7 kW) (LES-8)
	38.6 dBW (7.2 kW) (LES-9)
Receive G/T (BOL, maximum)	
UHF uplink	-20 dB/K (LES-8)
(∿35° beamwidth)	-20 dB/K (LES-9)
K-band horn uplink	- 8 dB/K (LES-8)
(9.5° beamwidth)	-7.3 dB/K (LES-9)
K-band dish uplink or	10.2 dB/K (LES-8)
crosslink (1.2° beamwidth)	10.5 dB/K (LES-9)

Status

Upon conclusion of the Joint Test Program in September, 1977, LES-8/9 began operational service. Four years after launch, both satellites continue in good health with every indication of providing useful services for many years to come.

References

"The Lincoln Experimental Satellites LES-8 and LES-9", F. W. Sarles, Jr., L. W. Bowles, L. P. Farnsworth, P. Waldron, EASCON-77 Record, p. 21-1A through 21-1U

"Developing, Testing, and Operating Lincoln Experimental Satellites 8 and 9 (LES-8/9)", W. W. Ward, Lincoln Laboratory TN 1979-3, 16 January 1979

LES 8 and 9 TECHNICAL PARAMETERS PART II

William W. Ward
IEEE Compendium

NATO I

(See IDCSP)

The NATO II satellites were designed and manufactured by Ford Aerospace and Communications Corp. under contracts administered by USAF Space Division for NATO. NATO II satellites were nearly identical to the SKYNET I satellites built by FACC for the United Kingdom. *The NATO I Communication Satellite Program utilized the IDCSP satellites also manufactured by FACC.*

NATO II

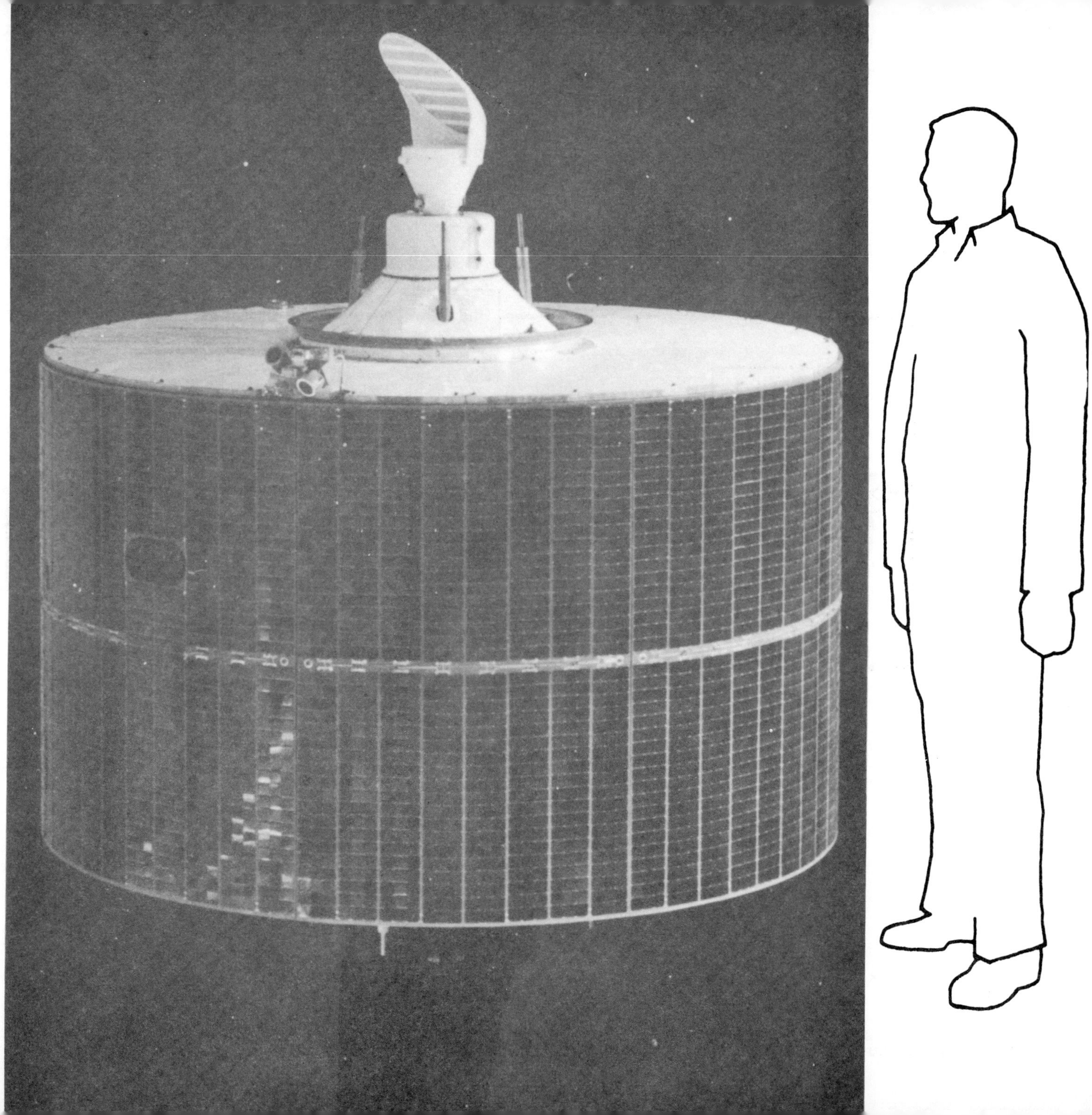

PHOTO COURTESY OF FORD AEROSPACE AND COMMUNICATIONS CORP.

Ray Berg
IEEE Compendium

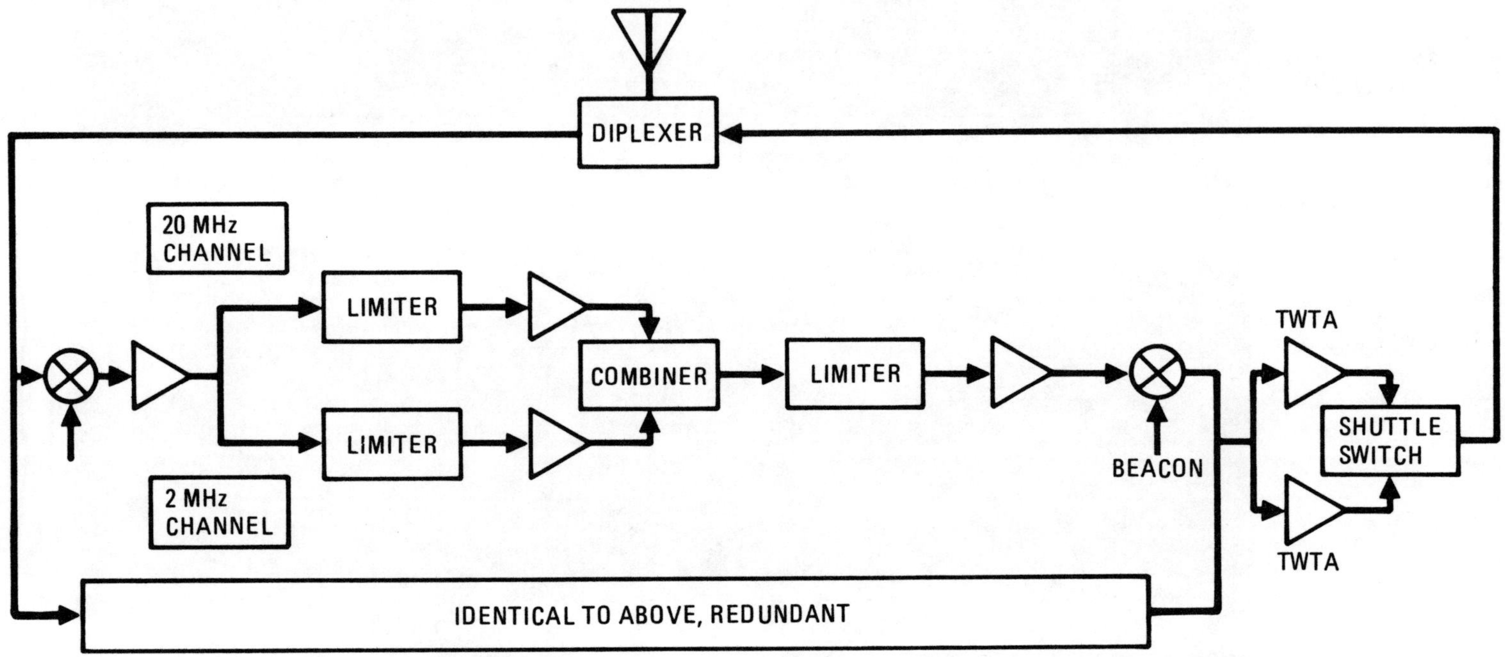

NATO II COMMUNICATION BLOCK DIAGRAM

Ray Berg
IEEE Compendium

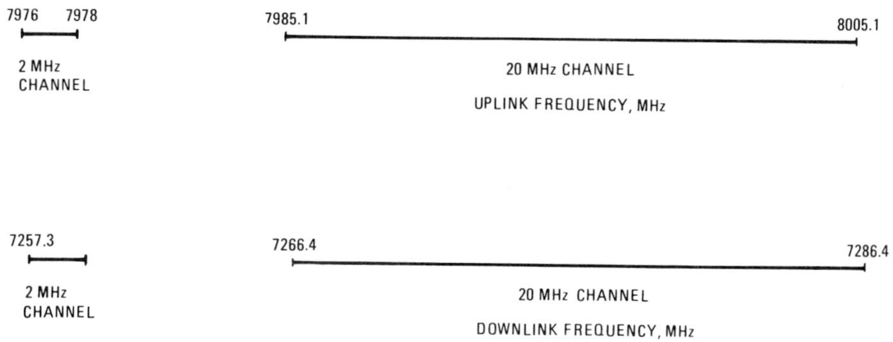

NATO II FREQUENCY PLAN

GENERAL	
NO. OF SATELLITES BUILT	2
DESIGN LIFE	5
LAUNCH VEHICLE	THOR DELTA
MAJOR CONTRACTOR	FACC
DATE OF FIRST LAUNCH	20 MAR 1970
PHYSICAL DESCRIPTION	
STABILIZATION	SPIN–EARTH ORIENTED DESPUN ANTENNAS
CYLINDER DIAM.	137.2 cm (54 in.)
CYLINDER HEIGHT	81.3 cm (32 in.)
OVERALL HEIGHT	157.5 cm (62 in.)
WEIGHT ON ORBIT	129.5 kg (285 lb)
ANTENNA CONFIGURATION	
MECHANICALLY DESPUN HORN – ALIGNED TO OPTIMIZE COVERAGE FOR NATO AREA.	
COMMUNICATION CAPABILITY	
TOTAL USEABLE BANDWIDTH	22 MHz
ASSIGNED FREQUENCY RANGE	
UPLINK	7976 – 8005.1 MHz
DOWNLINK	7257.3 – 7286.4 MHz
REPEATERS	DUAL CONVERSION
NO. ON LINE	1
REDUNDANT	1
TWTA	WJ251/EIMAC X1066
BANDWIDTH	300 MHz
RF POWER	3 Watts
AVAILABLE EIRP	
2 MHz CHANNEL	11 dBW
20 MHz CHANNEL	19.5 dBW
RECEIVER	
8.8 dB NOISE FIGURE	

NATO II TECHNICAL PARAMETERS

The NATO II satellites were designed and manufactured by Ford Aerospace and Communications Corp. under contracts administered by USAF Space Division for NATO. NATO II satellites were nearly identical to the SKYNET I satellites built by FACC for the United Kingdom. The NATO I Communication Satellite Program utilized the IDCSP satellites also manufactured by FACC.

Ray Berg
IEEE Compendium

NATO III

PHOTO COURTESY OF
FORD AEROSPACE AND
COMMUNICATIONS CORP.

Ray Berg
IEEE Compendium

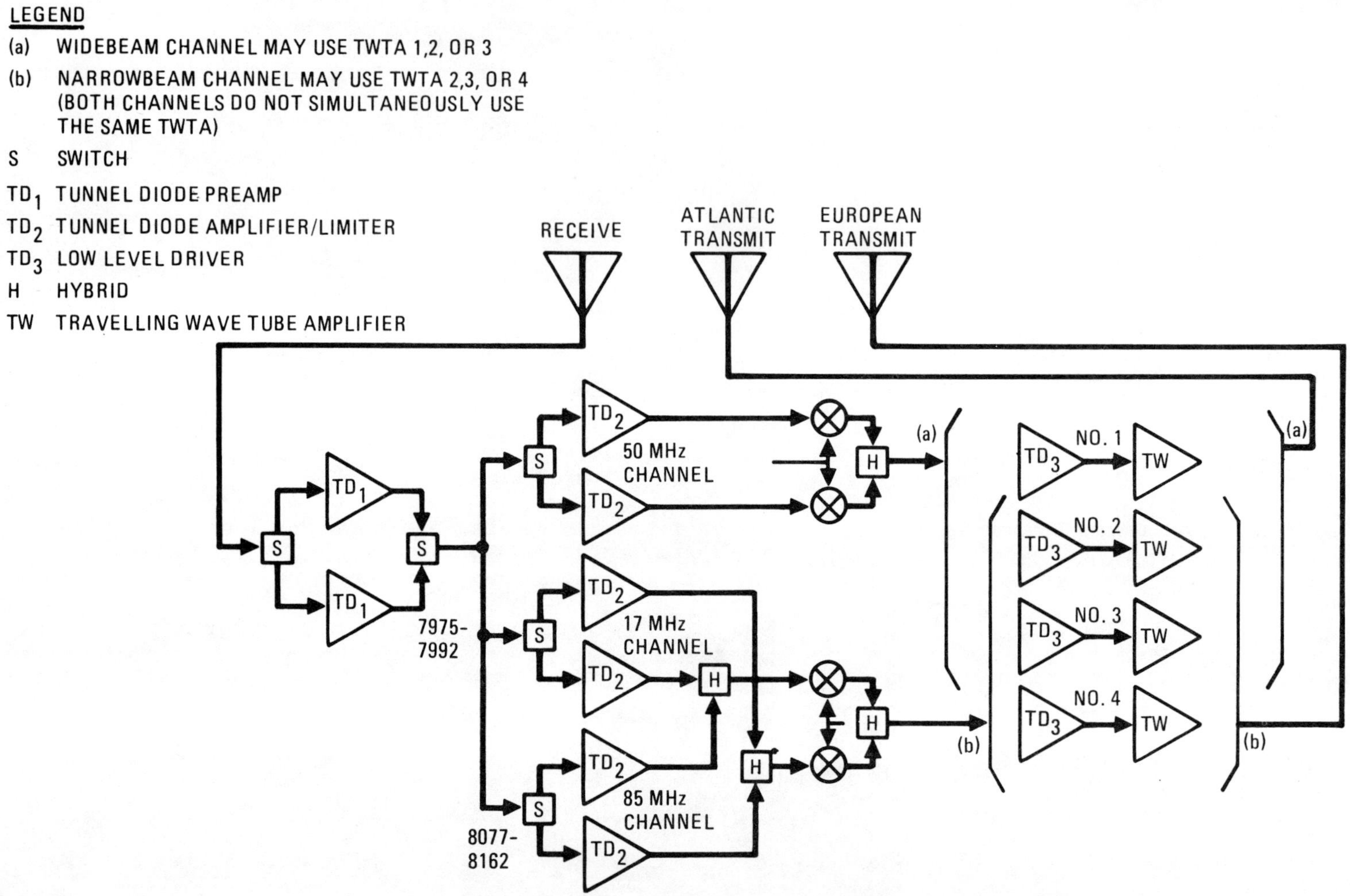

NATO III COMMUNICATION BLOCK DIAGRAM

Ray Berg
IEEE Compendium

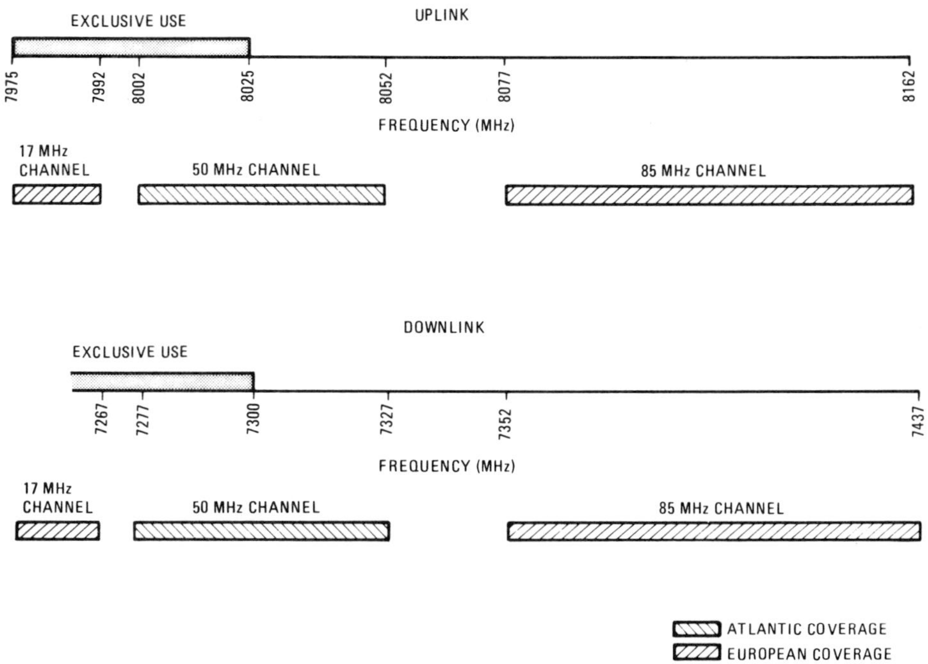

NATO III FREQUENCY PLAN

GENERAL	
NO. OF SATELLITES BUILT	3
DESIGN LIFE	7 yr
LAUNCH VEHICLE	DELTA 2914
MAJOR CONTRACTOR	FORD AEROSPACE AND COMM
DATE OF 1ST LAUNCH	22 APRIL 1976
PHYSICAL DESCRIPTION	
STABILIZATION	DUAL SPIN (EARTH ORIENTED DESPUN ANTENNAS)
CYLINDER DIAMETER	218.4 cm (86 in.)
CYLINDER HEIGHT	203.2 cm (80 in.)
OVERALL	309.9 cm (122 in.)
WEIGHT	
LAUNCH	695 kg (1530 lb)
ORBIT	336.4 kg (740 lb)
ANTENNA CONFIGURATION	
WIDEBEAM (9.2 X 16^{0} F.O.U.)	1 TRANSMIT 1 RECEIVE
1 NARROWBEAM (EUROPEAN SPOT)	1 TRANSMIT
COMMUNICATION CAPABILITY	
TOTAL USEABLE BANDWIDTH	152 MHz
FREQUENCY ASSIGNED	7975 – 8162 MHz UPLINK 7250 – 7437 MHz DOWNLINK
REPEATERS	SINGLE CONVERSION
ON LINE	2
REDUNDANT	2
TWTA	HUGHES 265 HA
BANDWIDTH	300 MHz
RF POWER	20 Watts
AVAILABLE EIRP	
NARROWBEAM	35 dBW
WIDEBEAM	29 dBW
RECEIVER	-80 dB/^{o}K G/ T

NATO III TECHNICAL PARAMETERS

The NATO III satellites were designed and manufactured by Ford Aerospace and Communications Corp. under contracts administered by USAF Space Division for the NATO Integrated Communication System Management Agency.

Ray Berg
IEEE Compendium

SKYNET I

PHOTO COURTESY OF
FORD AEROSPACE AND
COMMUNICATIONS CORP.

Richard W. Duesing
IEEE Compendium

SKYNET II

PHOTO COURTESY OF
FORD AEROSPACE AND
COMMUNICATIONS CORP.

Richard W. Duesing
IEEE Compendium

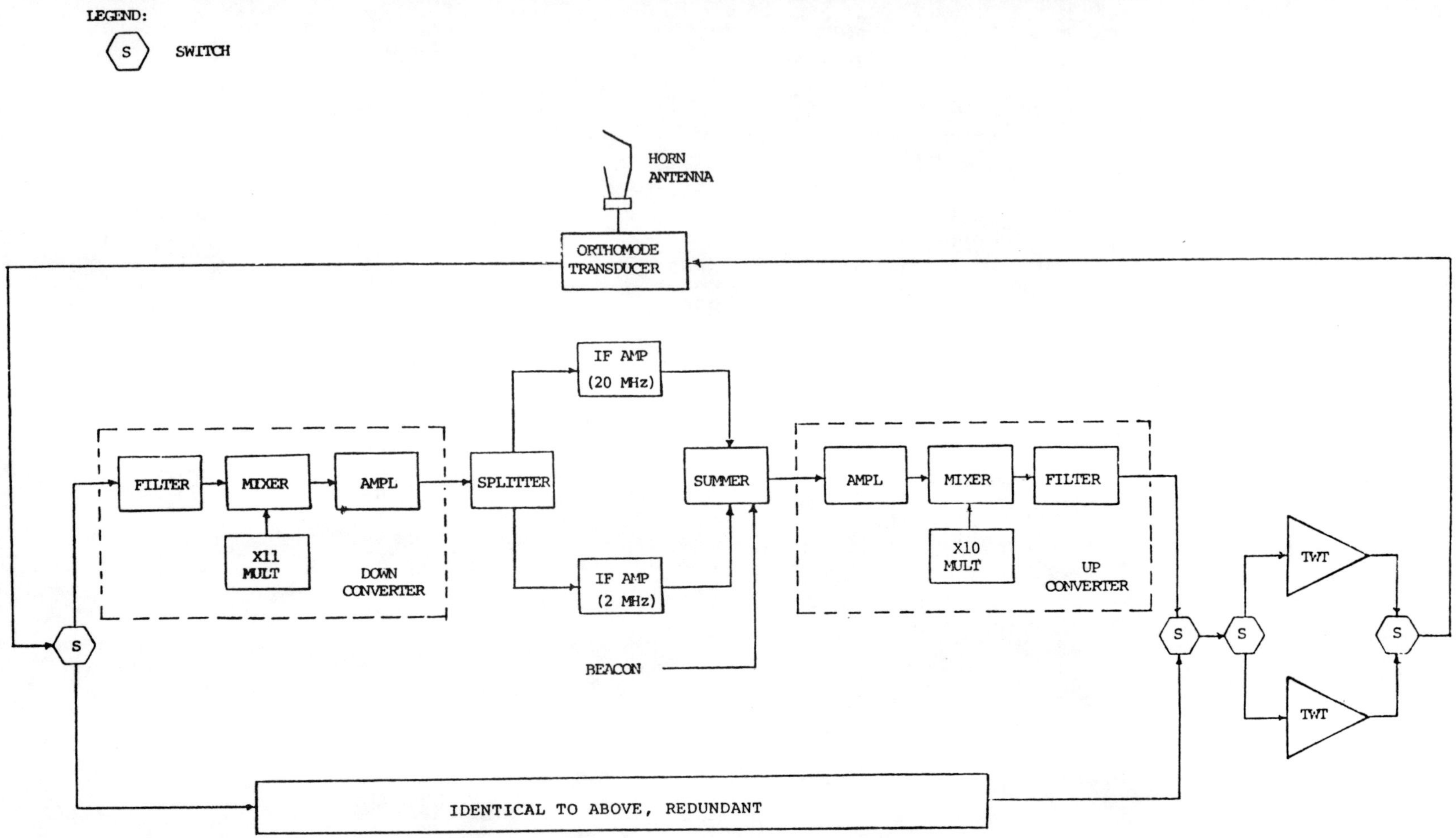

NOTE: SKYNET I TWT OUTPUT POWER = 3.5 WATTS
SKYNET II TWT OUTPUT POWER = 20 WATTS

SKYNET I and II
COMMUNICATION BLOCK DIAGRAM

Richard W. Duesing
IEEE Compendium

General

No. Satellites Built	2
Design Life	5 years
Launch Vehicle	Thor-Delta
Major Contractor	Philco-Ford
Date of Launch	
I-A:	21 November 1969
I-B:	19 August 1970 (Unsuccessful)

Physical Description

Stabilization	Spin
Cylinder Diameter	137 cm (54 in)
Cylinder Height	81.2 cm (32 in)
Height with Antenna	157.5 cm (62 in)
Weight	
Launch	240 kg (535 lbs)
Orbit (Initial)	126 kg (280 lbs)

Antenna Configuration

Earth Coverage - X-Band	1 Mechanically Despun Horn
TT&C - UHF	2 UHF Array

Polarization

Horn	Circular

Communication Capability

Total Usable Bandwidth	22 MHz
Assigned Frequency Range	
Up-Link (Receive)	7976.02-7978.02 MHz 7985.12-8005.12 MHz
Down-Link (Transmit)	7257.3-7259.3 MHz 7266.4-7286.4 MHz
Repeaters	
No. On-Line	1
No. Redundant	1
Type	Hard-Limiting
Bandwidth	20 MHz and 2 MHz
RF Power	3.5 Watts
Noise Figure	8.8 dB
Available e.i.r.p. (Nominal)	19.5 dBW per channel
Antenna Characteristics	
Gain	
X-Band	18.5 dBi
UHF	-0.7 dBi
Beamwidth	
X-Band	19°
UHF	Omnidirectional

Communication Capability

20 MHz Channel	Long distance strategic point-to-point digital communications utilizing SSMA techniques. Digital signals at medium speed rates of 75 x 2^n bauds.
2 MHz Channel	Tactical communications for mobil terminals using FDMA techniques.

SKYNET I TECHNICAL PARAMETERS

Richard W. Duesing
IEEE Compendium

General	
No. Satellites Built	2
Design Life	5 years
Launch Vehicle	Thor-Delta 2313
Major Contractor	Marconi Space Defence Systems/ Philco-Ford
Date of Launch	
II-A:	18 January 1974 (Unsuccessful)
II-B:	22 November 1974

Physical Description	
Stabilization	Spin
Cylinder Diameter	190.5 cm (75 in)
Cylinder Height	134.6 cm (53 in)
Height with Antenna	208.3 cm (82 in)
Weight	
Launch	432 kg (960 lbs)
Orbit (Initial)	233 kg (517 lbs)

Antenna Configuration	
Earth Coverage - X-Band	1 Mechanically Despun Horn
TT&C - S-Band	2 Toroidal Beam

Polarization	
Horn	Circular
Toroidal Beam	RHC

Communication Capability	
Total Usable Bandwidth	22 MHz
Assigned Frequency Range	
Up-Link (Receive)	7976.02-7978.02 MHz 7985.12-8005.12 MHz
Down-Link (Transmit)	7257.3-7259.3 MHz 7266.4-7286.4 MHz
Repeaters	
No. On-Line	1
No. Redundant	1
Type	Hard-Limiting
Bandwidth	20 MHz and 2 MHz
RF Power	20 Watts
Noise Figure	10 dB
Available e.i.r.p. (Nominal)	7 dBW (2 MHz Channel) 23 dBW (20 MHz Channel)
Antenna Characteristics	
Gain	
X-Band	17.0 dBi
S-Band	4.0 dBi
Beamwidth	
X-Band	19°
S-Band	Omnidirectional
Communication Capability	
20 MHz Channel	Long distance strategic point-to-point digital communications utilizing SSMA techniques. Digital signals at medium speed rates of 75 x 2^n bauds.
2 MHz Channel	Tactical communications for mobil terminals using FDMA techniques.

SKYNET II TECHNICAL PARAMETERS

Richard W. Duesing
IEEE Compendium

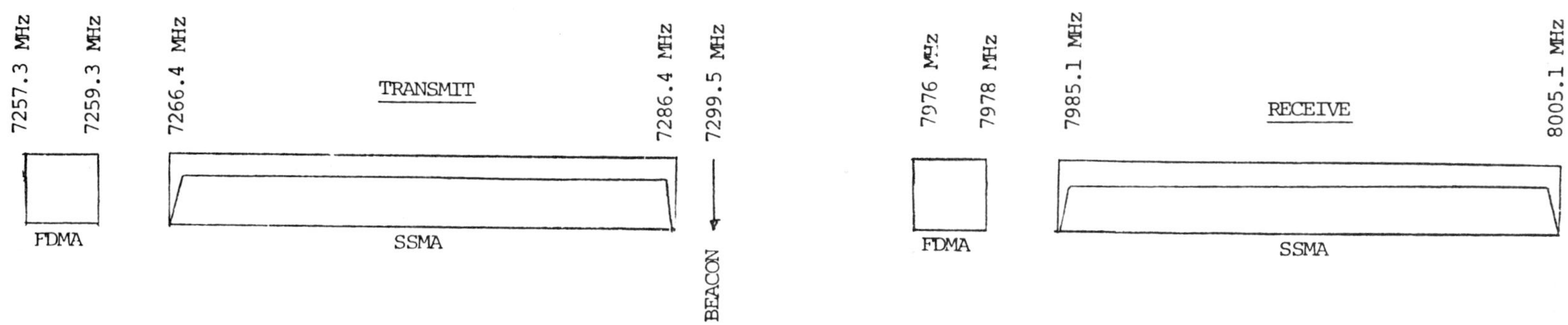

SKYNET I and II FREQUENCY PLAN

Richard W. Duesing
IEEE Compendium

TACSAT

PHOTO COURTESY OF
HUGHES AIRCRAFT CORP.

Martin P. Brown, Jr.
IEEE Compendium

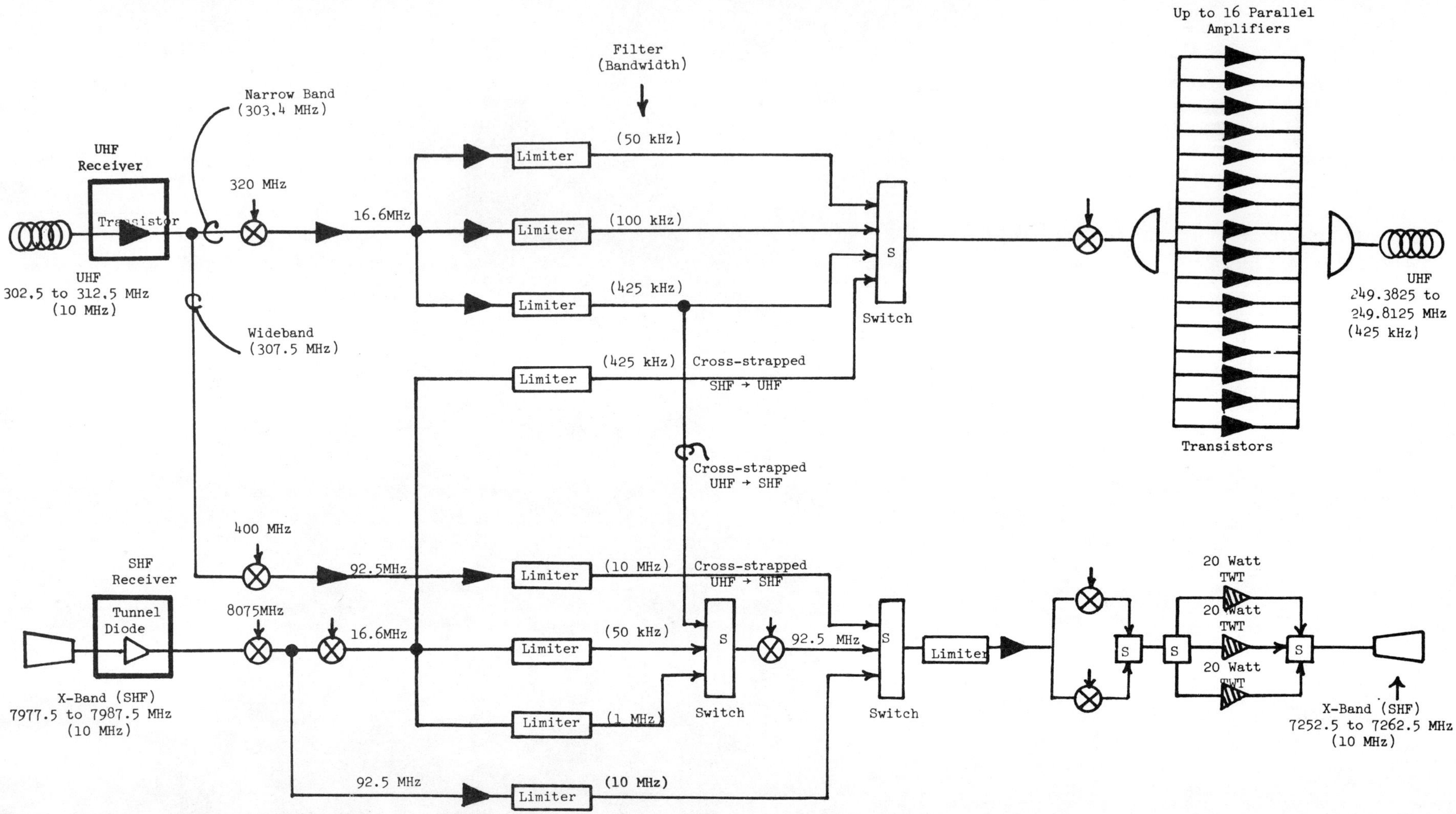

TACSAT COMMUNICATIONS BLOCK DIAGRAM

Martin P. Brown, Jr.
IEEE Compendium

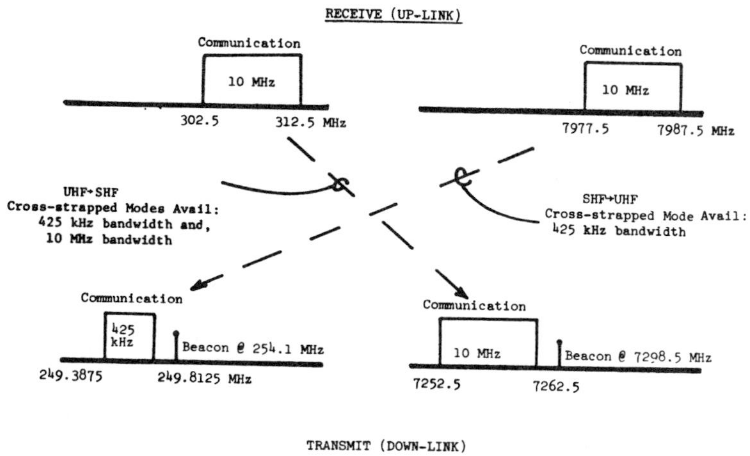

Communication Capability

UHF	40 vocoded voice or 700 teletype circuits
SHF (X-band)	Same as UHF
Modulation Techniques	FM/FDMA, MFSK and freq hopping, FSK, DPSK (288 kbps)/FDMA

TACSAT FREQUENCY PLAN

General

No. Satellites Built	One
Design Life	2.4 years
Launch Vehicle	Titan IIIC
Major Contractor	Hughes Aircraft Co
Sponsor	SAMSO
Date of Launch	9 February 1969 (operated until Dec 72)

Physical Description

Stabilization	Spin, (Spinning cylinder and despun platform)
Cylinder Diameter	2.7 meters(9 feet)
Cylinder Height	3.3 " (11 feet)
Height x/antenna mast	7.6 " (25 feet)
Weight (initial)	726 kg (1600 lbs)
in-orbit	645 kg (1424 lbs)

Antenna Configuration

Global Horns SHF (X-band)	
Transmit (one)	18.4 dBi, 19° Beamwidth
Receive (one)	19.3 dBi, 17° Beamwidth
Bifilar Helix, (five)	
Transmit	17.1 dBi, 19° Beamwidth
Receive	17.6 dBi, 19° Beamwidth
Biconical Horn	Telemetry and Command

Communication Parameters

Assigned Frequency Bands	
Uplink	
·UHF	302.5 to 312.5 MHz (10 MHz)
·SHF (X-band)	7977.5 to 7987.5 MHz (10 MHz)
Downlink	
·UHF	249.3875 to 249.8125 MHz (425 kHz)
·SHF (X-band)	7252.5 to 7262.5 MHz (10 MHz)
Command Selectable Modes	
UHF	50 kHz, 100 kHz, and 425 kHz bandwidths
SHF(X-band)	10 kHz, 1 MHz, and 10 MHz bandwidths
Cross-strapped	425 kHz (SHF→UHF) and 425kHz & 10MHz (UHF→SHF)
Beacon	
·UHF	254.1 MHz
·SHF (X-band)	7298.5 MHz
Repeaters	
UHF	16 parallel transistors (13 nominal on) 18.5 watts per amp 230 watts max from combiner
SHF (X-band)	3 TWT's (nominal 2 on) 20 watts per TWT 30 watts out of combiner
e.i.r.p. (nominal)	
UHF	40.7 dBW (with 16 amp)
SHF (X-band)	33.0 dBW (with 2 TWT)
Receiver	
UHF	3.7 dB Noise Figure (Transistor Amp)
SHF (X-band)	6.9 dB Noise Figure (Tunnel Diode)

TACSAT TECHNICAL PARAMETERS

SECTION VI

EXPERIMENTAL SATELLITES

- ATS 1 THROUGH 5
- ATS 6
- BS (BSE or "Yuri")
- COURIER
- CS ("Sakura")
- CTS ("Hermes")
- ECHO
- ECS (Japanese); "Ayame"
- ETS II ("Kiku II")
- OTS
- RELAY
- SCORE
- SIRIO
- SYMPHONIE
- SYNCOM
- TELSTAR
- WESTFORD

ATS 1 THROUGH 5

PHOTO COURTESY OF
HUGHES AIRCRAFT CORP.

Richard W. Duesing
IEEE Compendium

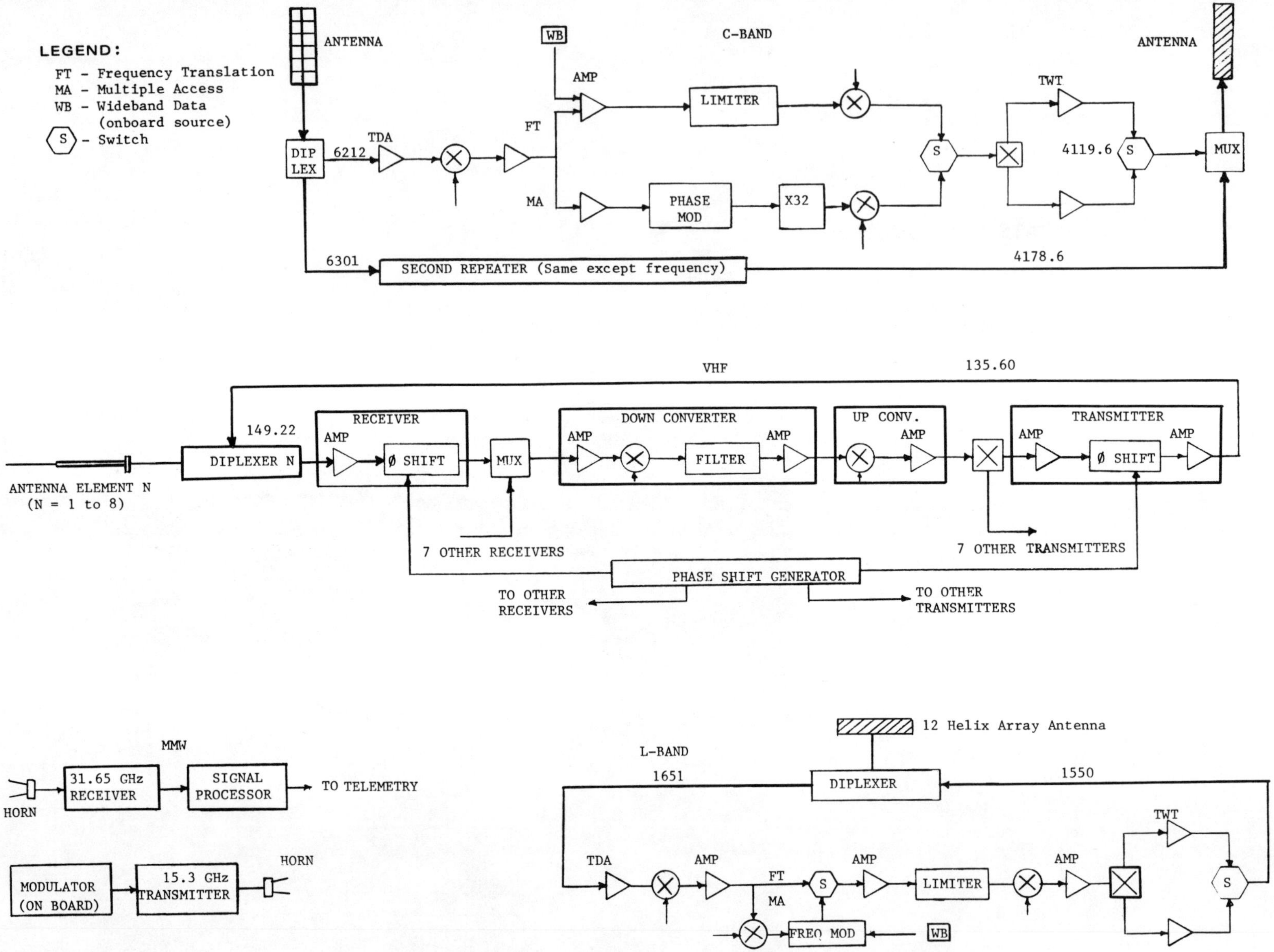

ATS 1 THROUGH 5 COMMUNICATIONS BLOCK DIAGRAM

Richard W. Duesing
IEEE Compendium

ATS 1 THROUGH 5 TECHNICAL PARAMETERS

Richard W. Duesing
IEEE Compendium

	ATS 1	ATS 2	ATS 3	ATS 4	ATS 5
General					
No. Satellites Built	1	1	1	1	1
Design Life	3 years	same	same	same	same
Launch Vehicle	Atlas-Agena	Atlas-Agena	Atlas-Agena	Atlas-Centaur	Atlas-Centaur
Major Contractor	Hughes Aircraft	same	same	same	same
Date of Launch	6 December 1966	5 April 1967	5 November 67	10 August 1968	12 August 1969
Date of Decay	N/A	2 September 69	N/A	17 October 1968	N/A
Orbit-Actual	Synchronous equatorial 150°W	100x6000 nmi	Synchronous equatorial 105°W	130x480 nmi	Synchronous equatorial 70°W
Orbit-Intended	As Achieved	circular (6000 nmi)	As Achieved	Synchronous equatorial	Achieved, however spin about axis normal to intended spin axis limits success of experiments
Status	In use as of 1/79	Decayed	In use as of 1/79	Decayed	
Physical Description					
Stabilization	Spin (electronically Despun Antennas)	Gravity Gradient	Spin (Mechanically Despun Antennas)	same as 2	same as 2
Cylinder Diameter	146cm (57.6 in)	142cm (56 in)	146cm(57.6in)	142cm (56 in)	142cm (56 in)
Cylinder Height	134cm (53 in)	183cm (72 in)	134cm (53in)	183cm (72 in)	183cm (72 in)
Weight					
Launch	733kg (1616 lbs)	318.4kg (702 lbs)	733kg (1616 lbs)	737kg (1625 lbs)	737kg (1625 lbs)
Orbit (initial)	351.5kg (775 lbs)	318.4kg (702 lbs)	351.5kg (775 lbs)	304kg (670 lbs)	304kg (670 lbs)
Communication Capabilities					
C-Band Communications					
Total Usable Bandwidth	50 MHz	same	same	same	same
Assigned Center Frequencies					
Up-Link (Receive)	6212,6301 MHz	same	same	same	same
Down-Link (Transmit)	4120,4179 MHz	same	same	same	same
Repeaters					
No.	2	2	2	2	2
Bandwidth	25 MHz	same	same	same	same
R F Power	4 Watts	4 Watts	4 Watts & 12 Watts	4 Watts	4 Watts
Receiver Noise Figure	6.2 dB	6.2 dB	6.2 dB	6.2 dB	6.2 dB
Available e.i.r.p.(Nominal)	19.5,22.0dBW (1,2 TWT)		22.0,25.0dBW (1,2 4-W TWT) 26.5(1 12-W TWT)		22.5,25.0dB (1,2 TWT)
Antenna Configuration					
Receive	Collinear Array	Horn	Mechanically Despun	Planar Array	Planar Array
Transmit	Phased Array	Horn	"	Planar Array	Planar Array
Polarization					
Receive	Linear	Linear	Linear	Linear	Linear
Transmit/Receive	Orthogonal	Orthogonal	Parallel	Orthogonal	Orthogonal
Gain					
Receive	7.8 dBi	10.5 dBi	18.0 dBi	16.5 dBi	16.5 dBi
Transmit	14.0 dBi	10.5 dBi	18.0 dBi	16.5 dBi	16.5 dBi
Capacity	1200 one-way Voice or 1 color TV channel	same	same	same	same

	ATS 1	ATS 2	ATS 3	ATS 4	ATS 5
VHF Communications		Not Applicable		Not Applicable	Not Applicable
Total Usable Bandwidth	100 kHz		100 kHz		
Assigned frequency					
Receive	149.22 MHz		149.22 MHz		
Transmit	135.60 MHz		135.60 MHz		
Repeater					
RF Power	5 W per element 40W Total		6.25W per element 50W Total		
Receiver Noise Figure	4.5 dB		4.0 dB		
Available e.i.r.p.	22.5 dBW		25.2 dBW		
Antenna Configuration	8-element(dipoles) phased array		same		
Polarization	Linear		Linear		
Gain					
Receive	8 dBi		8 dBi		
Transmit	9 dBi		10 dBi		
Capacity	1 duplex-voice channel between Earth terminal and aircraft		same		

ATS 1 THROUGH 5 TECHNICAL PARAMETERS (continued)

Richard W. Duesing
IEEE Compendium

ATS 1 THROUGH 5 TECHNICAL PARAMETERS (continued)

ATS 5	Millimeter Wave Propagation Experiment	L-Band Experiment					
		Frequency Translation	Narrow Band	Narrow-Narrow Band	Wideband	Cross-strapped L to C	Cross-strapped C to L
Total Usable Bandwidth		25 MHz	1.5 MHz	100 kHz	N/A	100 kHz	600 kHz
Assigned Frequency							
Receive	31.65 GHz	1651 MHz	1651 MHz	1651 MHz	N/A	1651MHz	6212 MHz
Transmit	15.3 GHz	1550 MHz	1550 MHz	1550 MHz	1550 MHz	4120MHz	1550 MHz
Repeater							
RF Power	200 mW	12.5 or 25W	12.5 or 25W	12.5 or 25W	12.5 or 25W	4 or 8W	12.5 or 25W
Receiver Noise Figure	15 dB	8 dB	8 dB	8 dB	N/A	8 dB	8 dB
Available e.i.r.p.		22.4,25.4dBW (1,2 TWT)	same	same	same	same	same
Antenna Configuration	2 Horns, 20° beamwidth	12 Helix Planar array, 24° beamwidth	same	same	same	same	same
Gain							
Receive	19 dBi	15 dBi	same	same	same	same	same
Transmit	19 dBi	14 dBi	same	same	same	same	same
Modulation	Phase,Freq of: none, 100kHz, 1MHz, 10MHz, or 50 MHz	FM	FM	FM	FM	FM	FM
Capability	Propagation Measurements	Ground/ Aircraft experiments	same	same	On-board Video only	same	same

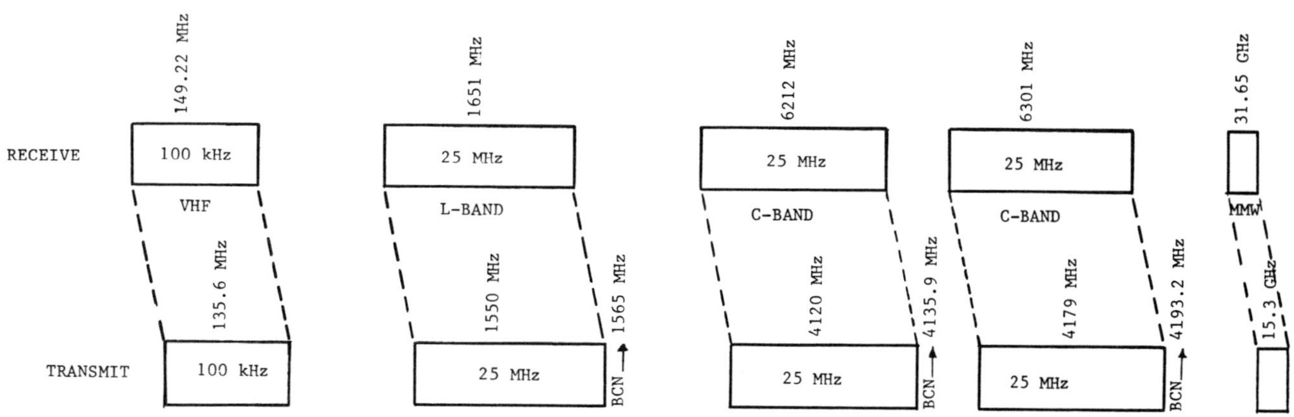

ATS 1 THROUGH 5 FREQUENCY PLAN

Richard W. Duesing
IEEE Compendium

PHOTO COURTESY OF FAIRCHILD

ATS 6

Richard W. Duesing
IEEE Compendium

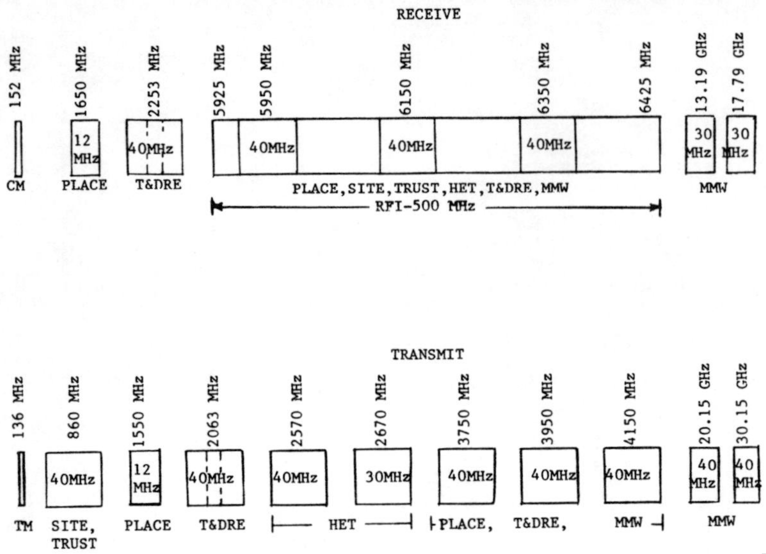

ATS 6 FREQUENCY PLAN

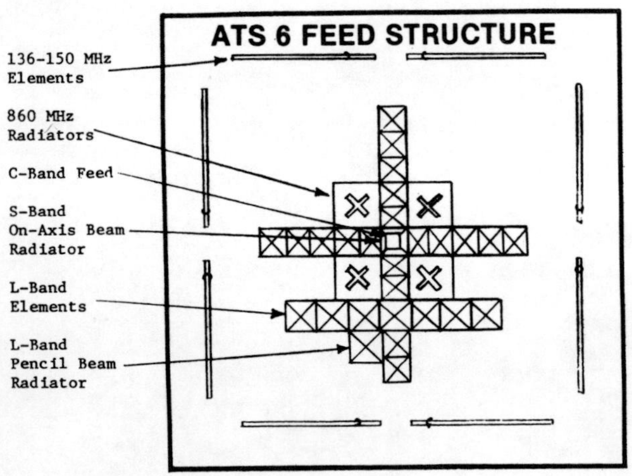

ATS 6 FEED STRUCTURE

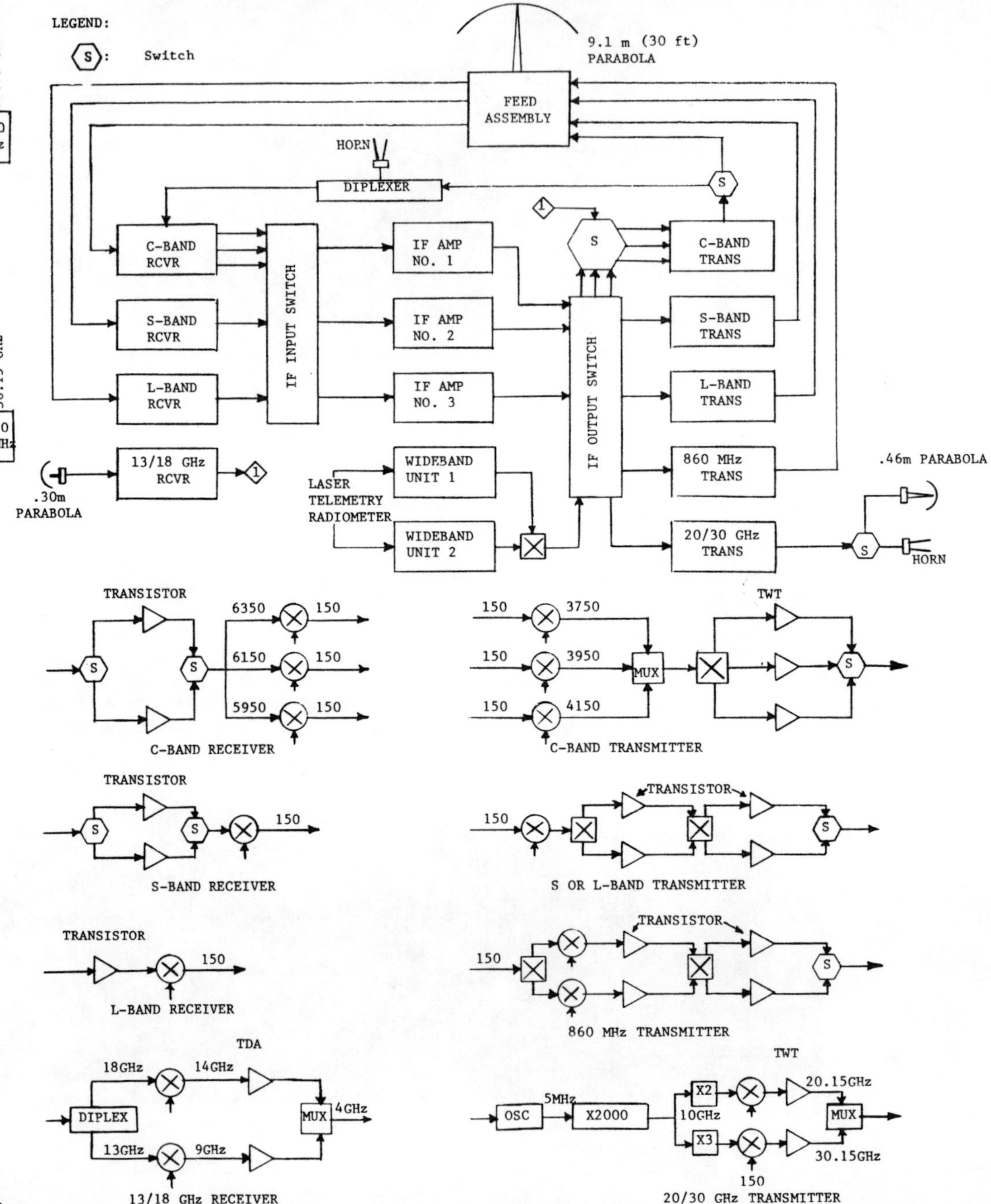

ATS 6 COMMUNICATIONS BLOCK DIAGRAM

Richard W. Duesing
IEEE Compendium

General

No. Satellites Built	1
Design Life	
Required	2 years
Goal	5 years
Launch Vehicle	Titan IIIC
Major Contractor	Fairchild
Date of Launch	30 May 1974

Physical Description

Stabilization	3-axis gravity gradient, 0.1° pointing accuracy
Maximum Height	838.2 cm (330 in)
Maximum Span	1574.8 cm (620 in)
Parabolic Reflector	914.4 cm (360 in) diameter
Earth-viewing Module	137x137x165 cm (54x54x65 in)
Two Solar Array Half Cylinders	
Radius	137 cm (54 in)
Length	238.8 cm (94 in)
Weight	
Launch	1406.7 kg (3101.8 lbs)
Orbit (initial)	1360 kg (2992 lbs)

Antenna Configuration

- Prime Focus Feed (PFF) illuminating the 9.1 m parabolic reflector for frequencies at VHF, UHF, L-band, S-band and C-band.
- Earth Coverage Horn (ECH) antennas for C-band interferometer and millimeter wave experiment.
- Parabolic (.46m) for millimeter wave experiment.
- Parabolic (.30m) for millimeter wave experiment.

Major Experiments

- Position Location and Aircraft Communication Experiment (PLACE)
- Satellite Instructional Television Experiment (SITE)
- Television Relay Using Small Terminals (TRUST)
- Health and Education Television Experiment (HET)
- Tracking and Data Relay Satellite Experiment (T&DRE)
- Radio Frequency Interference Experiment (RFI)
- Millimeter Wave Experiments (MMW)

Experiment/ Function	Center Frequency (MHZ)		Bandwidth (MHz)	Antenna Characteristics				Transmit Performance		Receive Performance		Capability
	Transmit	Receive		Antenna	Polarization (XMIT & RCV)	Field of View(deg)	Gain(dB)-Peak	RF Power (Watts)	e.i.r.p. (dBW)-Peak	G/T-peak (dB/K)	T_s (dBK)	
PLACE-ATS/ Aircraft	1550	1650	12	9.1m-fan 9.1m-pencil	RHCP RHCP	0.8x7.5 1.5	28-29 38.5	40 40	40.3 51.0	-4.4 5.5	31.1 31.1	Two way link through ATS-6 between ground & aircraft for voice & ranging.
PLACE-ATS/ Ground	3750 3950 4150	5950 6150 6350	40	ECH	Linear	13x20	16.5-16.3	12	28.0 on axis	-17.0	32.9	
SITE, TRUST-XMIT	860	-	40	9.1m	RHCP	2.8	30	80	51.0			TV for instruction to inexpensive ground terminals
SITE, TRUST-RCV	-	5950		ECH or 9.1m	Linear RHCP	13x20 0.4	16.3 48.4			-17.0 +13.7	32.9 32.9	
HET -XMIT chnl 1 -XMIT chnl 2 -RCV	2570 2670	5950	40 30	9.1m 9.1m ECH or 9.1m	RHCP RHCP Linear RHCP	1.0 1.0 13x20 0.4	41.5 41.5 16.3 48.4	15 15	53.0 53.0	-17.0 13.7	32.9 32.9	2 FM-TV carriers with audio subcarriers plus separate telephone carriers
T&DRE -ATS/Stle	2063	2253	Two chnls 12 or 40	9.1m	RHCP	13.2 switched array	36.4min	20	48.0min	7.0min	30.2	Two way link through ATS-6 between aground & low altitude satellite
-ATS/Grnd	3753 3953 4153	5938 6138 6338	12 or 40	ECH	Linear	13x20	16.5xmit 16.3rcv	12	28.0	-17.0	32.9	
RFI		5925-6425	500	9.1m ECH	Horizontal Vertical, RHCP linear	0.4 13x20	48.4 16.3			17.0 -17.0		Determine extent of interference between terrestrial and satellite services
MMW-NASA Propaga-tion Mode CW	(GHz) 20.0, 30.0			Horn	Linear	5x7	27	2	30			CW accommodated 60 dB fades, communication mode used digital rates up to 40 megabits per second
Multi-tone (9 tones)	same							.06/ tone	15/tone			
Communi-cations	20.15 (GHz)		40	.46m		2.4	37	2	40			
	30.15 (GHz)		40	.46m		1.6	39	2	42			
	3750 3950 4150	5950 6150 6350	40	ECH	Linear	13x20	16.5-16.3	12	28	-17.0	32.9	
				9.1m (RCV only)	RHCP	0.4	48.4			13.7	32.9	
MMW-COMSAT	4150	13.19 (GHz) 17.79 (GHz)	30	ECH(XMIT) .30m(RCV)	Linear Linear	13x20 4x8	16.5 26-28	0.2to 1.3mw per carrier	-13to-21	10 dB (Noise figure)		40 unmodulated carriers
Telemetry Command	136	152	2 .03	9.1m 9.1m	RHCP RHCP	15 15	17 17	2	10.2	-18.0	36.4	Telemetry/Command

ATS 6 TECHNICAL PARAMETERS

Richard W. Duesing
IEEE Compendium

BSE
("Yuri" or "Lily")

PHOTO COURTESY OF
TOSHIBA ELECTRIC CO/
GENERAL ELECTRIC CO.

Yoshitaka Kurihara
IEEE Compendium

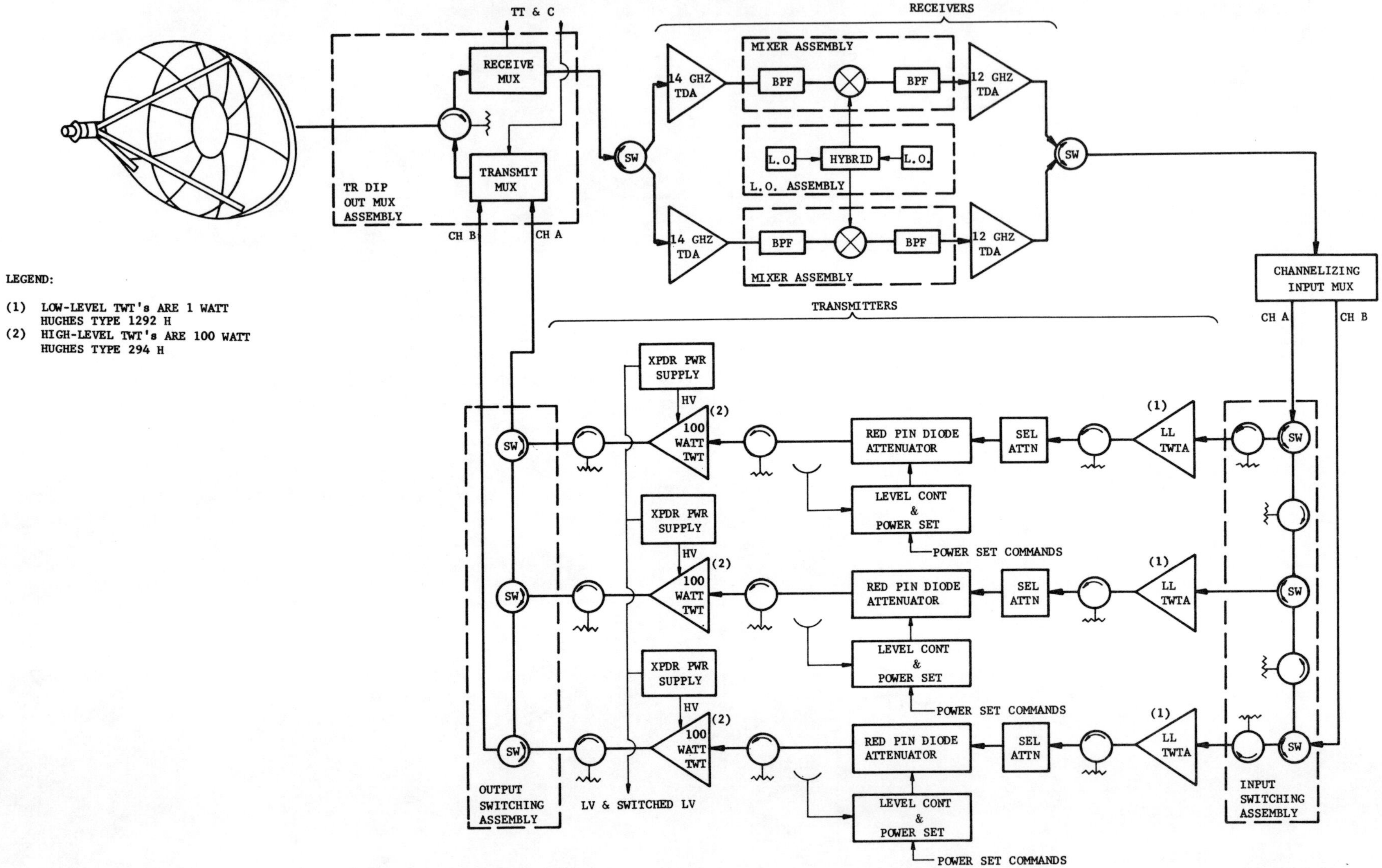

BSE COMMUNICATIONS BLOCK DIAGRAM

Yoshitaka Kurihara
IEEE Compendium

TELEMETRY

Television Channel

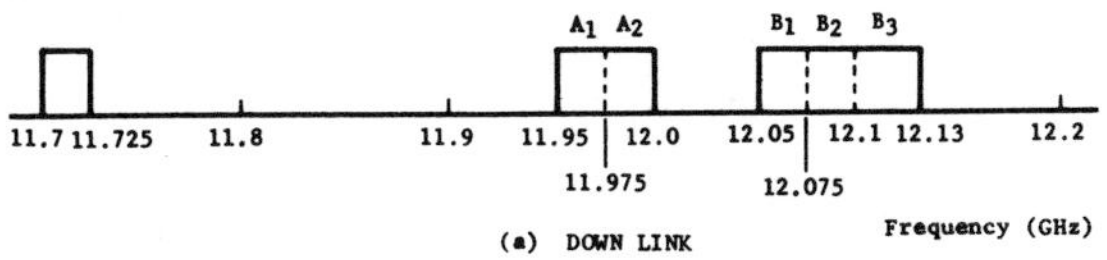

COMMAND

Television Channel

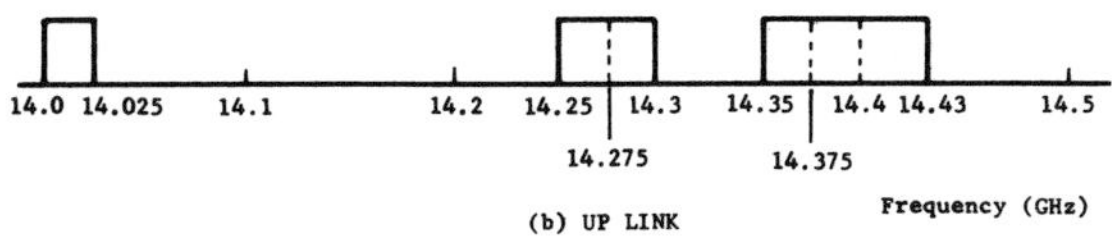

Carrier Frequency (MHz)

f_0A_1 = 11.9625/14.2625

f_0A_2 = 11.9875/14.2875

f_0B_1 = 12.0625/14.3625

f_0B_2 = 12.0875/14.3875

f_0B_3 = 12.1150/14.4150

BSE FREQUENCY PLAN

General

No. Satellite Built	: 1 of Flight Model, & 1 of Proto-flight Model
Design Life	: 3-years
Launch Vehicle	: Delta 2914
Major Contractor	: Toshiba Electric Co./ General Electric Co.
Date of Launch	: April 1978

Physical Description

Stabilization	: 3-axis Stabilized, Zero-momentum
Center Body Width	: 132.0 cm (52.0 in)
Length	: 119.9 cm (47.2 in)
Height	: 309.2 cm (121.7 in)
Solar Array Panel(Deployed)	
Total Length	: 894.8 cm (352.3 in)
Width	: 148.0 cm (58.3 in)
Weight	
Launch	: 673.4 Kg (1484.6 lbs)
Orbit(initial)	: 354.3 Kg (781.1 lbs)

Antenna Configuration

1 of center feed shaped beam parabolic antenna (transmit/receive) which has 3-feed horns

Polarization

Receive	: Linear (Orthogonal)
Transmit	: Linear (Orthogonal)

Communication Capability

Total Usable Bandwidth	: 130 MHz
Assigned Frequency Range	
Up-link (Receive)	: 14.25 to 14.43 GHz
Down-link (Transmit)	: 11.95 to 12.13 GHz
Repeaters	
No. on Line	: 2-color Television channels
No. of Redundant	: 1 for either channel (but not both)
Type	: Hughes 294-H TWT
Bandwidth	: 25 MHz
RF Power	: 100 W/channel
Available e.i.r.p.(Nominal)	
Japan Mainland	: 55.3 dBW/channel
Remote Islands	: 46.3 dBW/channel
Receiver Noise Figure	: less than 8.0 dB at receiver input point
Saturation Flux Density	: - 80 to -96 dBW/m^2
Gain Control	: over 16 dB range, automatically
Communication Capability	
FM Color Television	: 2 channels, simultaneously
Received Quality	
Television-Video	: S/N ≥ 45 dB at Rain loss 1 dB
-Sound	: S/N ≥ 50 dB

BSE TECHNICAL CHARACTERISTICS

COURIER

PHOTO COURTESY OF
NATIONAL AIR AND SPACE
MUSEUM (USA)

T. Michael Kelley
IEEE Compendium

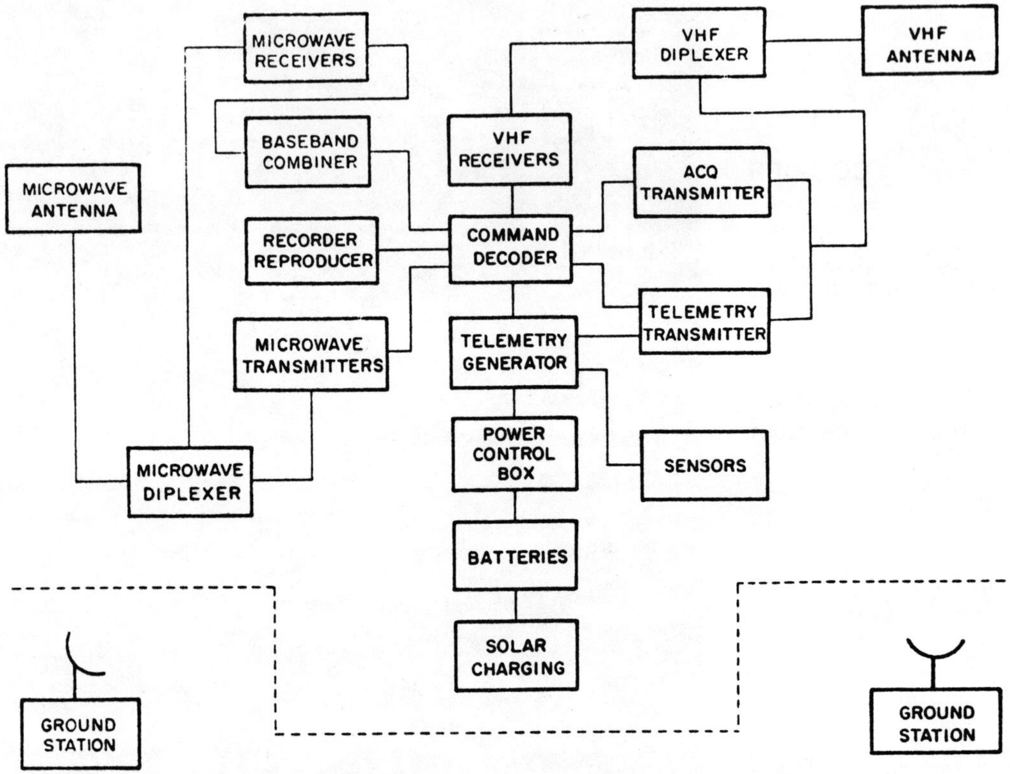

COURIER COMMUNICATION BLOCK DIAGRAM

T. Michael Kelley
IEEE Compendium

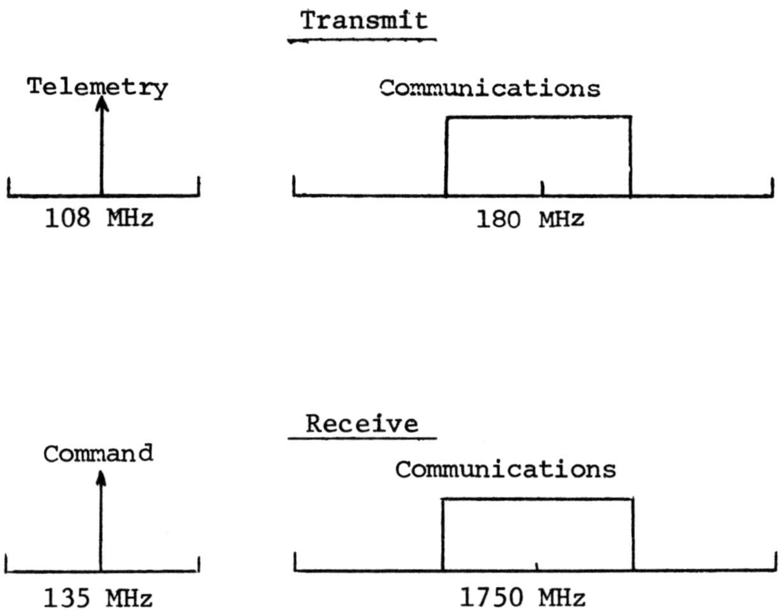

COURIER FREQUENCY PLAN

COURIER TECHNICAL CHARACTERISTICS

T. Michael Kelley
IEEE Compendium

General

No. of Satellites Built	2 (Courier 1A & 1B)
Launch Vehicle	Thor-Able Star
Launch Site	ETR (Cape Canaveral)
Sponsor/Manufacturer	U.S. Army Signal Corp/Philco-Ford Corp.
No. of Launches	2 18 August 1960 - Failure (1A) 4 October 1960 - Success (1B)

Physical Description

Stabilization	Spin-stabilized
Shape	Spherical
Size	1.3m in diameter
Weight	227 kg

Antenna

Communications (UHF) - 2 slotted-fin antennas 180° apart
TT&C (VHF) - 4 whip turnstile antennas
Both are approximately isotropic

Orbital Parameters

Perigee	942 km
Apogee	1200 km
Inclination	28.3°
Period	106.9 min

Communications Capability

Frequency Plan	
Transmit	
Communications	1800 MHz
Telemetry	108 MHz
Receive	
Communications	1750 MHz
Command	135 MHz
RF Power Output	1 Watt
Power Supply	
19152 Solar Cells	
Nickel-Cadmium Batteries	
Modulation	FM

Communications Capacity

Simultaneously transmits,receives, and stores 68,000 digitally encoded words per minute

Significance

Demonstrated the feasibility of using active repeater satellites for delayed message transmission

COMMUNICATIONS SATELLITE (CS)
("Sakura" or "Cherry Blossom")

PHOTO COURTESY OF
MITSUBISHI ELECTRIC CORP/
FORD AEROSPACE AND
COMMUNICATIONS CORP.

Yoshitaka Kurihara
IEEE Compendium

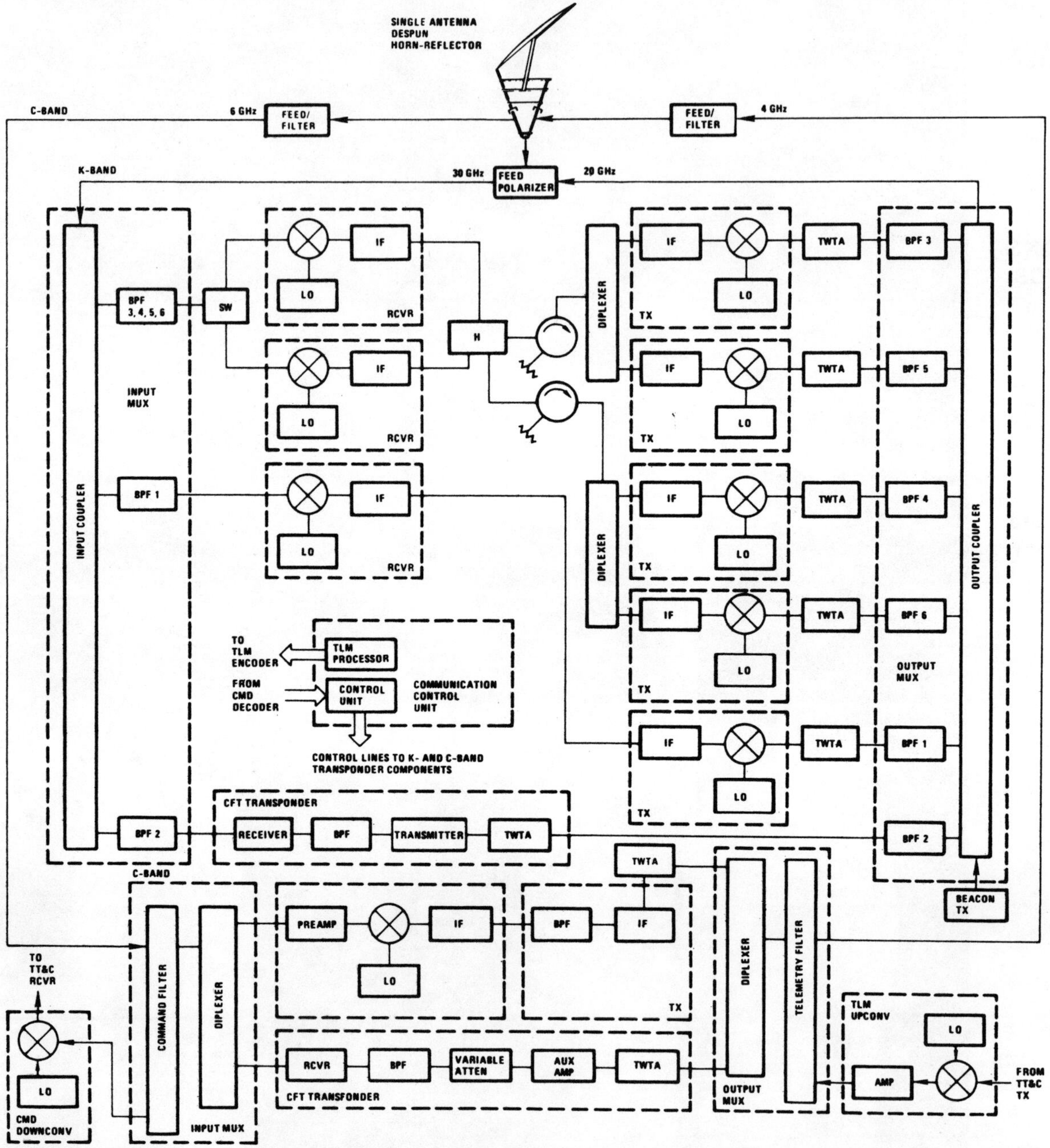

CS COMMUNICATIONS BLOCK DIAGRAM

Yoshitaka Kurihara
IEEE Compendium

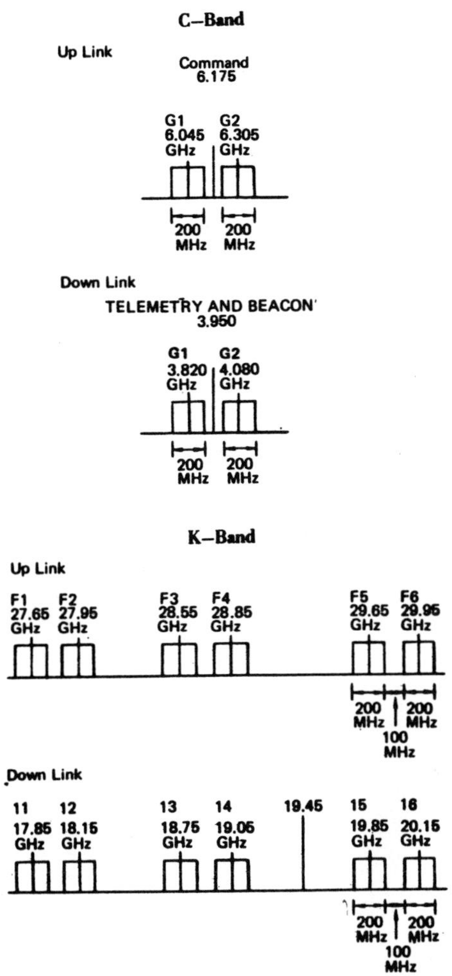

CS FREQUENCY PLAN

General	
No. Satellites Built	2 (FM and PFM)
Design Life	3 years
Launch Vehicle	Delta 2914
Major Contractor	Mitsubishi Electric Corp./ Ford Aerospace and Comm Corp.
Date of Launch	December 1977

Physical Description	
Stabilization	Spin (Despin Antenna Earth Oriented, Spinning Drum)
Cylinder Diameter	218 cm (86 in)
Cylinder Height	222 cm (87 in)
Height with Antenna	348 cm (137 in)
Weight	
Launch	696 kg (1490 lbs)
Orbit (initial)	340 kg (749 lbs)

Antenna Configuration	
	Multi-frequency Shaped Beam Horn-Reflector Mechanical Despun Antenna (Transmit and Receive)

Polarization	
Receive	LHC
Transmit	RHC

Communication Capability	
Total Usable Bandwidth	1600 MHz
Assigned Frequency Range	
Up-Link (Receive)	27.55-30.05 GHz (K-Band) 5.945-6.405 GHz (C-Band)
Down-Link (Transmit)	17.75-20.25 GHz (K-Band) 3.720-4.180 GHz (C-Band)
Repeaters	
No. on board	8 (6 K-Band and 2 C-Band)
Type	Hughes 292H (5 K-Band) NEC LD4098 (1 K-Band) Hughes 296HA(1 C-Band) NEC LD4147 (1 C-Band)
Bandwidth	200 MHz
RF Power	4 Watts (K-Band) 6 Watts (C-Band)
Available e.i.r.p. (Nominal)	
Beam Edge	39 dBW (K-Band) 30 dBW (C-Band)
Beam Peak	44 dBW (K-Band) 35 dBW (C-Band)
Receive G/T (Nominal)	
Beam Edge	+1.4 dB/ K (K-Band) -3.6 dB/ K (C-Band)
Beam Peak	+6.4 dB/ K (K-Band) +1.4 dB/ K (C-Band)
Flux Density at Beam Edge	-123 dBW/m^2 (K-Band) -132 dBW/m^2 (C-Band)
Communication Capability Per Repeater PCM/PM	100 mbps
Total Satellite Capacity 8 PCM/PM Repeaters	100 mbps x 8

CS TECHNICAL CHARACTERISTICS

CTS ("Hermes")

PHOTO COURTESY OF DEPARTMENT OF COMMUNICATIONS, GOVERNMENT OF CANADA

N. George Davies
IEEE Compendium

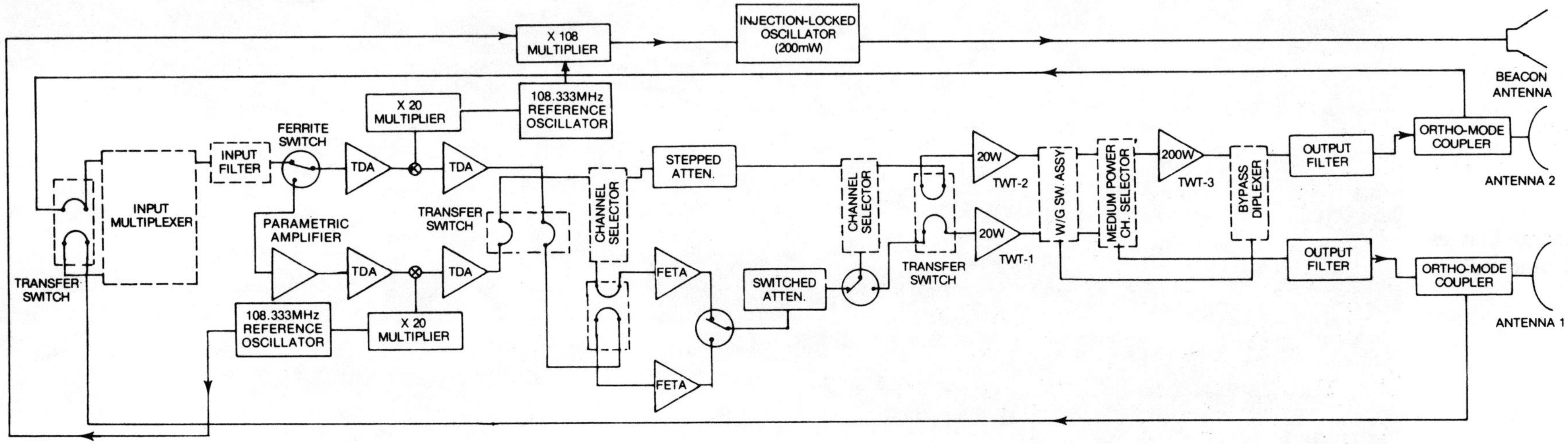

CTS COMMUNICATIONS BLOCK DIAGRAM

N. George Davies
IEEE Compendium

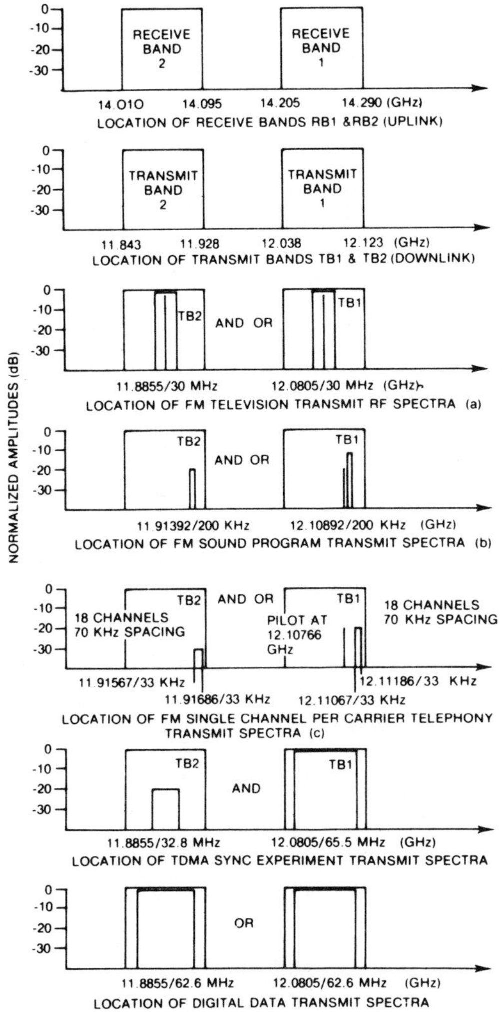

NOTES— 11.8855/30 MHz ≡ 11.8855 GHz CENTRE FREQUENCY/30 MHz RF BANDWIDTH

— SPECTRA OF RB1 & RB2 CORRESPONDS TO TB1 & TB2 RESPECTIVELY WITH APPROPRIATE FREQUENCY TRANSLATOR

— SPECTRA TYPE a, b, c OR d CAN BE CHOSEN INDEPENDENTLY FOR TB1/RB1 AND TB2/RB2

CTS FREQUENCY PLAN

General

No. Satellites Built	1
Design Life	2 years
Launch Vehicle	2914 Thor Delta
Design Authority	Communications Research Centre
Date of Launch	17 January 1976

Physical Description

Stabilization	3 Axis Stabilized, Body Stable with Forward Platform Earth Facing
Body - length x width	183 cm (72 in) x 167 cm (66 in)
Body Height	117 cm (46 in)
Height incl. antenna feed	188 cm (74 in)
Weight:	
Launch	676 kg (1490.5 lbs)
Orbit (initial)	362 kg (798.8 lbs)
Solar Array Tip to Tip	1600 cm (630 in)

Antenna Configuration

Transponder:	
2 Spot - steerable over Earth coverage	#1 Transmit/Receive #2 Transmit/Receive
- 2.5° beam to 3 dB points	
Beacon:	
Fixed conical	17.5° to 3 dB points

Polarization

Transmit Spot	Linear N-S
Receive Spot	Linear E-W
Beacon	R.H.C.

Beacon:

Power Output:	200 milliwatts CW
Frequency:	11.700 GHz

Communication Capability

Total Usable Bandwidth	2 x 85 MHz
Assigned Frequency Range	
Up-Link (Receive)	14.0-14.3 GHz
Down-Link (Transmit)	11.7-12.2 GHz
Repeaters	
No. on Line - 200W	1
No. on Line - 20W	1
No. Redundant - 20W	1
Type - 200W	Litton L-5394
Type - 20W	Thomson CSF
Bandwidth	85 MHz
RF Power - TB1	240 Watts
RF Power - TB2	20 Watts
Available e.i.r.p. (Nominal)	
Spot	
Beam Edge - TB1	56.5 dBW
Beam Centre - TB1	59.5 dBW
Beam Edge - TB2	45.4 dBW
Beam Centre - TB2	48.4 dBW
Receive G/T (Nominal)	
Beam Edge	+3.8 dB/K
Beam Centre	+6.8 dB/K
Saturation Flux Density at Beam Edge - TB1	
Min. Gain Step Attn setting	-81.4 dBW/m^2
Max. Gain Step Attn setting	-91.4 dBW/m^2
Nominal Gain Step Attn setting	-86.4 dBW/m^2
Saturation Flux Density at Beam Edge - TB2	
Min. Gain Step Attn setting	-84.1 dBW/m^2
Max. Gain Step Attn setting	-88.1 dBW/m^2

Communications Capability

Two of the following:

FM Television with up to 3 simultaneous voice channels
High quality 15 kHz sound program or radio broadcast
Two way, FM single channel per carrier, FDMA providing up to 18 telephone quality 3.1 kHz voice channels
Digital high speed data or TDMA at rates up to 65 Mb/s

Note: The location of the steerable spot beam antenna earth coverage determines the regions in which this capability can be provided at any given time.

CTS TECHNICAL PARAMETERS

ECHO

PHOTO COURTESY OF
NATIONAL AIR AND
SPACE MUSEUM (USA)

T. Michael Kelley
IEEE Compendium

ECHO I

PHOTO COURTESY OF NATIONAL AERONAUTICS AND SPACE ADMINISTRATION

T. Michael Kelley
IEEE Compendium

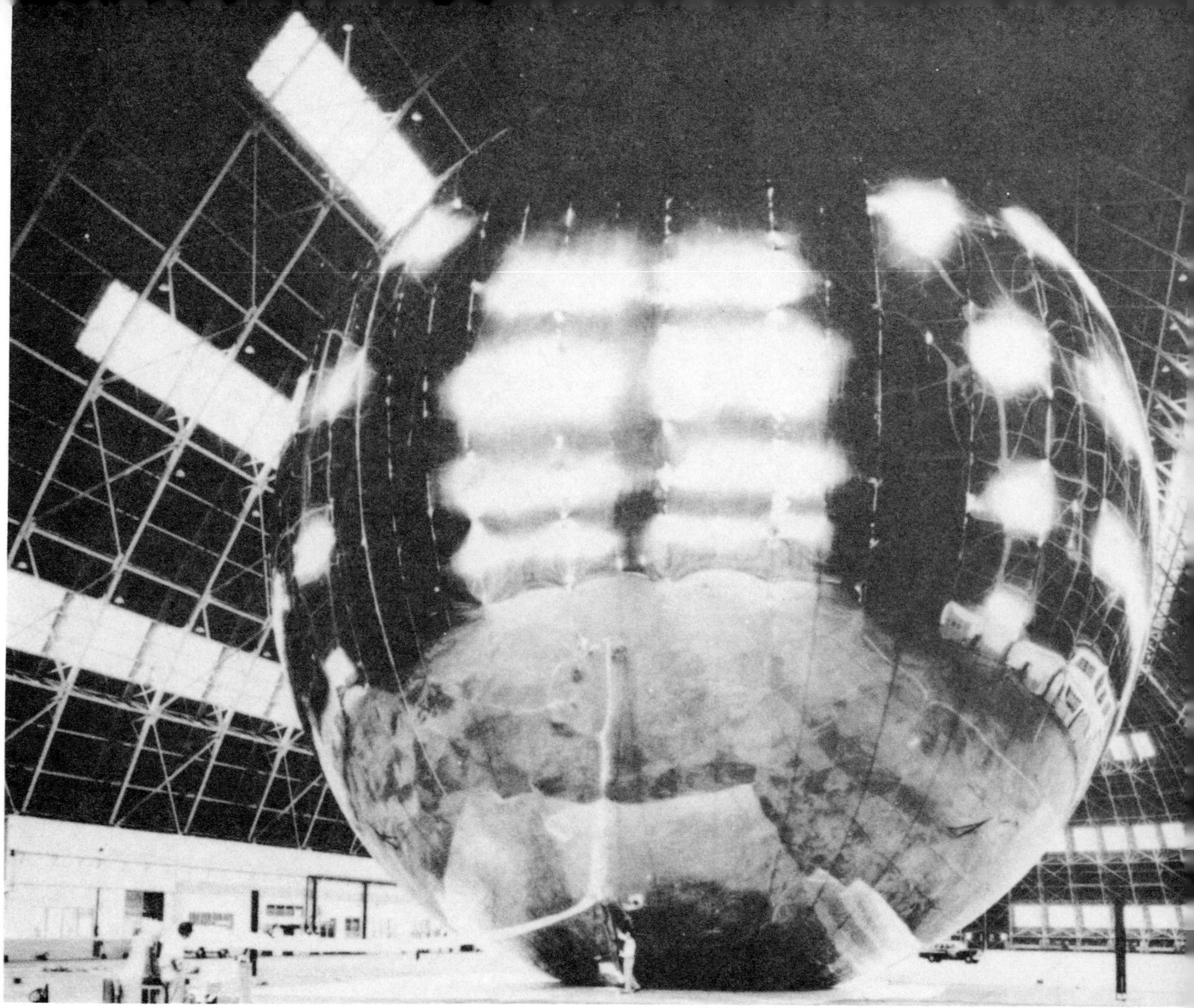

ECHO II

PHOTO COURTESY OF
NATIONAL AERONAUTICS AND
SPACE ADMINISTRATION

T. Michael Kelley
IEEE Compendium

ECHO 1 Frequency Plan

FROM	TO	FREQUENCY
Jet Propulsion Laboratory Goldstone, California (10 kW transmitter and 60' dish)	Naval Research Lab Stump Neck, Maryland (60' dish)	2390 MHz
Jet Propulsion Laboratory Goldstone, California (10 kW transmitter and 60' dish)	Bell Telephone Lab Holmdel, New Jersey (20'x20' horn antenna)	2390 MHz
Naval Research Lab Stump Neck, Maryland (60' dish)	Bell Telephone Lab Holmdel, New Jersey (20'x20' horn antenna)	2390 MHz
Bell Telephone Lab Holmdel, New Jersey (10 kW transmitter and 60' dish)	Jet Propulsion Lab Goldstone, California (85' dish)	960 MHz

ECHO 2 Frequency Plan

FROM	TO	FREQUENCY
Arapaho Station Dallas, Texas (60' dish)	Ohio State University Columbis, Ohio (4-30' dishes)	2260 MHz
Arapaho Station Dallas, Texas (60' dish)	Naval Research Lab Stump Neck, Maryland (60' dish)	2260 MHz 2380 MHz
Arapaho Station Dallas, Texas (28' dish)	Naval Research Lab Stump Neck, Maryland (60' dish)	2190 MHz

ECHO FREQUENCY PLAN

The ECHO satellite series consisted of passive reflector balloons only. No block diagram of a communications sub-system has been supplied.

ECHO COMMUNICATION BLOCK DIAGRAM

ECHO TECHNICAL CHARACTERISTICS

General

No of satellites built	3 (ECHO A-10, 1, & 2)
Launch Vehicles	A-10 & S/C 1 - Delta S/C 2 - Thor-Agena B
Launch Site	ETR (Cape Canaveral)
Sponsor/Manufacturer	NASA/G.T. Schieldahl
No. of launches	3 13 May, 1960 - Failure (A-10) 12 Aug, 1960 - Success (ECHO 1) 25 Jan, 1964 - Success (ECHO 2)

Physical Description

	ECHO 1	ECHO 2
Shape & Materials	Spherical pressurized balloon with an envelop of plastic mylar and aluminum	Same as ECHO 1
Diameter (Deployed)	30m	41m
Weight	75.5 kg	256 kg

Antenna

Quarter-wave monopole made of spring wire for tracking

Orbital Parameters

	ECHO 1	ECHO 2
Perigee (km)	1521	1299
Apogee (km)	1688	1341
Inclination	47.2°	81.5°
Period (min)	118.3	108.7

Communications Capability

Voice for ECHO 1 & 2	Baseband Bandwidth = 200 Hz to 3 kHz Frequency deviation = $\pm$30 MHz
Music for ECHO 1 & 2	Baseband Bandwidth = 30 Hz to 15 kHz Frequency deviation = $\pm$ 15 MHz
DIGITAL TRANSMISSION FOR ECHO 1 & 2	BER = 10^{-2} to 10^{-3}

Significance

Demonstrated the feasibility of passive reflector communication at high frequency

T. Michael Kelley
IEEE Compendium

EXPERIMENTAL COMMUNICATIONS SATELLITE (ECS)

("Ayame" or "Sweet Flower")

PHOTO COURTESY OF
MITSUBISHI ELECTRIC CORP/
FORD AEROSPACE AND
COMMUNICATIONS CORP.

Joji Tabata
IEEE Compendium

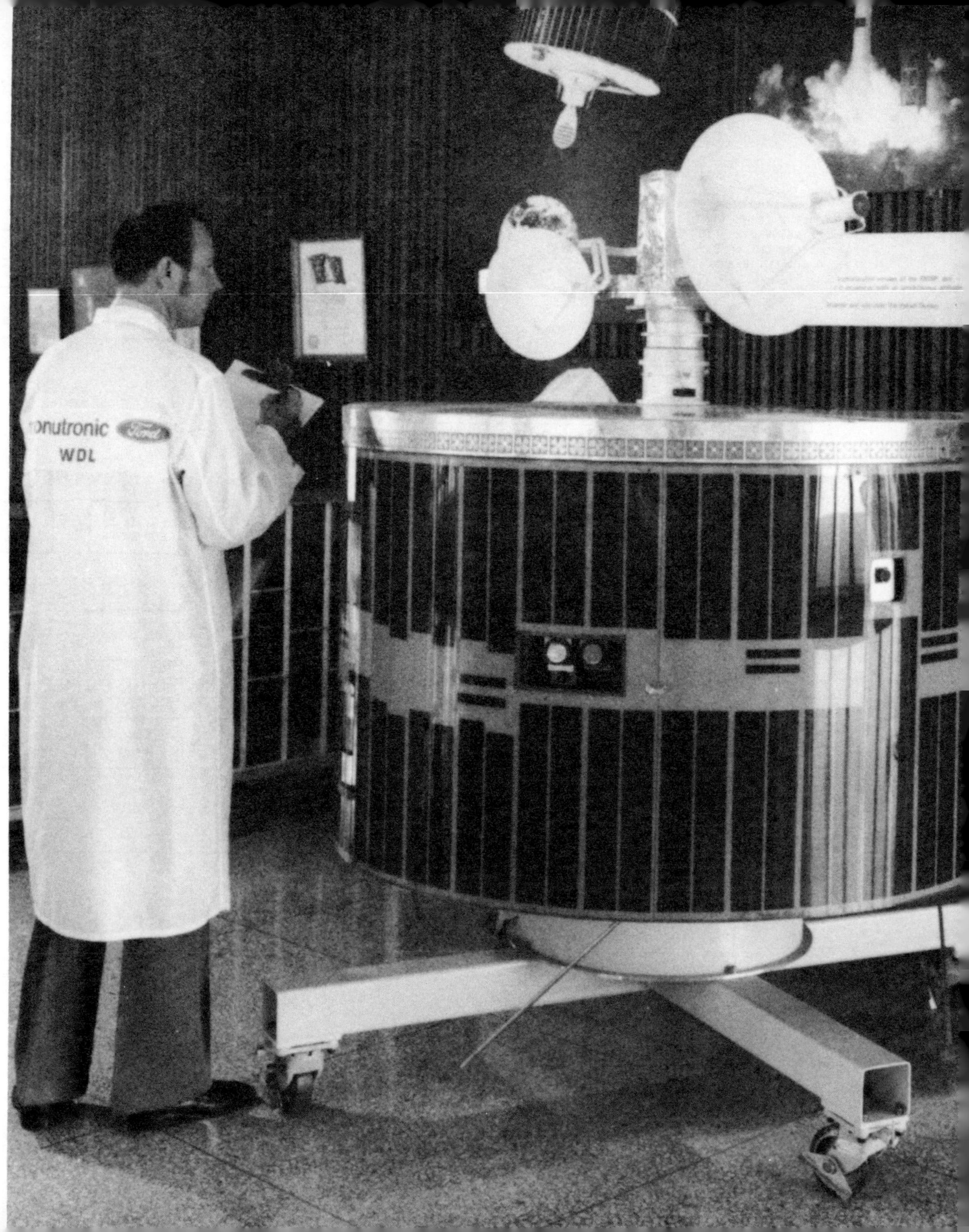

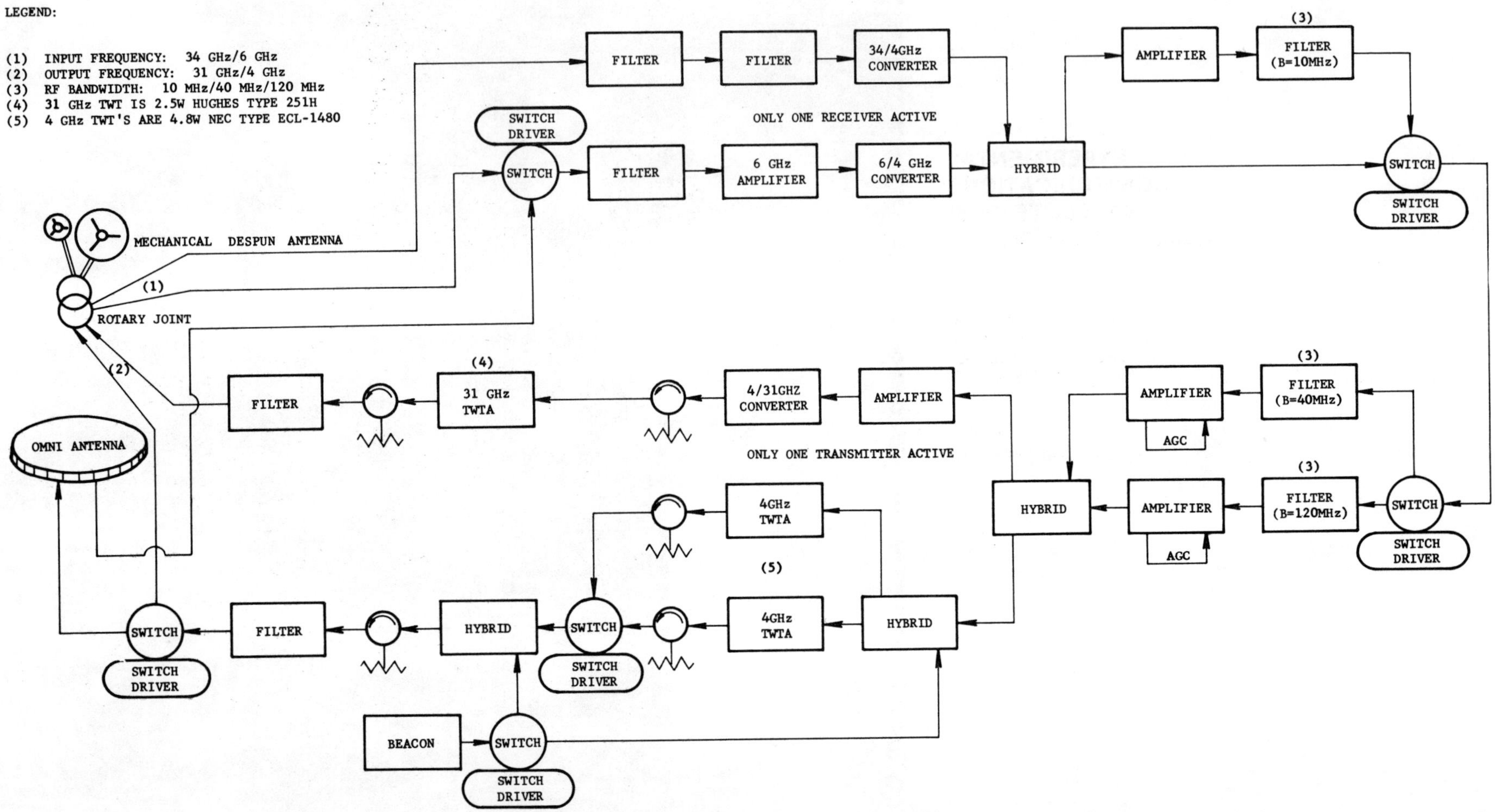

ECS COMMUNICATION BLOCK DIAGRAM

Joji Tabata
IEEE Compendium

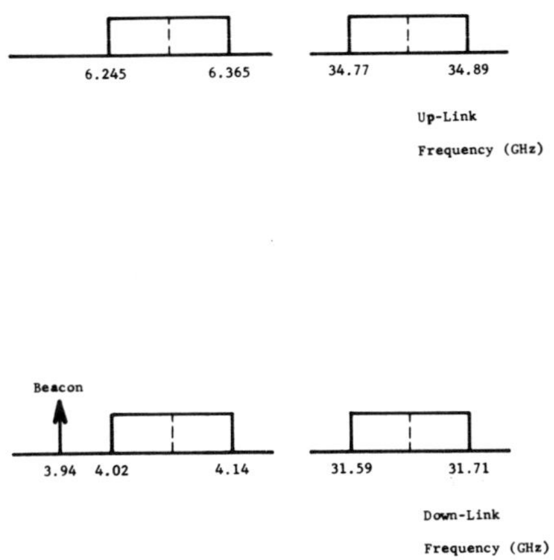

ECS FREQUENCY PLAN

General

No. Satellite Built	2
Design Life	1 year
Launch Vehicle	NLV-1
Major Contractor	Mitsubishi Electric Corp./ Aeronutronic Ford Corp., Nippon Electric Co.
Date of Launch	First Quarter of 1979

Physical Description

Stabilization	Spin
Cylinder Diameter	141.3 cm
Cylinder Height	93.7 cm
Height with Antenna	189.9 cm
Weight	
Launch	260 kg (Nominal)
Orbit (initial)	137 kg (Nominal)

Antenna Configuration

Despun Antenna	1 for 4 GHz & 6 GHz 1 for 31 GHz & 34 GHz
Omni Antenna	1 for 4 GHz & 6 GHz

Polarization

Receive	LHC
Transmit	RHC

Communication Capability

Total Usable Bandwidth	120 MHz
Assigned Frequency Range	
Up-Link	34,770-34,890 MHz 6,245- 6,365 MHz
Down-Link	31,590-31,710 MHz 4,020- 4,140 MHz 3,940 MHz (Beacon)
Repeaters	
No. on Line	1
Frequency Combination	34 GHz to 31 GHz (K to K) 6 GHz to 4 GHz (C to C) 34 GHz to 4 GHz (K to C) 6 GHz to 31 GHz (C to K)
Type	Hughes 251H (31 GHz) NEC ECL-1480 (4 GHz)
Bandwidth	120 MHz, 40 MHz or 10 MHz
RF Power	2.5 Watts for 31 GHz 4.8 Watts for 4 GHz
Available e.i.r.p. (Nominal)	
Despun Antenna	64.5 dBm for 31 GHz 53.2 dBm for 4 GHz
Omni Antenna	17.4 dBm for 4 GHz
Receive G/T (Nominal)	
Despun Antenna	-14.8 dB/ K for 34 GHz -11.8 dB/ K for 6 GHz
Saturation Flux Density at Beam Center	-90.7 dBW/m^2 for 34 GHz -89.6 dBW/m^2 for 6 GHz
Communication Capability (Nominal)	
FDM-FM	1224 channels
PCM-PSK	796 channels
FM TV	S/N = 57.4 dB

ECS TECHNICAL CHARACTERISTICS

Joji Tabata
IEEE Compendium

ETS II
("Kiku II" or "Chrysanthemum")

PHOTO COURTESY OF FORD AEROSPACE AND COMMUNICATIONS CORP.

Martin P. Brown, Jr. and
Kunihiko Niwa
IEEE Compendium

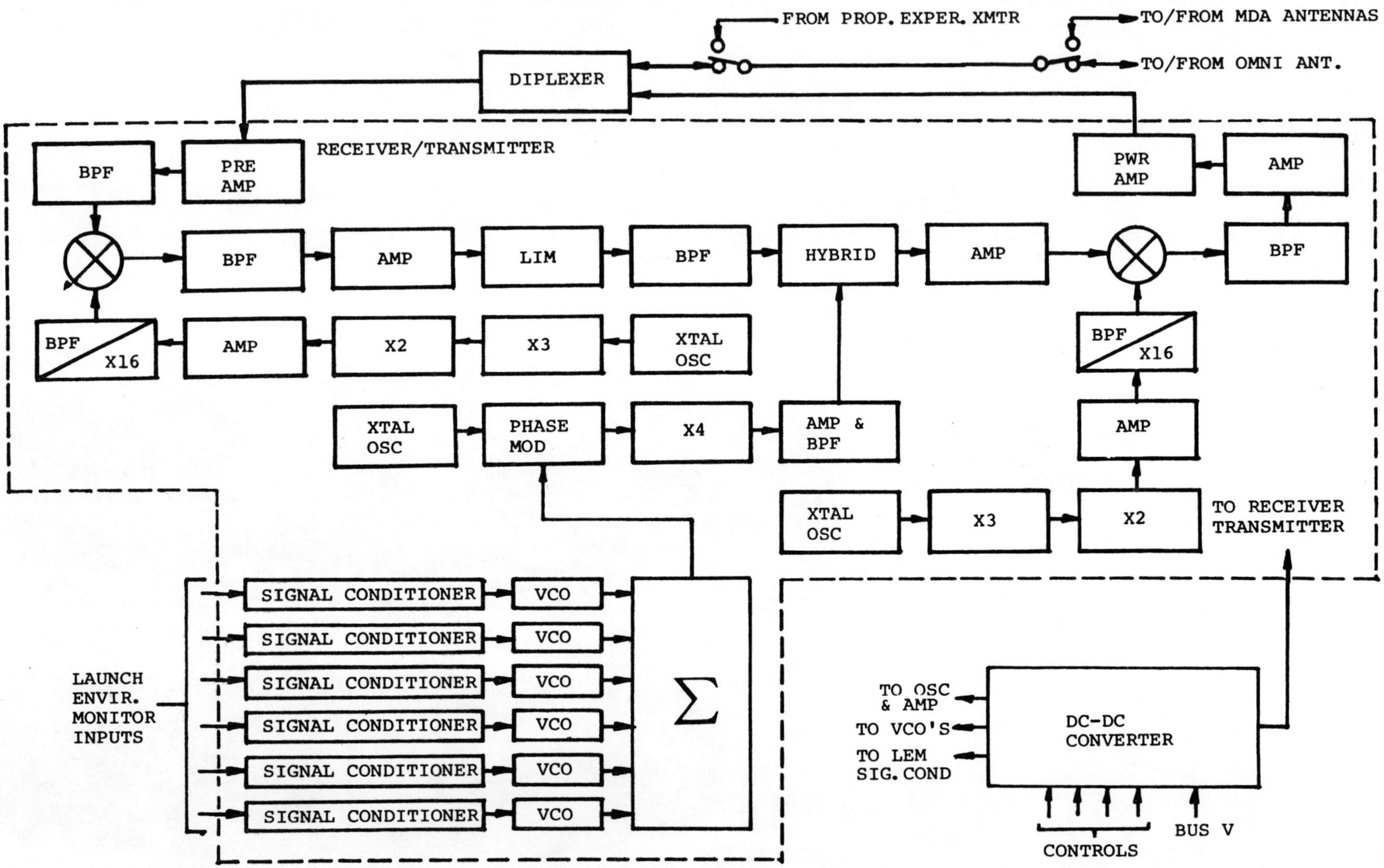

ETS II COMMUNICATION BLOCK DIAGRAM

Martin P. Brown, Jr. and
Kunihiko Niwa
IEEE Compendium

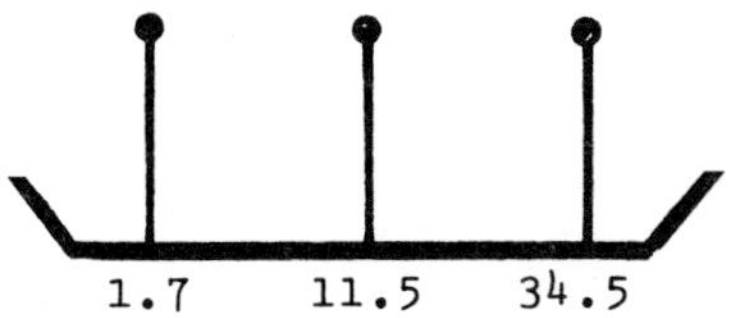

ETS II FREQUENCY PLAN

General

Operated by	NASDA (Japan)
Design Life	more than 1 year
Launch Vehicle	Japanese N rocket (No. 3)
Major Contractor	Philco Ford, Mitsubishi
Date of Launch	23 February 1977

Physcial Description

Stabilization	Spin stabilized (spin rate 100 ± 15 rpm)
Cylinder Diameter	141 cm (55.5 in)
Cylinder Height	93 cm (36.6 in)
Height w/antenna mast	158 cm (62.2 in)
Weight	254 kg (560.0 lbs)

Antenna Configuration

2 Parabolic reflectors
- one each for 11.5 and 34.5 GHz, 2.2° beamwidth at 34.5 GHz

1 end-fire antenna in a cavity for 1.7 GHz

Experimental Communication System

(1) Propagation exper. transmitter

1.705 GHz (S-band): CW or 100% AM by 300Hz square wave, 6 dBW measured ERP
11.50875 GHz (X-band): CW, 20 dBW measured ERP
34.52625 GHz (K-band): CW or 100% AM by 300Hz square wave, 24 dBW measured ERP

(2) S-band transponder

Beacon	Carrier frequency	1.7 GHz
	Spurious < -40dB	
Ranging	Transponder bandwidth	8.2 MHz
	Up link	2.1 GHz
	Down link	1.7 GHz

(3) Launch environment monitor

Carrier frequency	1.7 GHz
FM/PM	

Telemetry/Command System

(1) VHF transponder

Transmitter power 2W	
Transmitter frequency	136 MHz
Receiver frequency	148 MHz

(2) Telemetry system

Modulation	PM/PCM
Bit rate	250 BPS

Capacity 64 words/mainframe

o Analog data	48 words
o Digital data	5
o Analog sub-frame	3
o Digital sub-frame	2
o Serial data	6

(3) Command system

Modulation	PCM/FSK/AM/PM
Bit rate	128 BPS

ETS II TECHNICAL PARAMETERS

Martin P. Brown, Jr. and
Kunihiko Niwa
IEEE Compendium

OTS

PHOTO COURTESY OF MESH

Barry G. Evans
IEEE Compendium

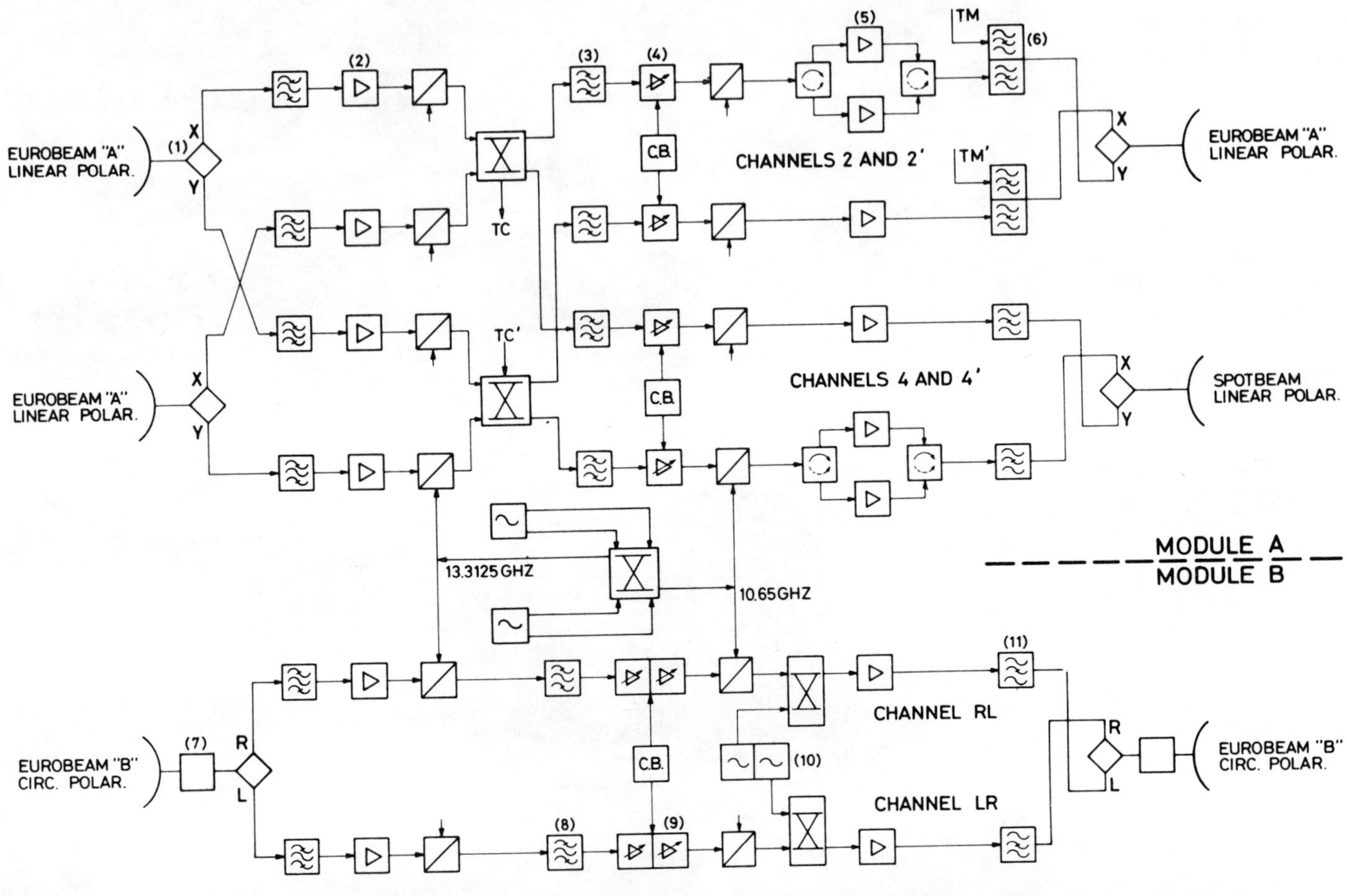

OTS COMMUNICATION BLOCK DIAGRAM

Legend

(1) Orthomode waveguide transitions for dual linear polarisation

(2) GaAs varactor diode parametric amplifiers. Gain : 13 dB, NF : 3.3 dB

(3) 14-cavity linear-phase IF filters. Bandwidth : 36 and 108 MHz.

(4) IF main amplifiers. Gain variable over a 14-dB range in 16 steps.

(5) Dual-collector TWTAs. Output power : 20 W.

(6) Output filters. Bandwidth : 36 and 108 MHz.

(7) Circular polariser

(8) Chebychef IF filters. Bandwidth : 12 MHz.

(9) IF main amplifiers. Gain variable over a 32-dB range in 16 steps.

(10) Redundant impatt-diode 12 GHz beacon generator.

(11) Output filters. Bandwidth : 20 MHz

Barry G. Evans
IEEE Compendium

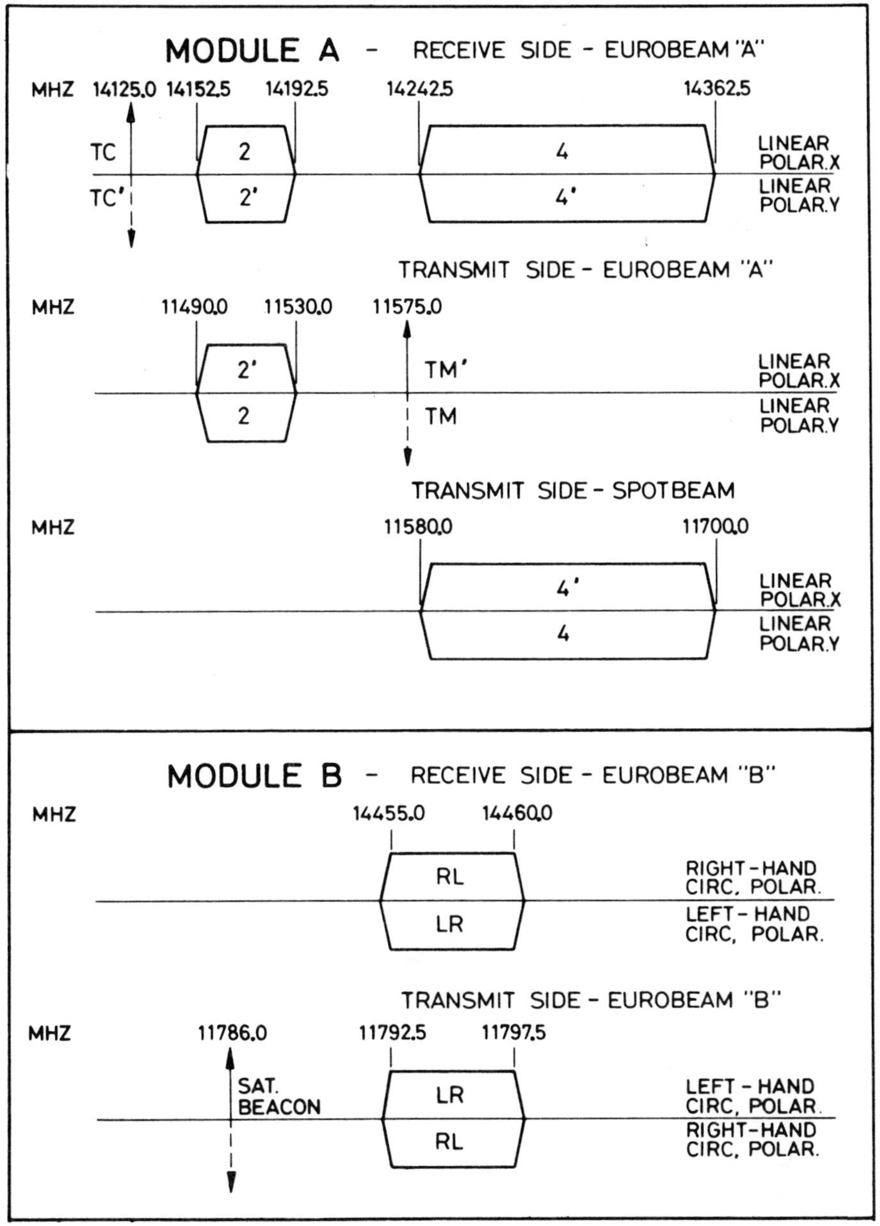

OTS FREQUENCY PLAN

General

No of satellites	2
Design life	5 years
Launch vehicle	Delta 3914
Major contractor	MESH Consortium
Date of launch	**13 Sep 1977**

Physical Description

Stabilisation	3-axis momentum wheel
Spacecraft size	
Diametrical envelope	2.18 m
Height	2.1 m
Deployed arrays	8.62 m
Weight at launch	902 kg
in orbit	443 kg (initial)

Antennas

Receive	2 Eurobeam "A" (dual linear polar)
	1 Eurobeam "B" (dual circular polar)
Transmit	1 Eurobeam "A" 1 Spotbeam (dual linear polar)
	1 Eurobeam "B" (dual circular polar)

Communication Payload Characteristics

Total usable bandwidth	330 MHz
Assigned freq. range	
Up-link (Receive)	14125-14460 MHz
Down-link (Transmit)	11490-11797.5 MHz
Repeater chains	
Number/Bandwidth	2 x 40 MHz
	2 x 120 MHz
	2 x 5 MHz
RF power output	20 W
Available EIRP (Nominal)	
Eurobeam "A"	38.0 dBW (centre)
	32.0 dBW (edge)
Spotbeam	47.3 dBW (centre)
	41.3 dBW (edge)
Eurobeam "B"	41.0 dBW (centre)
	35.0 dBW (edge)
Saturation flux density at beam edge	
Eurobeam "A" chains	
Min. gain setting	-63.5 dB (W/m^2)
Max. gain setting	-77.5 dB (W/m^2)
Spotbeam chains	
Min. gain setting	-61.0 dB (W/m^2)
Max. gain setting	-75.0 dB (W/m^2)
Eurobeam "B" chains	
Min. gain setting	-69.7 dB (W/m^2)
Max. gain setting	-101.7 dB (W/m^2)

Communication Capability

Per Eurobeam "A" chain	
PSK/TDMA 60 Mbit/s	900 channels
1 TV	S/N = 55 dB
Per Spotbeam chain	
PSK/TDMA 180 Mbit/s	2700 channels
Per Eurobeam "B" chain	Data transmission at 2 Mbit/s
Total sat. capacity	7200 channels on 5400 ch. and 2 TV

OTS TECHNICAL CHARACTERISTICS

Barry G. Evans
IEEE Compendium

RELAY

PHOTO COURTESY OF
NATIONAL AIR AND
SPACE MUSEUM (USA)

T. Michael Kelley
IEEE Compendium

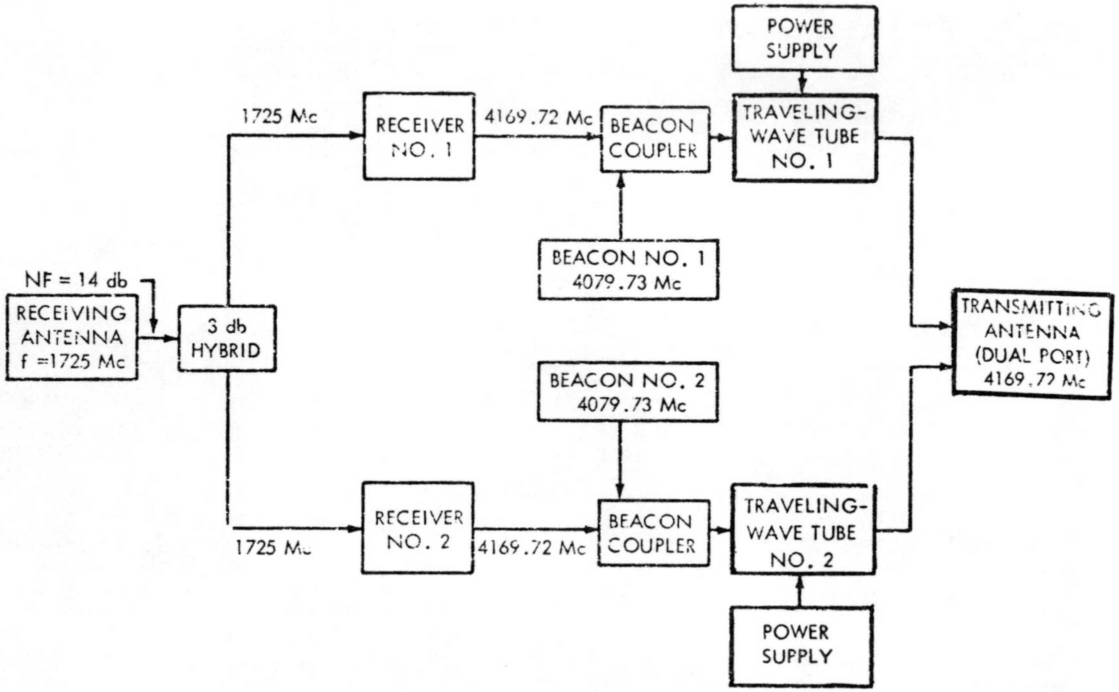

RELAY COMMUNICATIONS BLOCK DIAGRAM

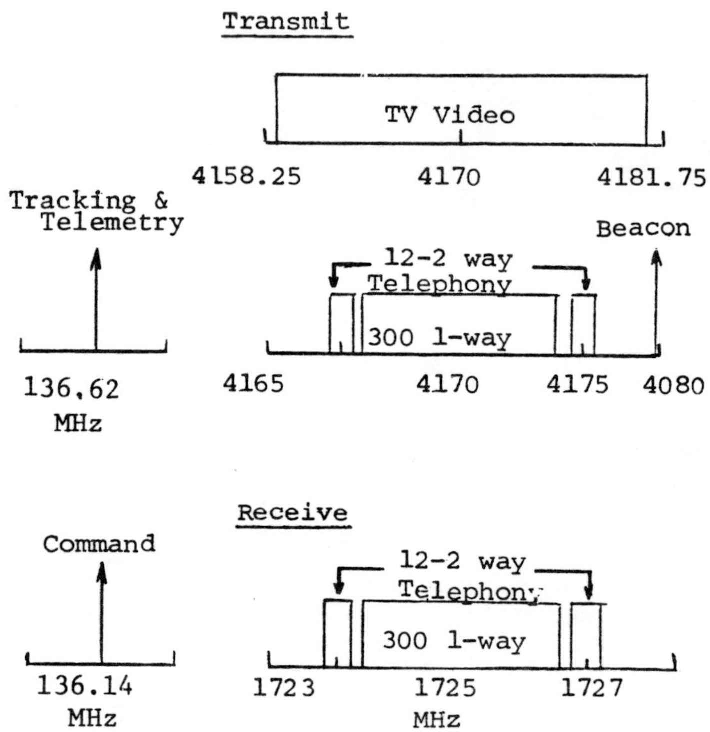

RELAY FREQUENCY PLAN

T. Michael Kelley
IEEE Compendium

General

No. of Satellites Built	2 (Relay 1 & 2)
Lanuch Vehicle	Thor-Delta
Launch Site	ETR (Cape Canaveral)
Sponsor/Manufacturer	NASA/RCA Astro-Electronics Div.
No. of Launches	2
	13 December, 1962 - Success
	21 January, 1964 - Success

Physical Description

	Relay 1	Relay 2
Stabilization	Spin-stablized	Spin-stabilized
Shape & Size	8-sided Prism	8-sided Prism
Diameter	.71 km	.74 km
Height	.83 km	.84 km
Weight	78kg	78kg

Antenna

Approximately Isotropic

Polarization

Transmit	Lefthand Circular
Receive	Righthand Circular
Beacon	Linear

Orbital Parameters

	Relay 1	Relay 2
Perigee (km)	1320	2089
Apogee (km)	7420	7413
Inclination	47.5°	46.3°
Period (Min)	185.09	194.73

Communications Capability

Total Usable Bandwidth	25 MHz
Frequency Plan	
Receive	
All Communication Experiments	1725 MHz
Command	136.140 MHz
Transmit	
Tracking Beacon	4080 MHz
Tracking & Telemetry	136.620 MHz
Wideband FM (TV)	4170 MHz (10 W)
Narrowband FM (Telephone)	4165 & 4175 MHz (4W)
RF Power Output	10 W (Saturated)
Power Supply	8215 Solar Cells Nickel-Cadmium Batteries
Modulation	FM

Communications Capacity

1 Television Program or

12 2-way telephone/data/facsimile channels or

300 1-way telephone/data/facsimile channels

Significance

First communications satellite designed to link North America, South America, and Europe

RELAY TECHNICAL CHARACTERISTICS

T. Michael Kelley
IEEE Compendium

SCORE SATELLITE EQUIPMENT

PHOTO COURTESY OF US ARMY THROUGH THE EFFORTS OF SAMUEL BROWN.

T. Michael Kelley
IEEE Compendium

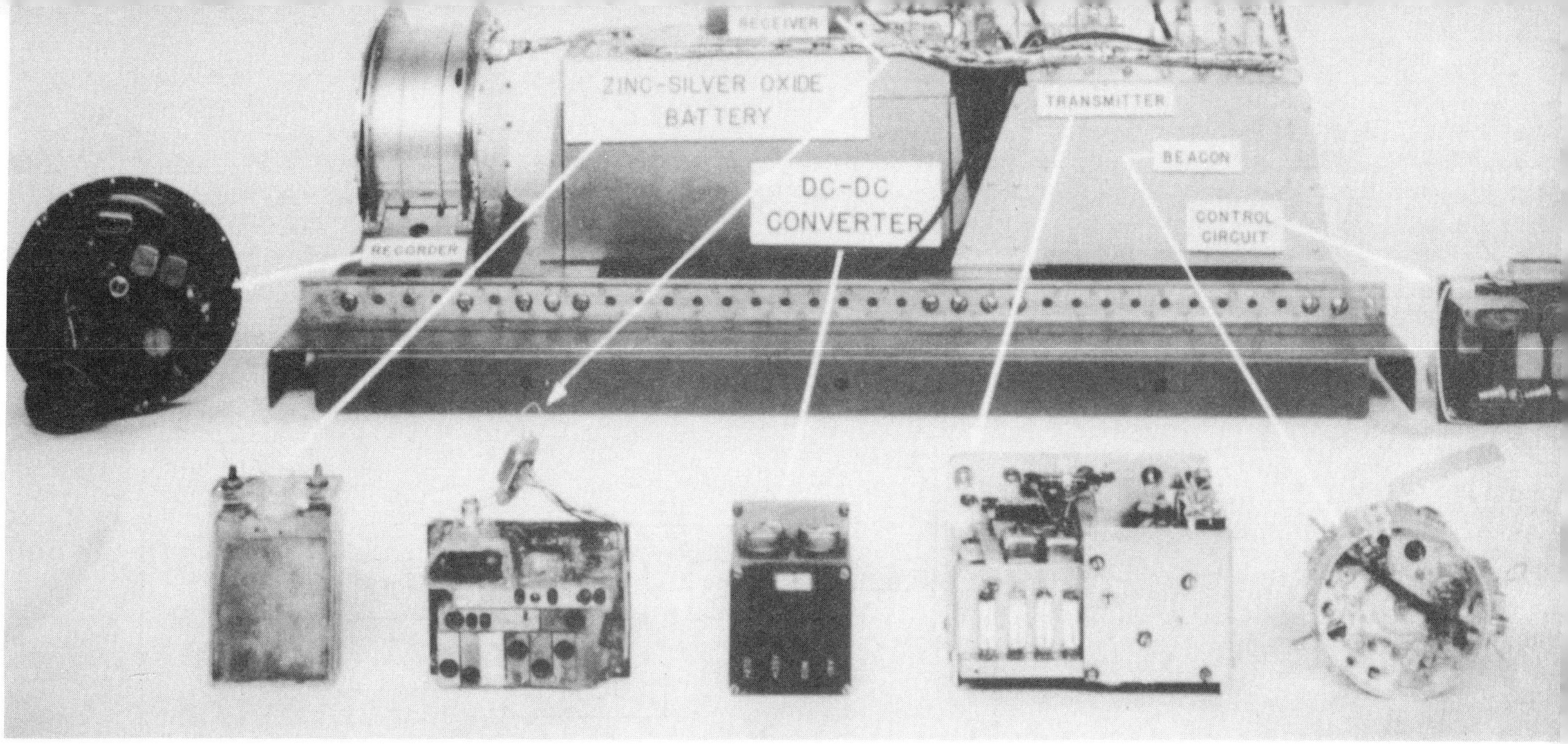

SIGFM/EL- 59-182 FORT MONMOUTH, N. J.

"SCORE" SATELLITE EQUIPMENT . (Lab. Design.-Constr.)
Used with Atlas Intercontinental Ballistic Missile
Overall Front View . Showing Components as Indicated
12 Feb 59

U. S. ARMY SIGNAL RESEARCH AND DEVELOPMENT LABORATORY

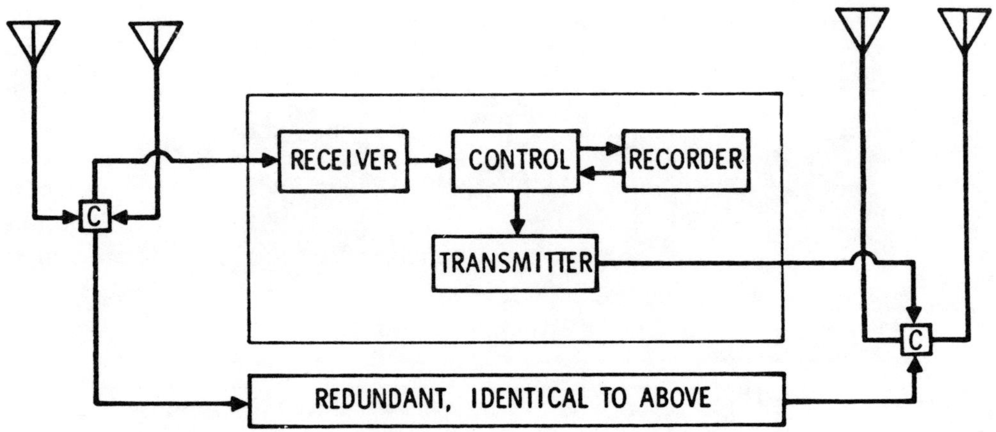

SCORE COMMUNICATION BLOCK DIAGRAM

T. Michael Kelley
IEEE Compendium

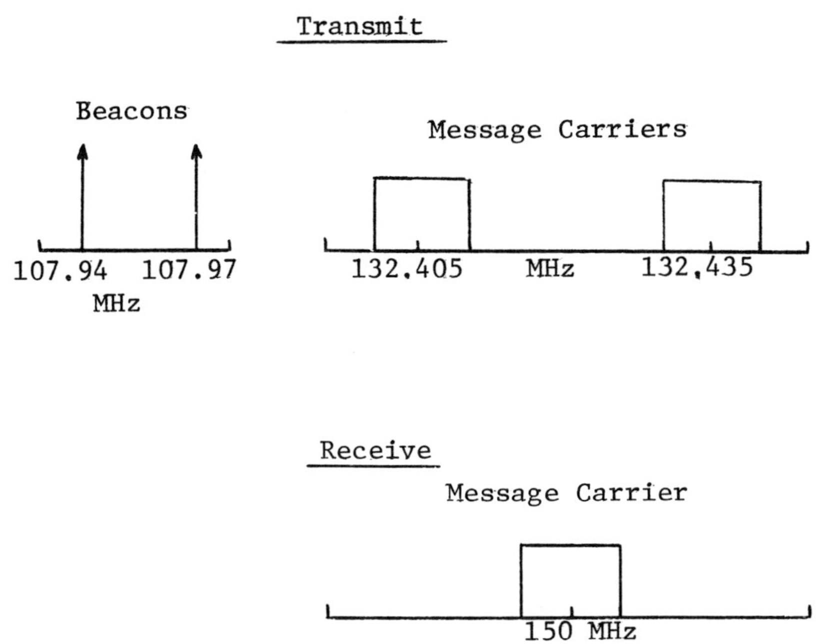

General

No. of Satellites Built	1
Launch Vehicle	Atlas 10-B
Launch Site	ETR (Cape Canaveral)
Sponsor/Manufacturer	ARPA/U.S. Army
No. of Launches	1
	18 December, 1958 -- Success

SCORE FREQUENCY PLAN

Physical Description

Instrumentation in Nose of an Atlas 10-B Rocket

Weight = 68 kg

Antenna	Wire monopoles protruding through slots in the pod cover of the instrument package

Orbital Parameters

Perigee	185 km
Apogee	1470 km
Inclination	32.3°
Period	101.5 min.

Communications Capability

Receive Messages	150 MHz
Transmit Messages (Delayed Repeater)	132.435 MHz or 132.905 MHz
Beacons	107.94 MHz or 107.97 MHz
Power Supply	Zinc - Silver Oxide Battery
RF Power Output	8 Watts (Max.)
Modulation	Narrowband FM

Communications Capacity

1 Voice Channel

or

7 60 Wpm Teletype Channel

Significance

- First Communications Satellite
- Demonstrated the feasibility of the delayed repeater technique

SCORE TECHNICAL CHARACTERISTICS

SIRIO

PHOTO COURTESY OF
COMPAGNIA NATIONALE
AEROSPAZIALE

Pier L. Bargellini
IEEE Compendium

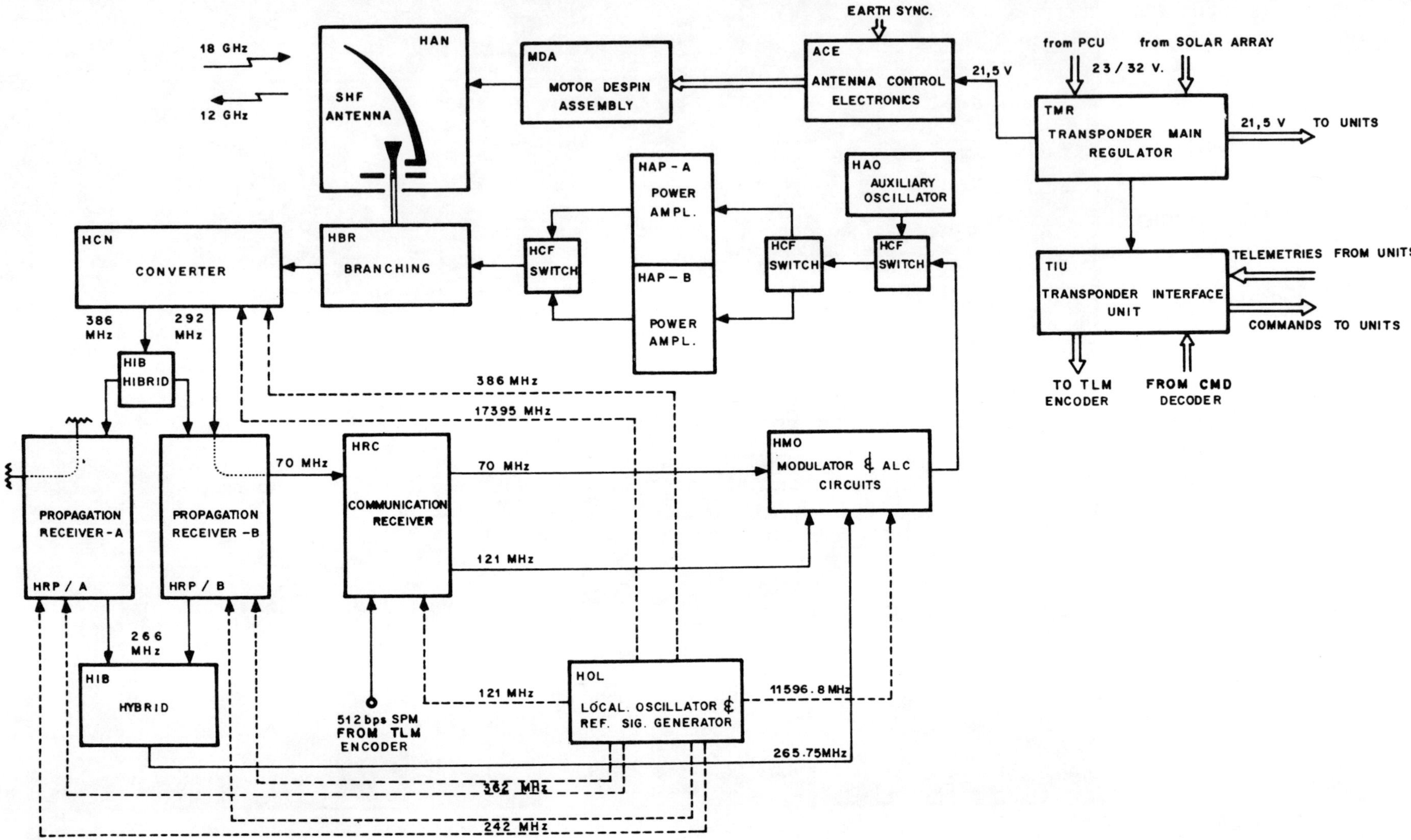

SIRIO COMMUNICATION BLOCK DIAGRAM

Pier L. Bargellini
IEEE Compendium

Propagation Experiments

Absolute attenuation at 17 or 17.8 GHz

Differential attenuation between 17 and 17.8 GHz

Absolute attenuation at 11.6 GHz

Differential attenuation between 17 and 17.8 GHz

Phase distortion in the band 11.3-11.9 GHz

Propagation Experiment Spectra

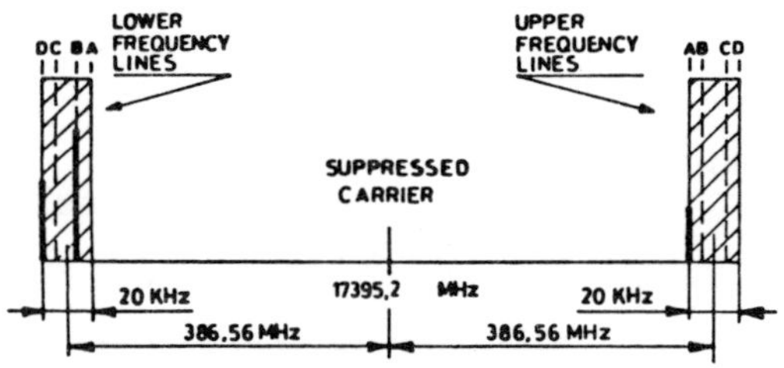

RECEIVED SPECTRUM (UP LINK)

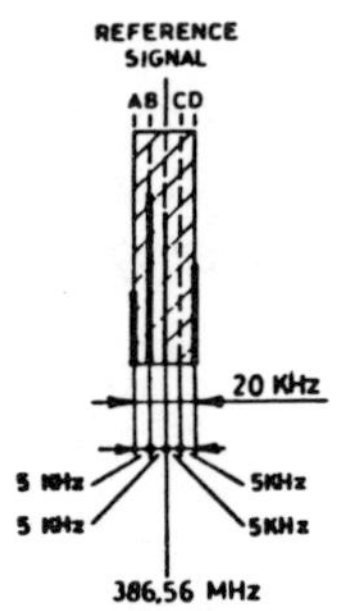

INTERMEDIATE FREQUENCY SPECTRUM

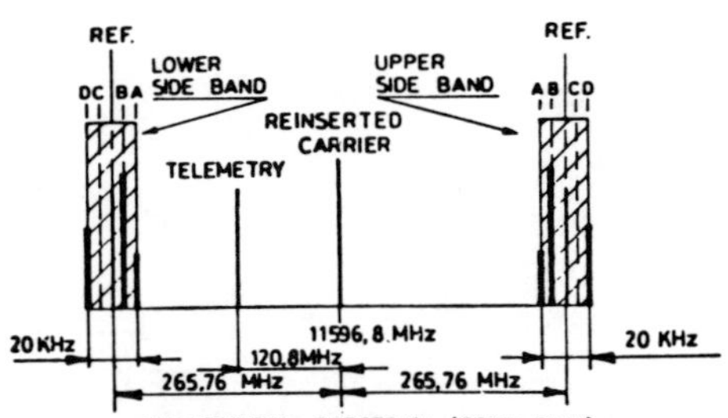

TRANSMITTED SPECTRUM (DOWN LINK)

Communications Experiments

Narrowband mode:

FM/FDMA telephony

PCM/PSK/FDMA digital transmissions

Wideband mode:

1 or 2 color TV channels

Experimental TV transmissions with sound in synch

Digital TV transmissions

High-speed digital transmissions up to 60 Mbit/s

Communications Experiment Spectra

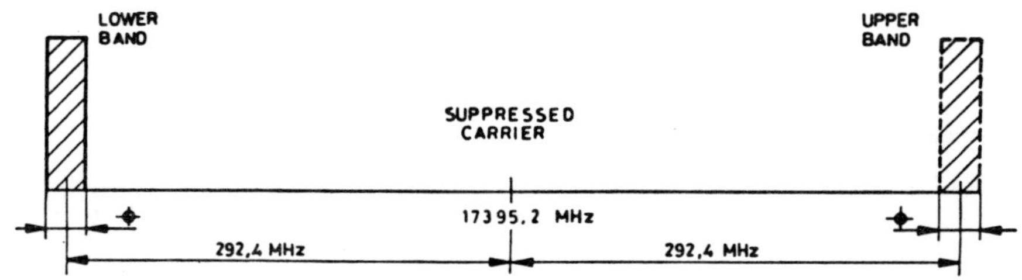

RECEIVED SPECTRUM (UP LINK)

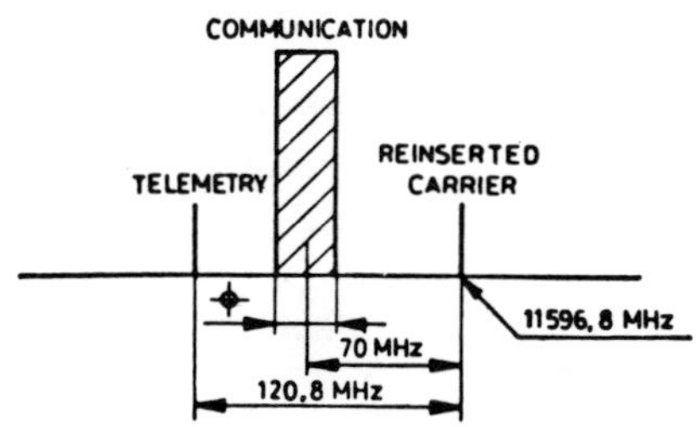

TRANSMITTED SPECTRUM (DOWN LINK)

		TRANS. COMM. PWR. @ ANTENNA INPUT
NARROW BAND EXPERIMENT	1,5 MHz	2,5 W. MIN.
WIDE BAND EXPERIMENT	26 MHz	4,5 W. MIN.

SIRIO FREQUENCY PLAN

Pier L. Bargellini
IEEE Compendium

General

No. of Satellites Built	1 flight unit + 1 qualification prototype
Design Life	2 yr
Launch Vehicle	Thor-Delta 2313
Major Contractor	Compagnia Nationale Aerospaziale
Date of Launch	August 25, 1977
Orbital Position	15°W

Physical Description

Stabilization	Spinning drum (despun earth oriented antenna)
Cylinder Diameter	1.438 m (56.61 in.)
Cylinder Height	0.954 m (37.56 in.)
Height (with antenna and apogee motor nozzle)	1.999 m (78.70 in.)
Weight	
At Launch	398.0 kg (877.6 lb)
On Orbit	185.6 kg (409.2 lb)

Antenna

Configuration	Paraboloid with offset feed
Beamwidth	5° x 10°
Coverage	Europe or W. Europe and East Coast U.S.
Polarization	
Receive (18 GHz)	LHC
Transmit (12 GHz)	RHC

Propagation and Communications Experiment Data

	Propagation	Communications
Bandwidth	2 x 20 kHz (with a freq. separation of 673 MHz)	1.5 MHz (narrowband) 26.0 MHz (wideband)
Nominal Frequency Bands		
Up-Link (receive)	17009-17782 MHz	17090-17200 MHz
Down-Link (transmit)	11331-11863 MHz	11476-11596.8 MHz
Repeaters	One on line, one redundant (power output: 2.5-4.5 W)	

	Coverage	
Satellite e.i.r.p.	Europe	Europe/U.S.
Sidebands	15.1 dBW	11.4 dBW
Carrier	15.5 dBW	11.7 dBW
Wideband Communications Signal	29.6 dBW	25.8 dBW
Satellite G/T	-10.3 dB/K	-16.1 dB/K
Flux Density Range at the Satellite		
Propagation	-59 to -16 dBW/m^2	
Communications		
Wideband	-33 to -15 dBW/m^2	
Narrowband	-43 to -18 dBW/m^2	

SIRIO TECHNICAL PARAMETERS

Pier L. Bargellini
IEEE Compendium

SYMPHONIE

PHOTO COURTESY OF
FRANCO-GERMAN
CONSORTIUM CIFAS

Marcel R. E. Bichara
IEEE Compendium

1 Receive antenna
2 Receive diplexer
3 Tunnel diode amplifier
4 Down converter
5 3-dB coupler
6 Local oscillator
7 IF bandpass filter
8 IF amplifier
9 Equalizing network
10 Filter chain
11 Up converter
12 Transmit amplifier
13 Coaxial switch
14 Transmit diplexer
15 Transmit antenna

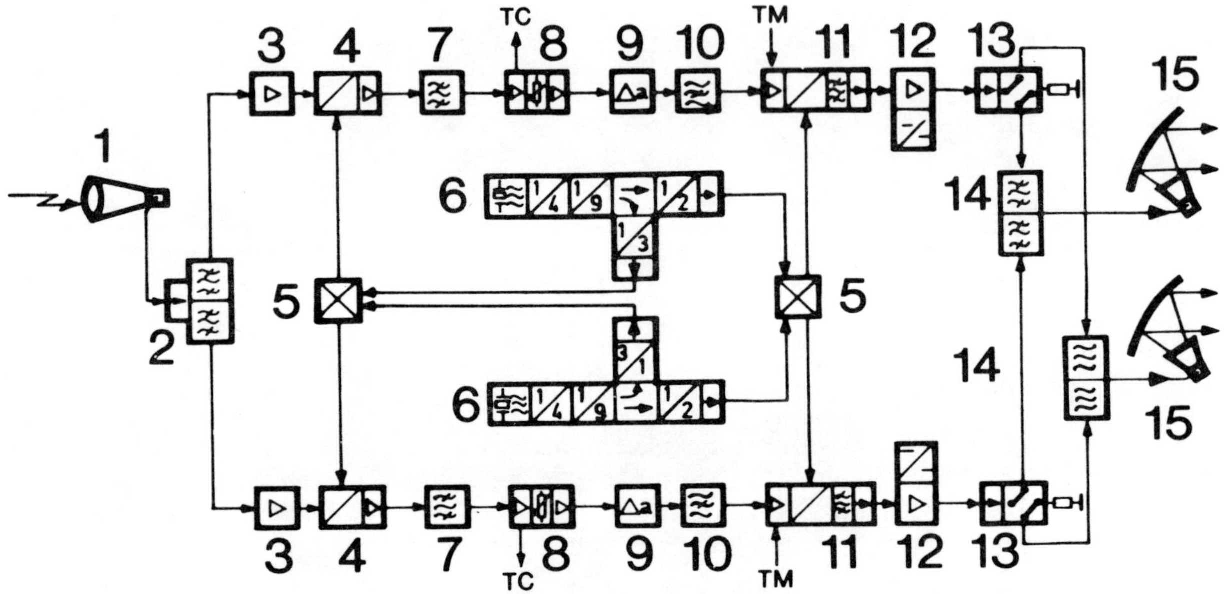

SYMPHONIE COMMUNICATION BLOCK DIAGRAM

Marcel R. E. Bichara
IEEE Compendium

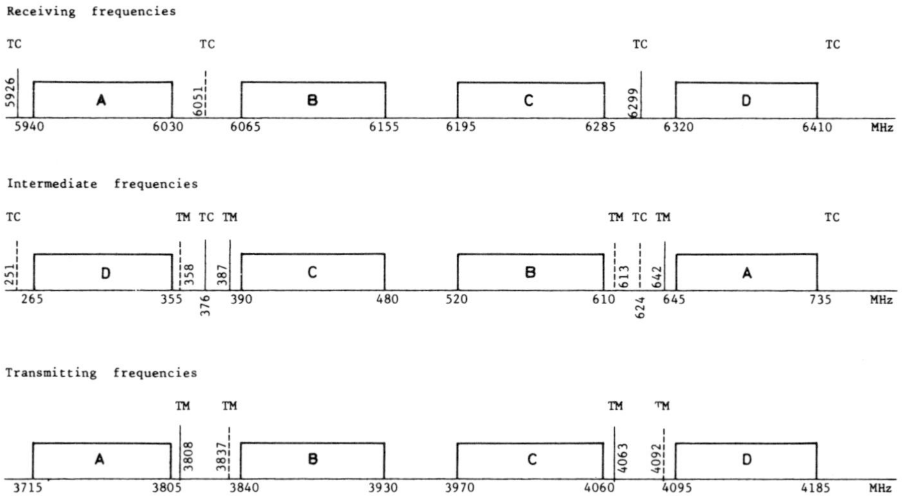

b/d flight model no. 1
a/c flight model no. 2

SYMPHONIE FREQUENCY PLAN

General

No. satellite built	2
Type	experimental geostationary telecommunications satellite
Position	11.5° W
Design life	5 years
Launch vehicle	Thor Delta 2914
Main contractor	French-German Consortium CIFAS
Launch dates	
Flight model no. 1	19 Dec 1974
Flight model no. 2	27 Aug 1975

Physical description

Stabilization	three-axis stabilized by flywheel and gas ejection
Central body diameter	1.7 m
Central body height	0.5 m
Diameter with deployed solar arrays	6.8 m
Overall height	2.0 m
Weight	
Launch	402 kg (including 176.5 kg apogee motor)
Orbit (initial)	230±5 kg

Antenna configuration

Receiving section	
Receiving zone	global
Antenna	1 horn antenna
Beam aperture	18.5°
Polarization	RHC
Transmitting section	
Transmitting zone	2 spots: Europe, Africa and America
Antenna	2 parabolic reflectors
Beam aperture	14.4° x 9.3°
Polarization	LHC

Communication capability

Total usable bandwidth	360 MHz
Assigned frequency range	
Up-link (receive)	5940-6410 MHz
Down-link (transmit)	3715-4185 MHz
Transponder	
Total number	4 (2 per flight model)
No. redundant	none
Bandwidth	90 MHz
RF power	13 watts

Available e.i.r.p. (nominal)	
Beam edge	29.5 dBW
Beam center	34.5 dBW
Receive G/T (nominal)	
Beam edge	- 14.0 dB/°K
Beam center	- 9.5 dB/°K
Saturation flux density at beam edge	
Low gain	- 77 dBW/m^2
Med. gain	- 79 dBW/m^2
High gain	- 87 dBW/m^2
Transponder capacity (involving earth stations with 16 m diameter antenna)	2 TV with sound or 800 telephone channels (FDM) or a corresponding capacity of other information.

SYMPHONIE TECHNICAL CHARACTERISTICS

SYNCOM

PHOTO COURTESY OF
NATIONAL AIR AND
SPACE MUSEUM (USA)

T. Michael Kelley
IEEE Compendium

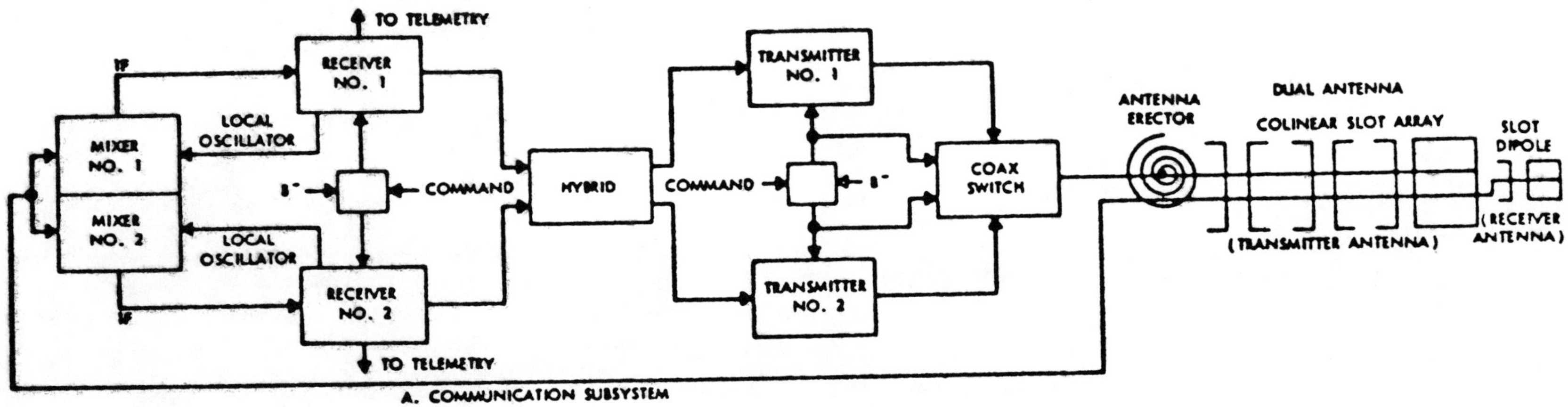

SYNCOM COMMUNICATIONS BLOCK DIAGRAM

T. Michael Kelley
IEEE Compendium

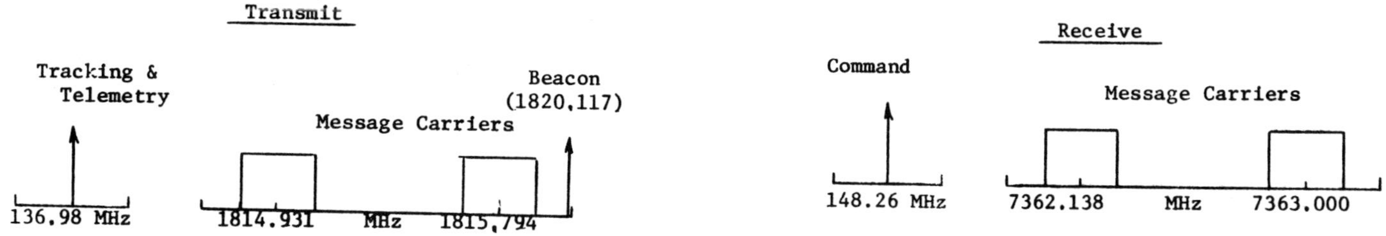

SYNCOM FREQUENCY PLAN

General

No. of Satellites Built	3 (Syncom 1,2, & 3)
Launch Vehicle	Syncom 1 & 2 - Delta Syncom 3 - Thrust-Augmented Delta
Launch Site	ETR (Cape Canaveral)
Sponsor/Manufacturer	NASA/Hughes Aircraft Company
No. of Launches	3
	14 February, 1963 - Launch successful Failed to perform in orbit
	26 July, 1963 - Success
	19 August, 1964 - Success

Physical Description

Stabilization	Spin Stablized with Axis Perpendicular to orbital plane
Shape & Size	
Diameter	.71 m
Height	.394 m
Weight	39 kg

Antenna

UHF Transmit - Colinear Array of Slot Dipoles
SHF Receive - Slot Dipole
VHF TT&C - 4 Whip Turnstile

Oribital Parameters

	Syncom 1	Syncom 2	Syncom 3
Perigee	34228	35505	36662
Apogee	36974	36612	36900
Inclination	33.5°	33.1°	0.31°
Period (Min)	1426.5	1454	1423

Significance

First communications satell*te designed to provide communications services while operating at synchronous altitude

Communications Capability

	Syncom 1 & 2	Syncom 3
Bandwidth per channel	0.5 MHz	0.5 MHz
Frequency		
Receive		
Command	148.260 MHz	148.260 MHz
Communications	7361.275 MHz 7362.582 MHz 7363,000 MHz	7362.138 MHz 7363.000 MHz
Transmit		
Telemetry	136.470 MHz	136.980 MHz
Beacon	1820.117 MHz	1820.117 MHz
Communications	1814.969 MHz 1815.794 MHz	1814.931 MHz 1815.794 MHz
RF Power Output	2 Watts (Nom.)	2 Watts (Nom.)
Power Supply	3840 Solar Cells Nickel-Cadmium Batteries	
Modulation	FDM/FM for Duplex Operation PSK for certain tests	

TYPE EXPERIMENT	NATURE OF RESULTS OBTAINED
1. Telephony	Single channel S+N/N of 35 dB readily attained with maximum FM deviation ratio employed. Multichannel operation demonstrated on Syncom III at reduction in per channel performance.
2. Data Transmission	Using vestigial sideband modems operating into 4 kHz baseband input to normal FM terminal equipment, 3 kbps rates at low error rates were possible. Using PSK RF modulation, data rates as high as 50 kbps were possible.
3. Teletype	For single channnel operation 4 kHz baseband input to normal FM terminal equipment, error rates of 0.1% were readily attained. For 1 of 16 channels into 4 kHz baseband, error rates less than 1% were attained.
4. Facsimile	Operating into standard FM terminal equipment, overall picture quality numerically rated 7 on a 0 to 10 scale was commonly obtained. Factors degrading quality included bandwidth limitations, phase delay distortion, and satellite spin rate modulation.
5. Television	Fort Dix to Andover through Syncom II realized 26 db p—p signal to rms noise ratio. Kashima to Point Mugu through Syncom III realized 34 db signal to rms noise ratio.
6. Direct-to-Aircraft TTY	Pan American scheduled aircraft to Camp Roberts through Syncom III VHF command receiver and telemetry transmitter realized under proper conditions, up to 60 wpm TTY.

SYNCOM TECHNICAL CHARACTERISTICS

T. Michael Kelley
IEEE Compendium

TELSTAR

PHOTO COURTESY OF
NATIONAL AIR AND
SPACE MUSEUM (USA)

T. Michael Kelley
IEEE Compendium

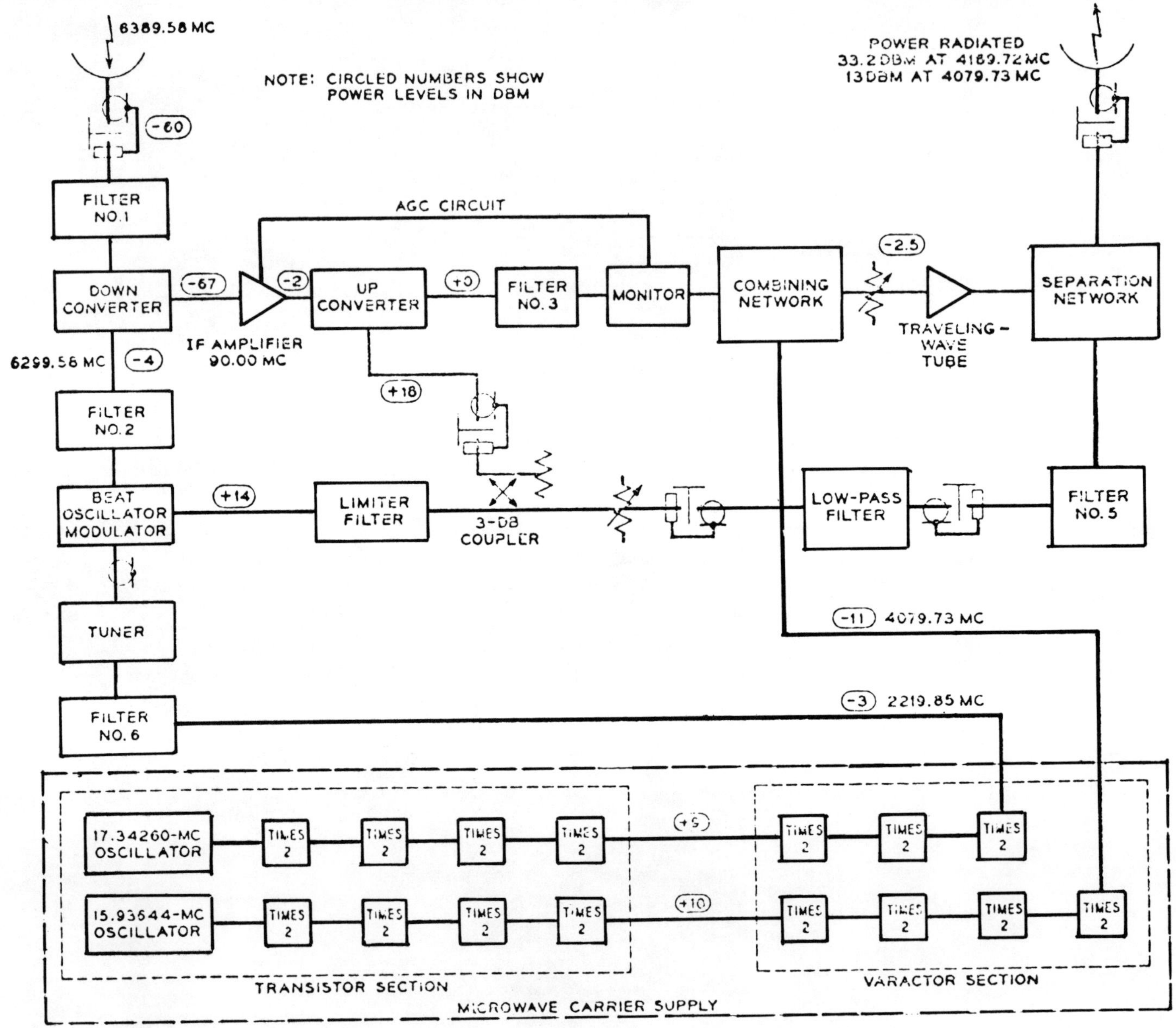

TELSTAR COMMUNICATIONS BLOCK DIAGRAM

T. Michael Kelley
IEEE Compendium

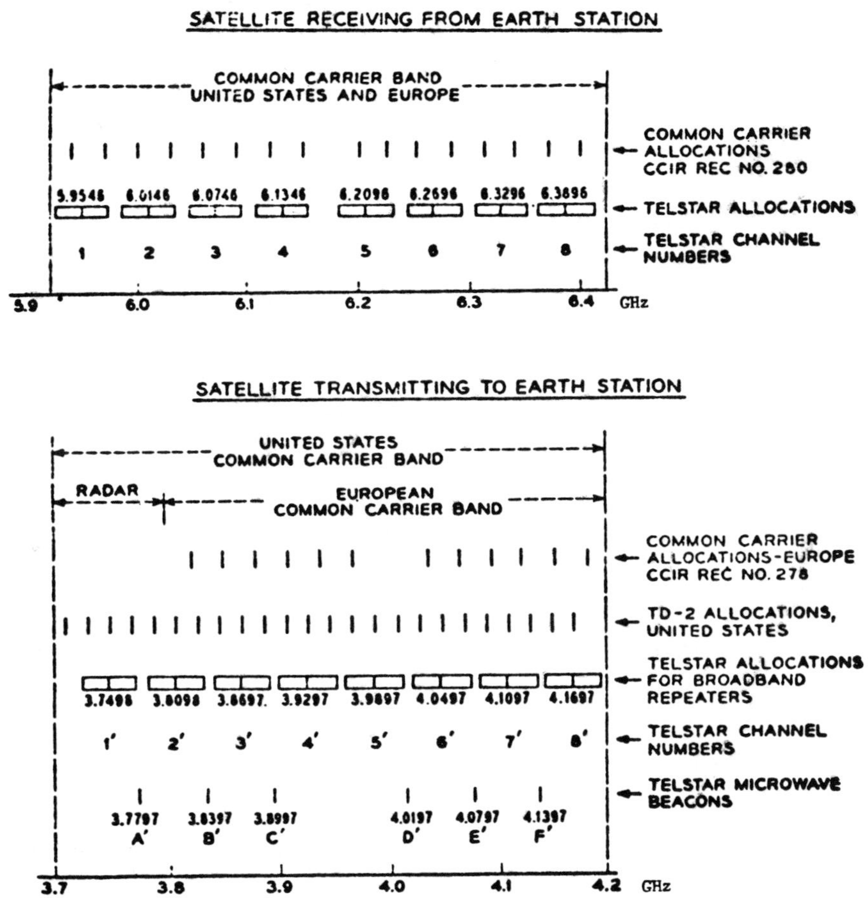

TELSTAR FREQUENCY PLAN

TELSTAR TECHNICAL CHARACTERISTICS

General

No of satellites built	2 (TELSTAR 1 & 2)
Launch Vehicle	Delta
Launch Site	ETR (Cape Canaveral)
Major Contractor	AT&T
No of launches	2 10 July, 1962 - Success 21 January, 1964 - Success

Physical Description

Stabilization: Spin stabilized with axis normal to plane of ecliptic

Shape & Size	TELSTAR 1	TELSTAR 2
Equatorial Diameter	.85m	.86m
Axial Diameter	.85m	.94m
Weight	77kg	79kg

Antenna

Approximately isotropic

Polarization

Microwave channels	Circular
VHF beacon	Linear
Command	Circular

Orbital Parameters

Orbital Parameters	TELSTAR 1	TELSTAR 2
Perigee	955 km	971 km
Apogee	6238 km	10508 km
Inclination	44.8°	42.7°
Period	157.8 min	221.5 min

Communications Capability

Total Usable Bandwidth: 25 MHz

Frequency Plan

Receive

Ground Command	123.00 MHz
TV/Telephony	6389.48 MHz

Transmit

Tracking Beacon	4079.73 MHz
Telemetry Beacon	136.05 MHz
TV/Telephony	4169.72 MHz

RF Power Output: 2 Watts

Power Supply: 3600 Silicon Solar Cells
Ni-Cd storage battery

Modulation: FM

Communications Capacity

Television	One channel
Telephone Channels	600 simplex channels
Telephone Channels	12 duplex channels

Significance

First active repeater for satellite communications

First trans-Atlantic transmission of TV and voice communication using a communications satellite

T. Michael Kelley
IEEE Compendium

PROJECT WEST FORD

PHOTO COURTESY OF
LINCOLN LABORATORY

William W. Ward
IEEE Compendium

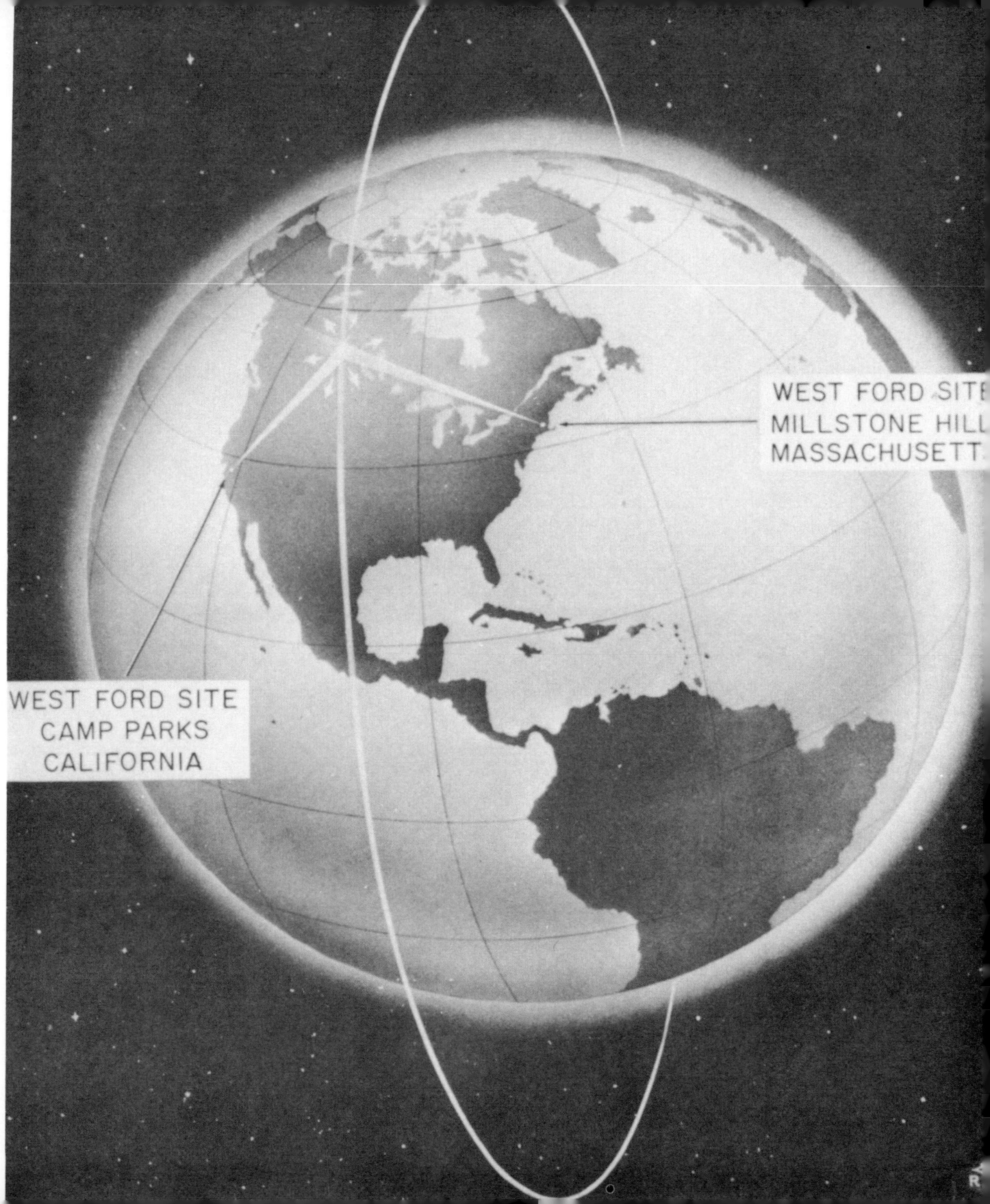

WEST FORD DIPOLES

PHOTO COURTESY OF
LINCOLN LABORATORY

William W. Ward
IEEE Compendium

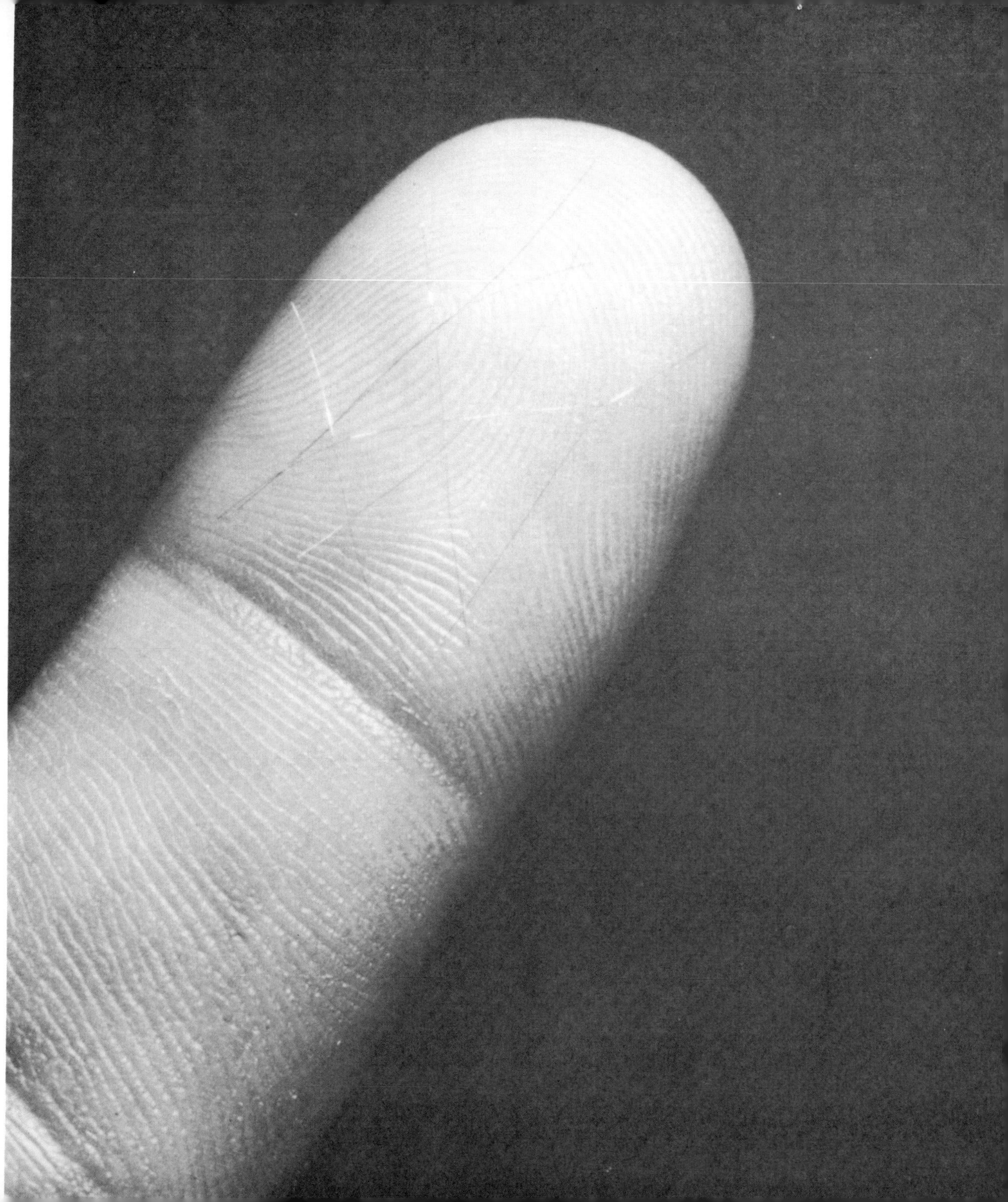

General

Number of packages launched	2
Design life	3 to 6 years (for experimental belt)
Launch vehicle	Atlas-Agena B
Major contractor	MIT Lincoln Laboratory
Date of first launch	21 October 1961
Orbit	Circular, 3500-km (1900-nm) altitude, 96° inclination
COSPAR designation	Piggyback on 1961-AΔ1
First payload failed to dispense its dipoles	
Date of second launch	8 May 1963
Orbit	Circular, 3650-km (2000-nm) altitude, 87° inclination
COSPAR designation	Piggyback on 1963-14A
Second payload dispensed its dipoles satisfactorily	

Physical Description

An individual dipole (made from AWG-53 copper wire) in second payload	Length 1.78 cm (0.7 in.), ∿ half-wave at 8 GHz Diameter 0.00178 cm (0.0007 in.) Weight 40 μgm (9×10^{-8} lb)
Number of dipoles in second payload	4.8×10^{8}
Weight of second payload	19 kg (42 lb)

Terminals

Number built	2, at Westford, Massachusetts, and Camp Parks, Pleasanton, California
Antenna	18.3 m (60 ft) cassegrain paraboloid ∿ 60 dBi gain ∿ 0.15° half-power beamwidth
Modulation/demodulation	MFSK
Signalling bandwidth	4 MHz
Transmit center frequency	7750 MHz (Westford) 8350 MHz (Camp Parks)
Receive center frequency	7750 MHz (Westford radar) 8350 MHz (Westford communications) 8350 MHz (Camp Parks radar) 7750 MHz (Camp Parks communications)
Polarization	Transmit RHCP Receive LHCP (radar) Receive LHCP (communications)
Transmitter power (nominal)	43 dBW (20 kW) at Westford 46 dBW (40 kW) at Camp Parks
EIRP on beam nose (nominal)	103 dBW (20 GW) at Westford 106 dBW (40 GW) at Camp Parks
System noise temperature (nominal)	60 K (communications) 120 K (radar)
Receive G/T on beam nose (nominal)	42 dB (K^{-1}) (communications) 39 dB (K^{-1}) (radar)

Status

The orbiting dipole belt from the second payload was gone by 1966 because of atmospheric drag. The mean orbit of the belt had been arranged so that the pressure of solar radiation would cause reduction of its perigee. Upon conclusion of the test program, the terminals were adapted to other uses.

References

"Project West Ford Issue", Proc IEEE, 52(1964)449, 452-606 (preface and 12 papers)

"The Last of the West Ford Dipoles", I. I. Shapiro, Science, 154(1966)1445-1448

WEST FORD TRANSMISSION PARAMETERS

William W. Ward
IEEE Compendium

SECTION VII

AMATEUR SATELLITES

- OSCAR 1
- OSCAR 2
- OSCAR 3
- OSCAR 4
- OSCAR 5
- OSCAR 6
- OSCAR 7
- OSCAR 8
- OSCAR 9
- RS 1 & 2

* OSCAR 9

VITAL STATISTICS AND TECHNICAL CHARACTERISTICS	OSCAR 1	OSCAR 2	OSCAR 3	OSCAR 4	OSCAR 5	OSCAR 6	OSCAR 7	OSCAR 8	USSR SATELLITE RS1 & RS2
Launch Date	12/12/61	2/6/62	9/3/65	21/12/65	23/1/70	15/10/72	15/11/74	5/3/78	26/OCT/78
Operated Until	1/1/62	20/6/62	24/3/65	16/3/66	15/3/70	21/6/77	Operational as of March 1980	Operational as of March 1980	Operational as of March 1980
Size	0.30x0.35x 0.15m	0.30x0.35x 0.15m	0.17x0.30x 0.43m	0.48x0.50x 0.50m	0.30x0.43x 0.15m	0.165x0.31x 0.44m	0.434x0.36x 0.424m	0.38x0.33x 0.38m	0.420x0.390m
Type of Radio Equipment Carried	Telemetry Beacon	Telemetry Beacon	Translator	Repeater	Telemetry Beacon	Transponder	Transponder	Transponder	**Transponder**
Transmit Frequency (MHz)	144.983	144.990	145.875-145.925 145.85(telem-beacon)	431.933-431.943 431.928(telem-beacon	29.45 HF 144.05 VHF	29.45-29.55 29.45 (telemetry & 435.10 beacon)	MODE A 29.400-29.500 29.502(beacon) MODE B 145.975-145.925 145.972(beacon)	MODE A 29.400-29.500 29.402(beacon) MODE J 435.100-435.200 435.095(beacon)	29.360-29.400 29.401(telem-beacon)
Transmitter Power	100mW e.i.r.p.	140mW e.i.r.p.	1W e.i.r.p 50mW(telem-beacon)	3W e.i.r.p.	250mW e.i.r.p. HF 50mW e.i.r.p. VHF	1 to 1.3W e.i.r.p. 0.2W(beacon)	1 to 2W e.i.r.p. 2W(telem-beacon	1 to 2We.i.r.p 1 to 2W(telem-beacon)	1 to 2W e.i.r.p.
Receive Frequency (MHz)	None	None	144.075-144.125	144.095-144.105	None	145.90-146.00	MODE A 145.850-145.950 MODE B 432.125-432.175	MODE A 145.850-145.950 MODE J 145.900-146.000	145.880-145.920
Antennas	Single Monopole ¼ λ	Single Monopole ¼ λ	Four Independent Monopoles	Four Independent Monopoles	Monopole ¼λ VHF Dipole Hf	Dipole(29MHz) Monopole (146MHz)	Dipole(29MHz) Monopole (146MHz)	Dipole(29MHz) Turnstile (146MHz) Monopole(432 MHz)	Inverted "V" (144 MHz) Whip ¼λ (29 MHz)
Bandwidth (kHz)	Beacon	Beacon	50	10	Beacon	100	Mode A:100 Mode B:50	Mode A:100 Mode J:100	40
Apogee	416 km	385 km	938 km	33544 km (elliptical)	1478 km	1452 km	1455 km	910.372 km	1724 km
Perigee	235 km	207 km	904 km	163 km	1432 km	1443 km	1450 km	898.259 km	1688 km
Altitude	250km	250 km	900 km	200-33500km	1444 km	1460 km	1460 km	880 km	1706 km
Period (mins.)	91.1	90.5	103.53	587.5	115	114.95	114.94513	103.227	120.38943
Inclination (at equator)	81.2°	74.26°	70.08° (polar)	26.8°	102°	101.73°	101.5966°	98.99°	82.66°

OSCAR 1 to 8 and RADIO 1 & 2

*See section on OSCAR 9 for details.

AMATEUR SATELLITE

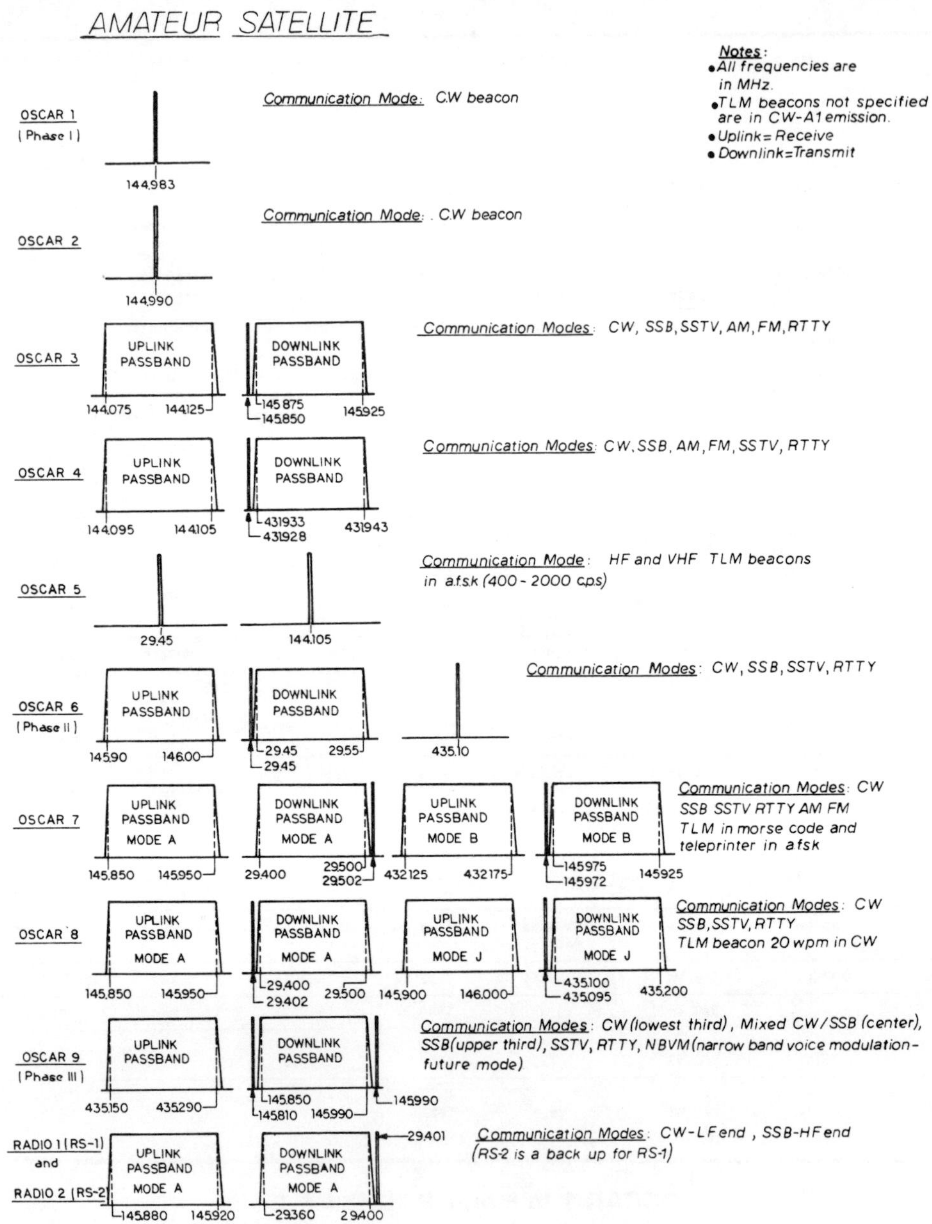

Gustavo D. Moreno
IEEE Compendium

FREQUENCY PLANS

OSCAR 1 & 2

PHOTO COURTESY OF
TELECOMMUNICATION JOURNAL

Gustavo D. Moreno
IEEE Compendium

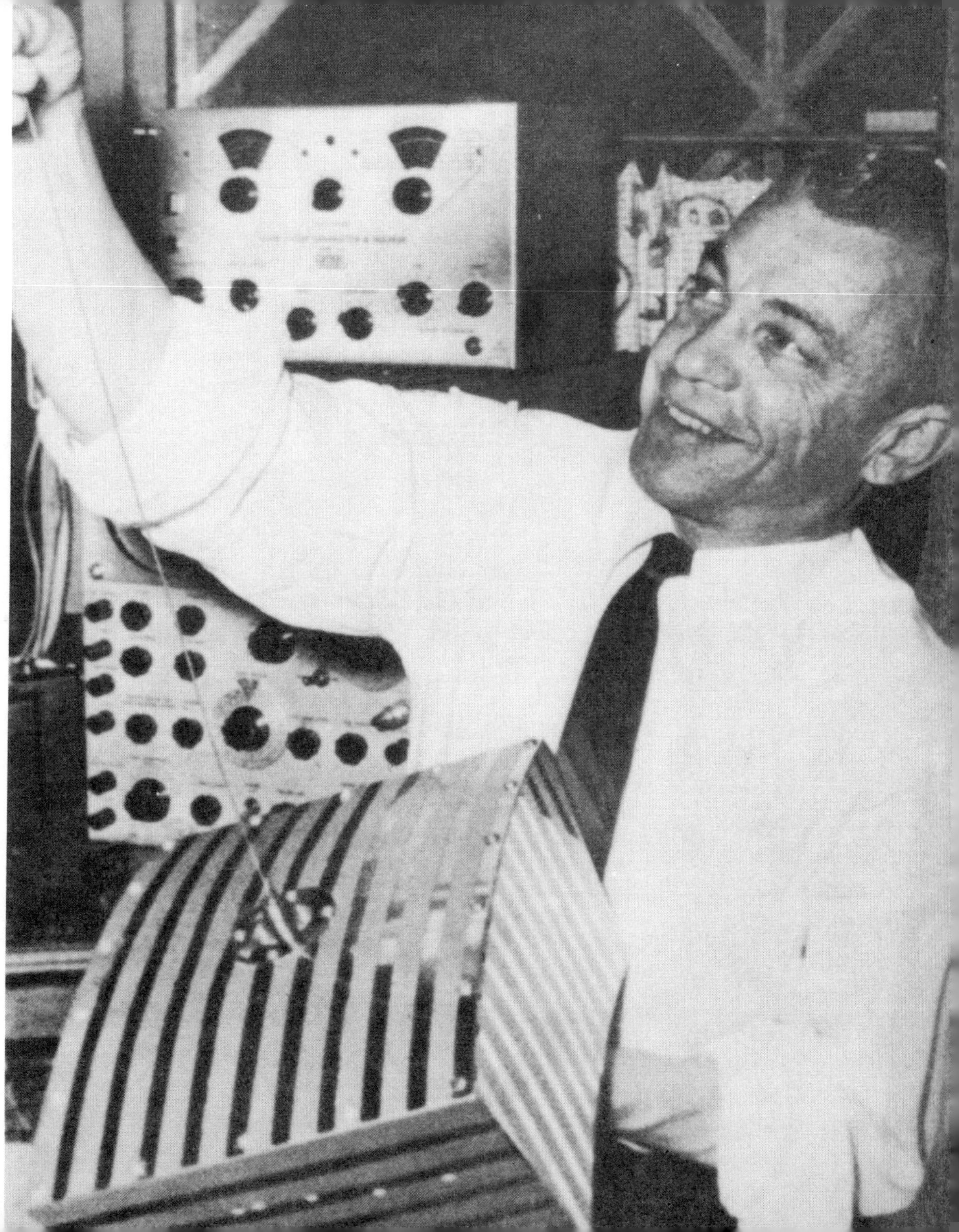

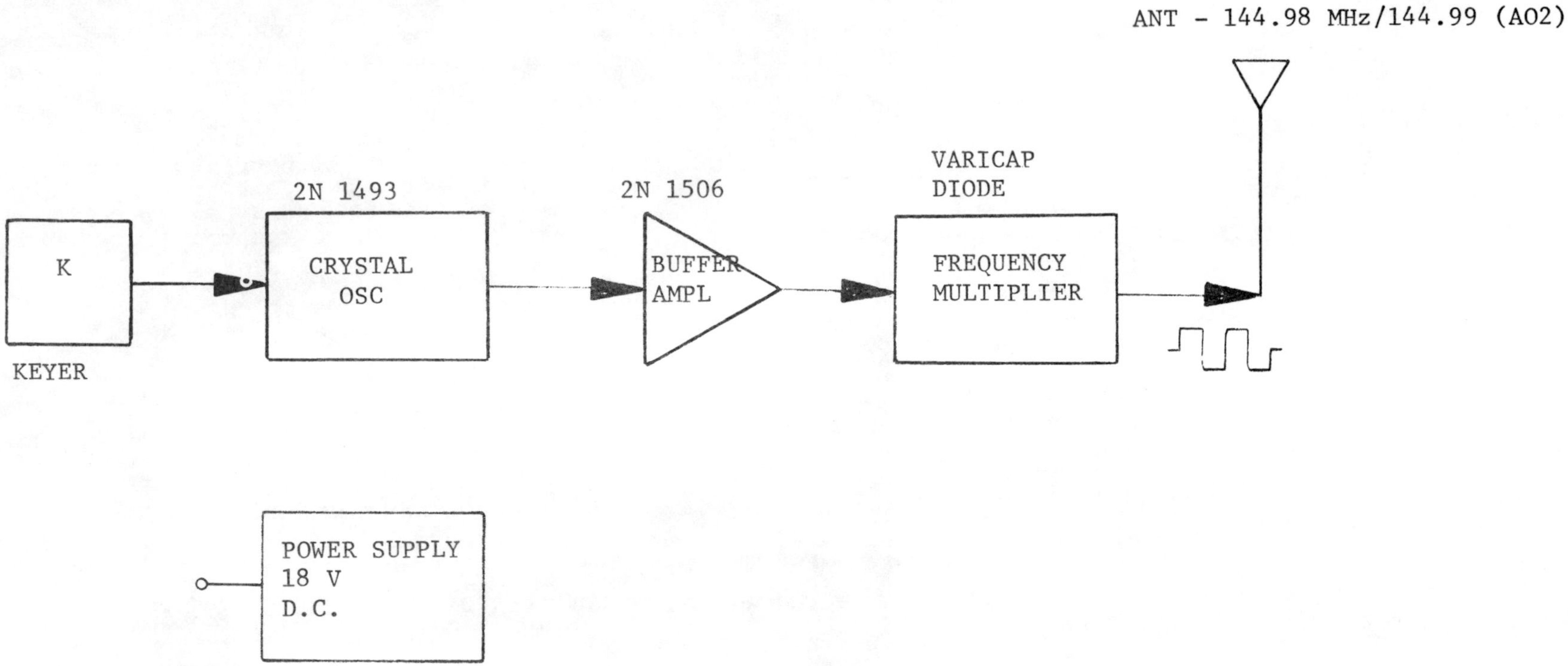

OSCAR 1 & 2

LEGEND: The R.F. unit consisted of a 2N 1493 crystal - controlled oscillator operating on the fifth overtone of the crystal to produce a 72.5 MHz signal source. The signal was amplified by a 2N1506 buffer stage which was base driven. The R.F level of the buffer output was 180 mW. A varicap diode doubler stage delivered 140 mW at 145 MHz. The output tank circuit was tapped at the proper point to provide a match to the 50 Ω coax line which fed the antenna. The transmitter keyer made use of digital circuits.

OSCAR 1 & 2
COMMUNICATION BLOCK DIAGRAM

Gustavo D. Moreno
IEEE Compendium

OSCAR 3

PHOTO COURTESY OF
AVIATION WEEK MAGAZINE

Gustavo D. Moreno
IEEE Compendium

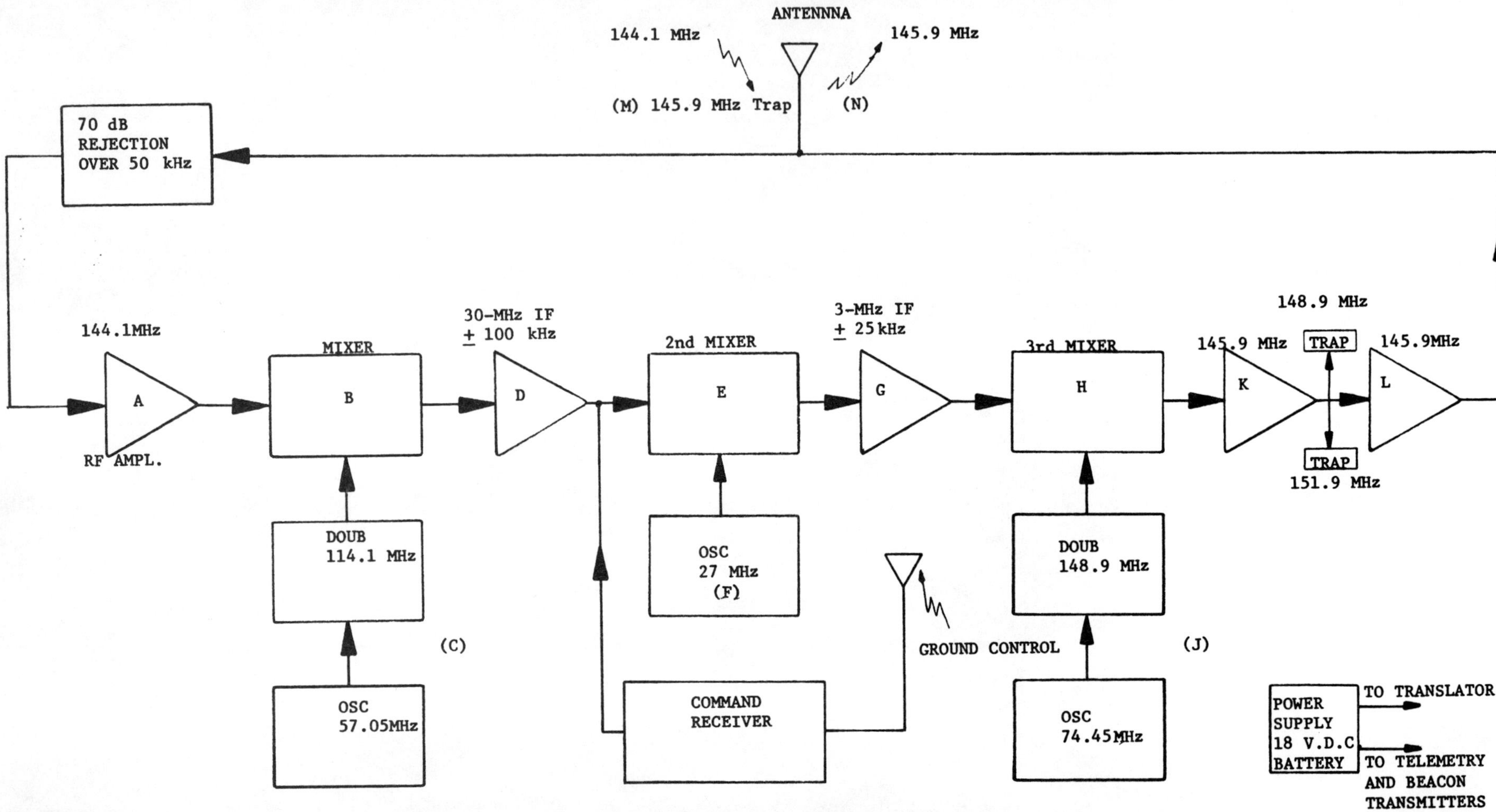

LEGEND:

Signal passes through the rejection filter (M), is impressed upon an RF ampl. stage (A) and converted to the first intermediate ampl. freq. by the mixer (B). The input signal of 144.105 Mc is now one of 30.005 Mc. A57.050 Mc. cry.osc and doubler(C) supply a 114.100 Mc. Local injection signal to accomplish this conversion. A broadband, low-gain 30 Mc first if ampl (D) drives a second mixer (E) which converts the signal to the second intermediate freq of 3.005 Mc by virtue of mixing with the second local osc (F) working on 27 Mc. The high-gain 3 Mc IF strip (G) has a total passband of 50 Kc which determines the oper passband of the translator. From here, the signal is mixed higher in freq (H) by a beating signal of 148.900 MHz supplied by a 74.450 Mc Osc douber chain (J). Successives linear ampl stages (K&L) further amplify the 145.895 resulting signal and feed it back to the translator antenna (N).

OSCAR 3
COMMUNICATION BLOCK DIAGRAM

Gustavo D. Moreno
IEEE Compendium

OSCAR 4

PHOTO COURTESY OF
QST MAGAZINE

Gustavo D. Moreno
IEEE Compendium

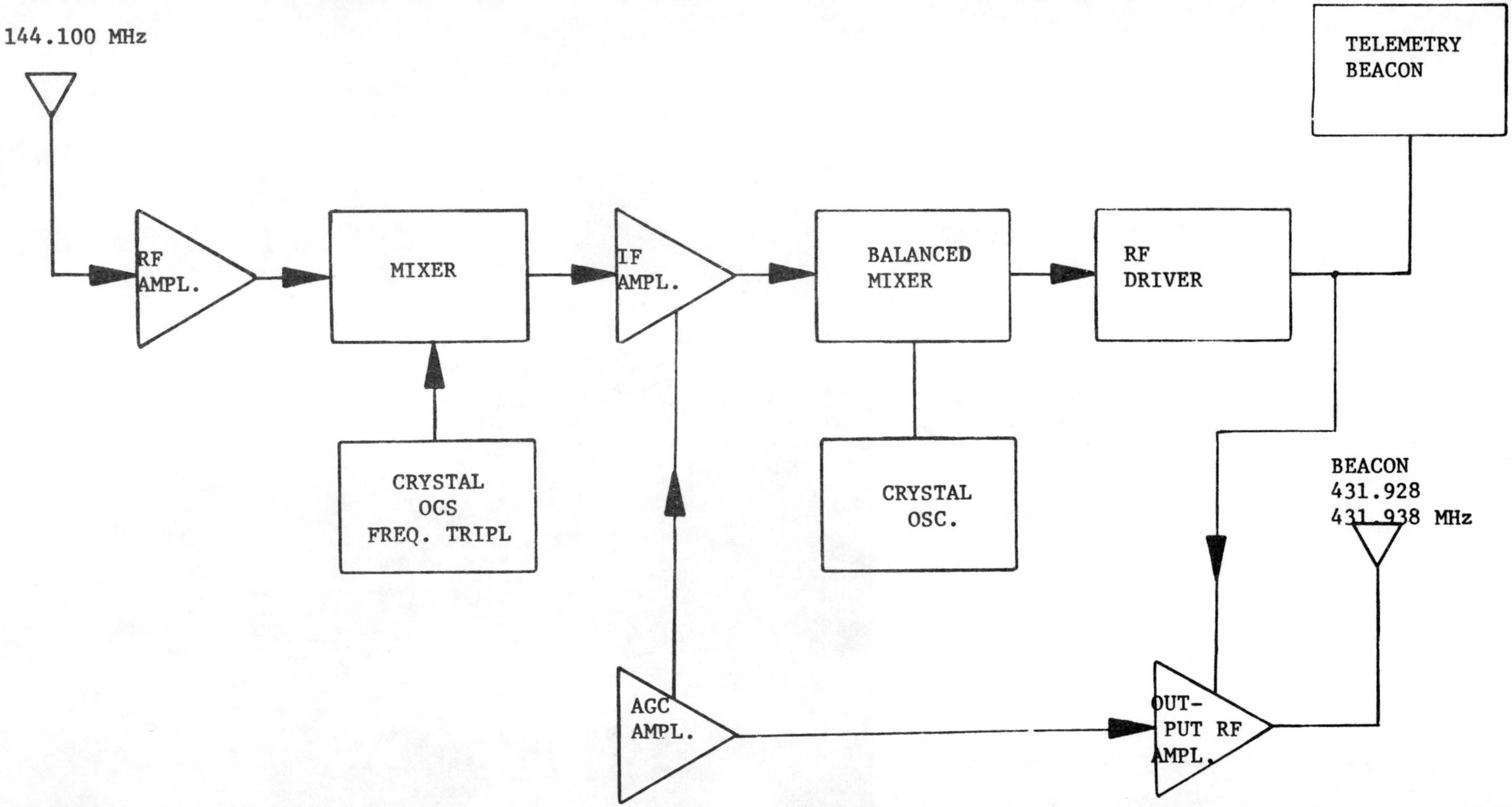

OSCAR 4: Linear translator consisting of a receiver and transmitter section. It also had a small module with the "digital logic" circuitry designed to gate the translator and transmit the beacon signal, and a command decoder unit. Three antennas for 144 MHz input signals and a single monopole for the 432 MHz transmitter powered completely with solar cells mounted on all sides of its outer surface. OSCAR 4 never attained its planned synchronous equatorial orbit and instead it travelled along an elliptical one.

OSCAR 4
COMMUNICATION BLOCK DIAGRAM

Gustavo D. Moreno
IEEE Compendium

OSCAR 5

PHOTO COURTESY OF
TELECOMMUNICATION JOURNAL

Gustavo D. Moreno
IEEE Compendium

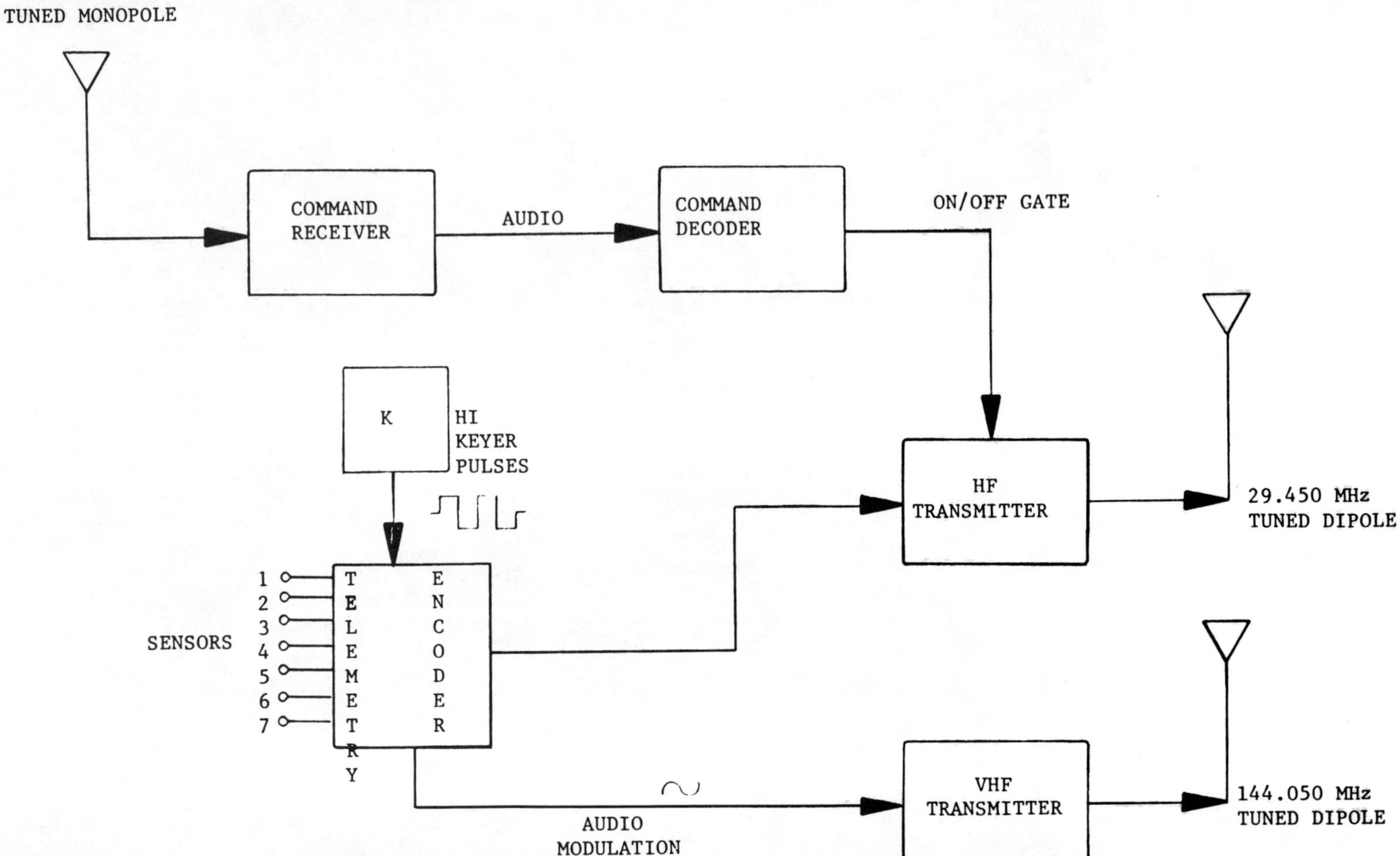

LEGEND

SENSOR - TELEMETRY CHANNELS

CHANNEL 0	HI IDENTICATION	CHANNEL 3 BATTERY VOLTAGE	CHANNEL 6 Z AXIS STABILIZATION
CHANNEL 1	BATTERY CURRENT DRAIN	CHANNEL 4 Y AXIS STABILIZATION	CHANNEL 7 SKIN TEMPERATURE OF PACKAGE
CHANNEL 2	X AXIS STABILIZATION	CHANNEL 5 INTERNAL TEMPERATURE	

The telemetry system aboard converted impulses produced by 7 channels into audio tones. These tones modulated the ten and two meter transmitters which relayed the information back to earth.

OSCAR 5
COMMUNICATION BLOCK DIAGRAM

Gustavo D. Moreno
IEEE Compendium

OSCAR 6

PHOTO COURTESY OF
CQ MAGAZINE

Gustavo D. Moreno
IEEE Compendium

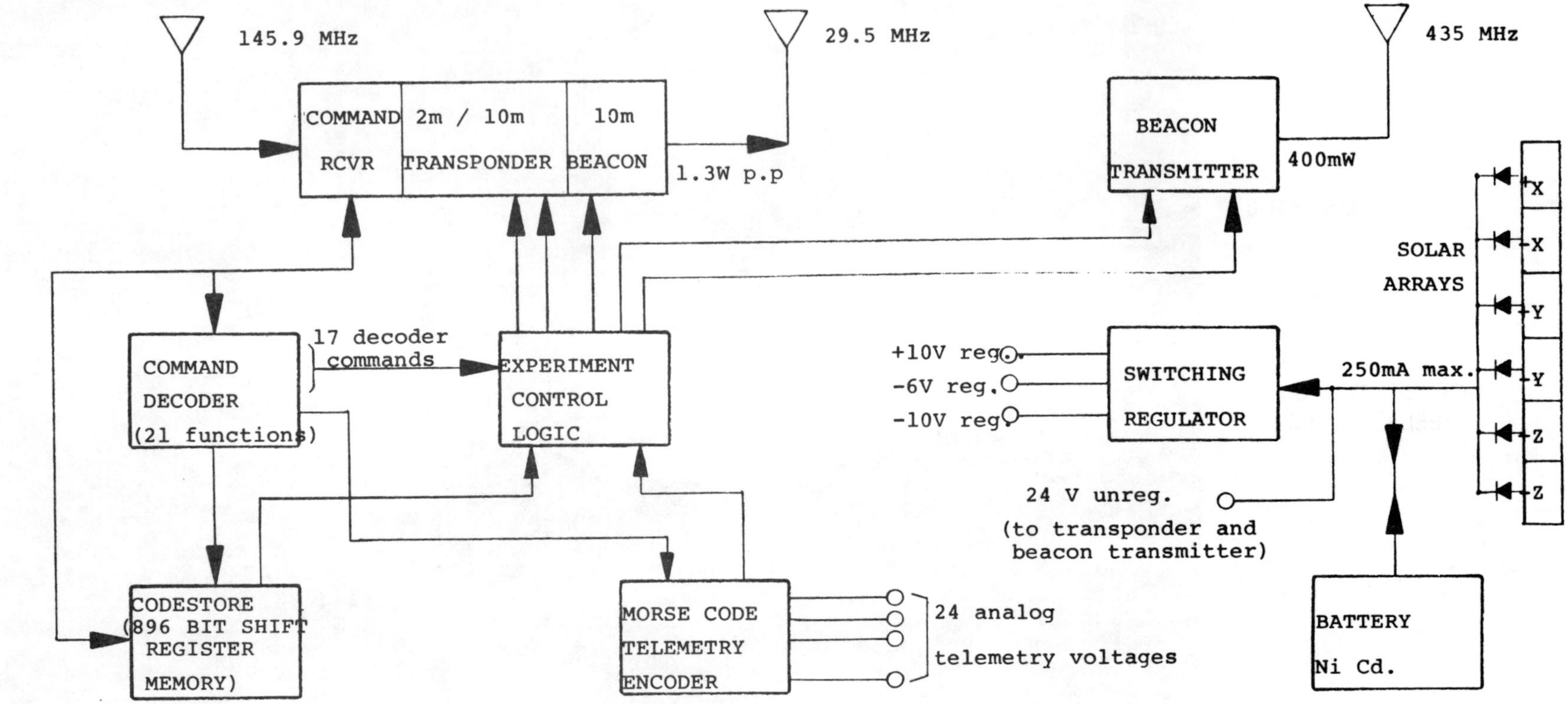

LEGEND: The beacon was keyed with 24 channels of telemetry data sent one channel at atime as three digit numbers in morse code . First time morse code telemetry used with an OSCAR satellite, also a morse code message storage unit was new aboard, called codestore. The battery was rechargable by banks of solar cells mounted on the satellite's outer surface. The experiment control logic module was responsible for accepting decoded pulsed commands from the command decoder and converting them to level commands which turn on and off the 2/ 10 m repeater and the 435.1 MHz beacon transmitter.

OSCAR 6
COMMUNICATION BLOCK DIAGRAM

Gustavo D. Moreno
IEEE Compendium

OSCAR 7

PHOTO COURTESY OF
PERRY KLEIN
(Pres. AMSAT)

Gustavo D. Moreno
IEEE Compendium

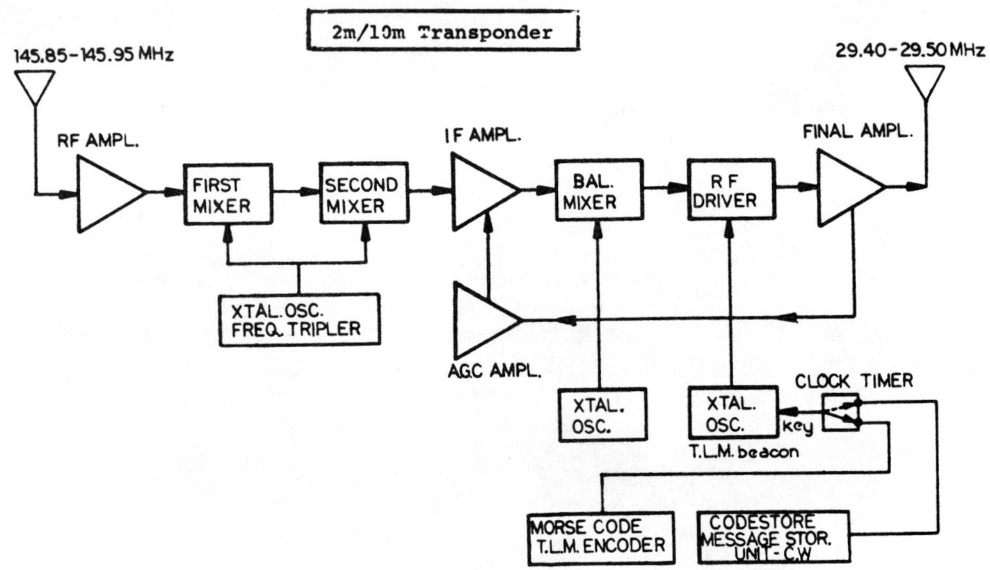

OSCAR 7
COMMUNICATION BLOCK DIAGRAM

Gustavo D. Moreno
IEEE Compendium

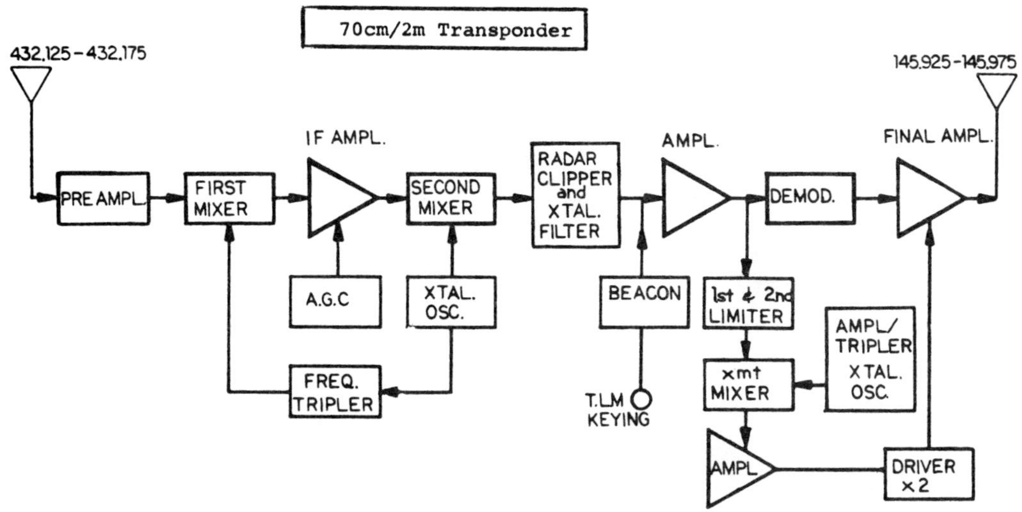

LEGEND:

OSCAR 7 contains two repeaters and two auxiliary beacons, and both morse and teletype telemetry encoders .
Ground control of the spacecraft is achieved by means of command receivers in each repeater, redundant command decoders and an experimental control logic sub-system.
Downlink telemetry and stored messagedata are generated by the morse code telemetry encoder, or the Codestore unit.
The primary power source consists of eight solar cell arrays supplying 2.2A at 6.4 V when illuminated by the sun
The spacecraft will normally alternate between Modes A and B. An internal timer generates a pulse every 24 hours.

OSCAR 7 (continued)
COMMUNICATION BLOCK DIAGRAM

OSCAR 8

PHOTO COURTESY OF
PERRY KLEIN
(Pres. AMSAT)

Gustavo D. Moreno
IEEE Compendium

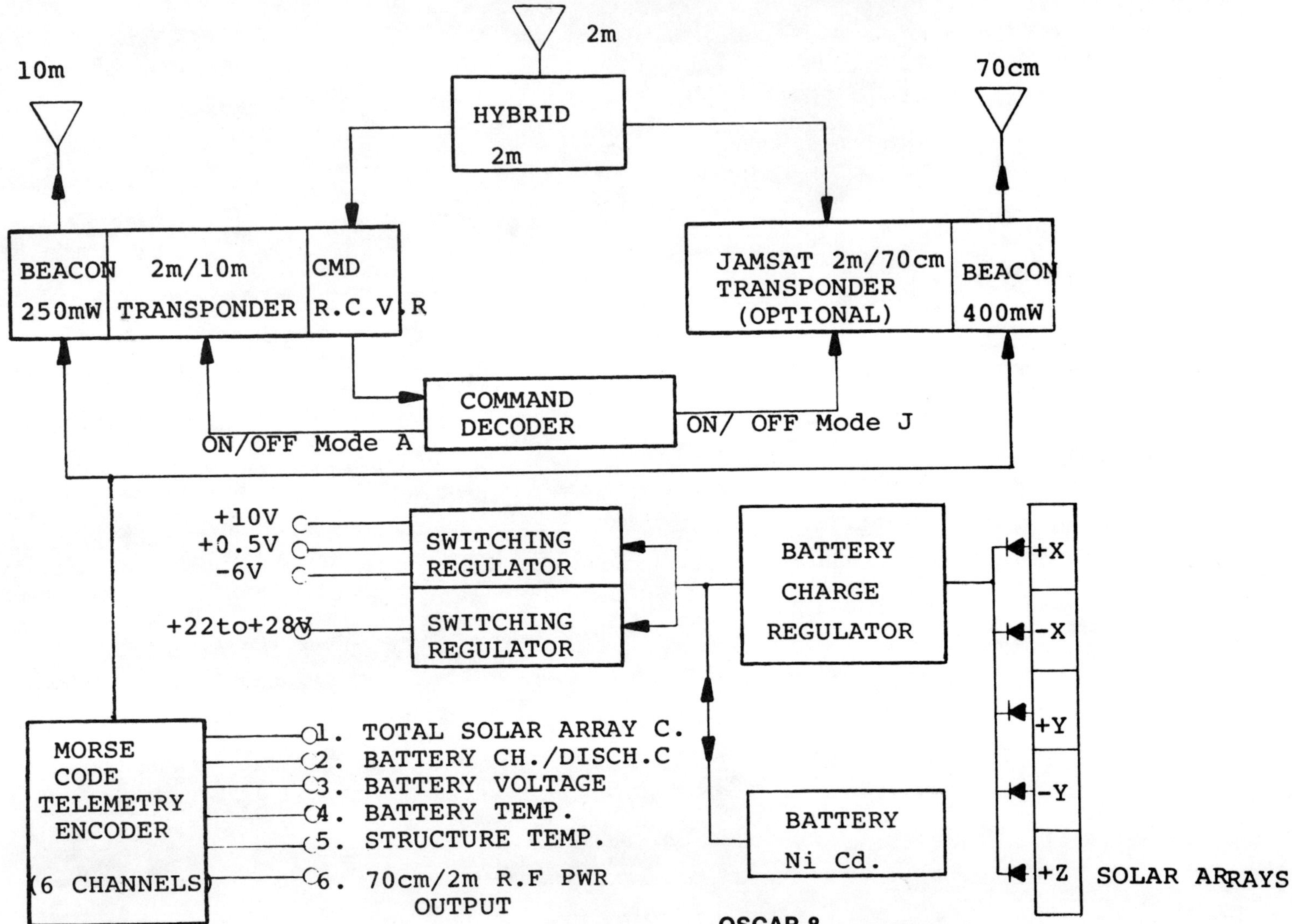

OSCAR 8
COMMUNICATION BLOCK DIAGRAM

LEGEND:
In Transponder Mode J the output passband is inverted. Mode A and Mode J Transponders use the same receiving antenna. The five function telecommand system are: Mode A select 2 to 10 m Transp. ON, Mode J select 2 to 70 cm. ON, Mode D select recharge mode, both transp. OFF—10 m antenna deployment—10 m antenna reset. TELEMETRY: Transmit sequentially in morse code at 20 w.p.m.

Gustavo D. Moreno
IEEE Compendium

OSCAR 9

PHOTO COURTESY OF
PERRY KLEIN
(Pres. AMSAT)

Gustavo D. Moreno
IEEE Compendium

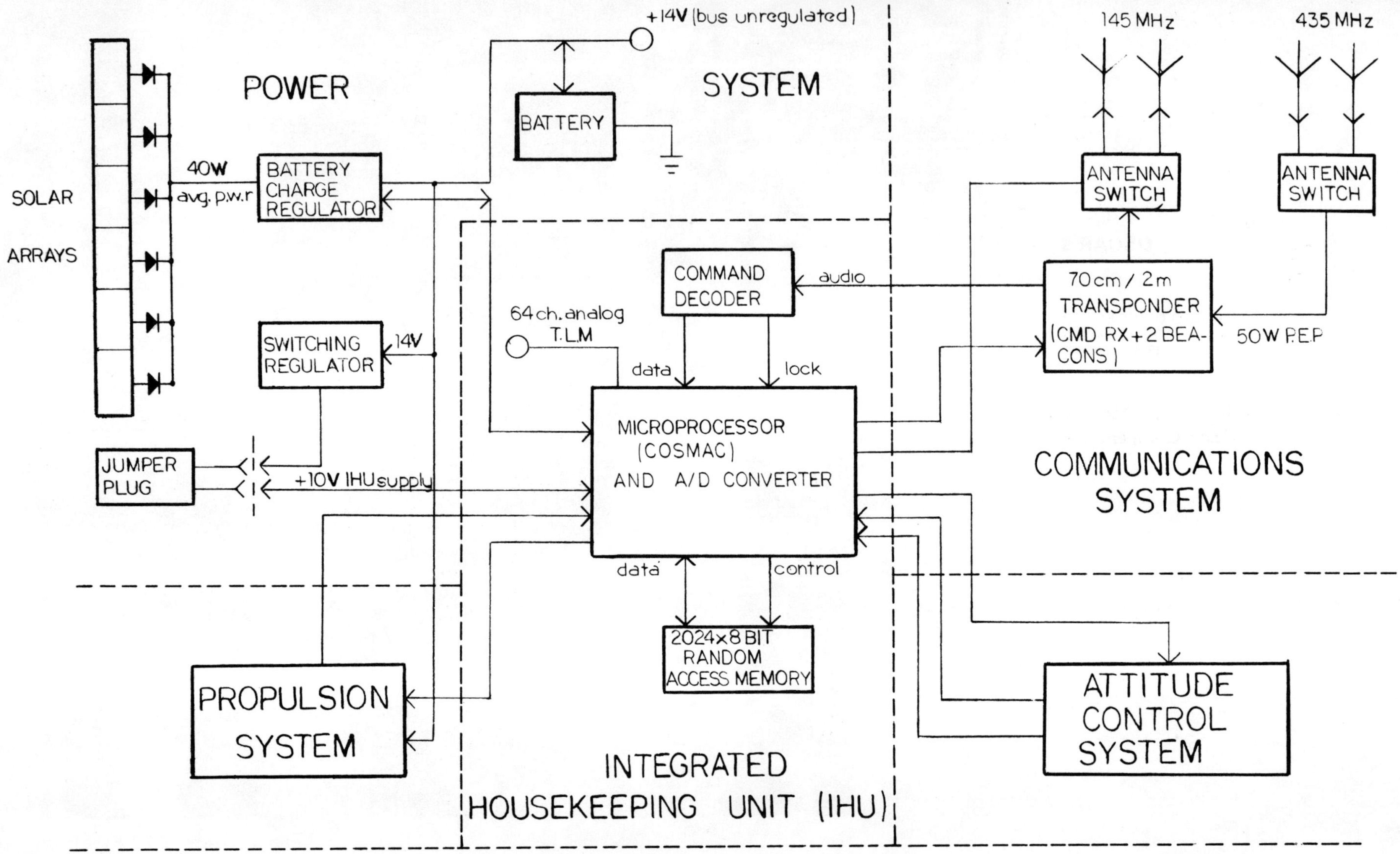

OBSERVATIONS: The passbands are inverted. USB on the uplink will become LSB on the downlink; signals at the high end of the uplink passband will be found on the low end of the downlink passband and vice versa. AMSAT PHASE III is the first of a series of long-life satellites intended for high-altitude elliptical orbit and offers a significant communications advantage over the previous low-orbiting OSCAR's of providing improved coverage for hours at a time over trans-continental distances.

OSCAR 9
COMMUNICATION BLOCK DIAGRAM

Gustavo D. Moreno
IEEE Compendium

OSCAR 9
TECHNICAL PARAMETERS

General

• Launch Site:	Kourou, French Giuana
• Date of Launch:	28 May 1980, Launch Failure
• Launch Vehicle:	Ariane

Frequency Bands

Transmit	145.850 - 145.990 MHz
Receive	435.150 - 435.290 MHz
Beacons	145.810 and 145.990

Transmitter Power

Transponder output	50 watts
e.i.r.p	500 to 1000 watts

Antennas

Omni	146/435 MHz
Dipoles	435 MHz
Monopoles	144 MHz on three arms

Orbit Parameters (if successful launch)

Apogee	35,786 km at 24.8 N latitude
Perigee	1500 km
Period	10 1/2 hours per orbit/2.3 orbits a day
Inclination	57°

RADIO 1 AND 2 (IDENTICAL)

PHOTO COURTESY OF
RADIO MAGAZINE

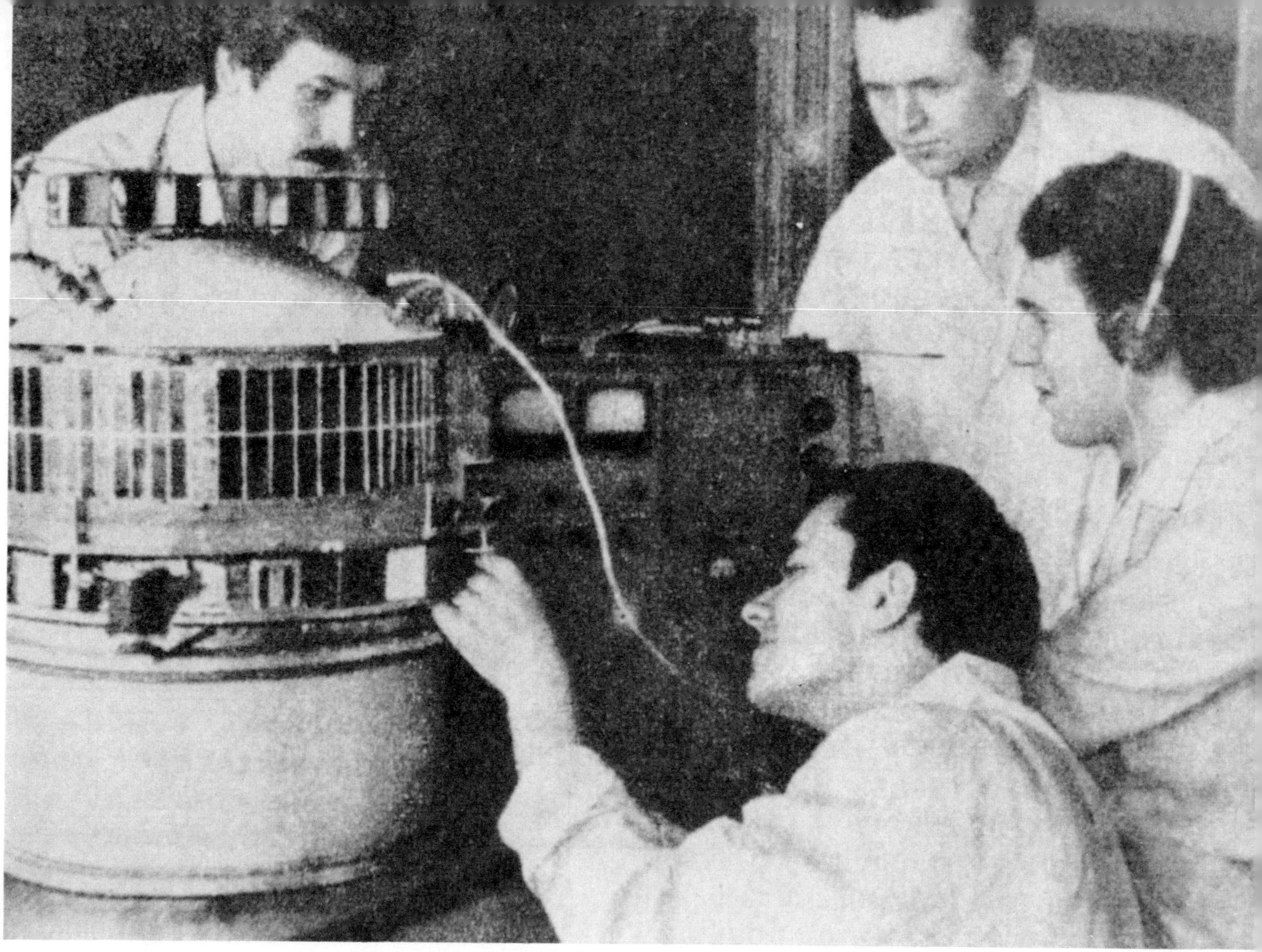

Gustavo D. Moreno
IEEE Compendium

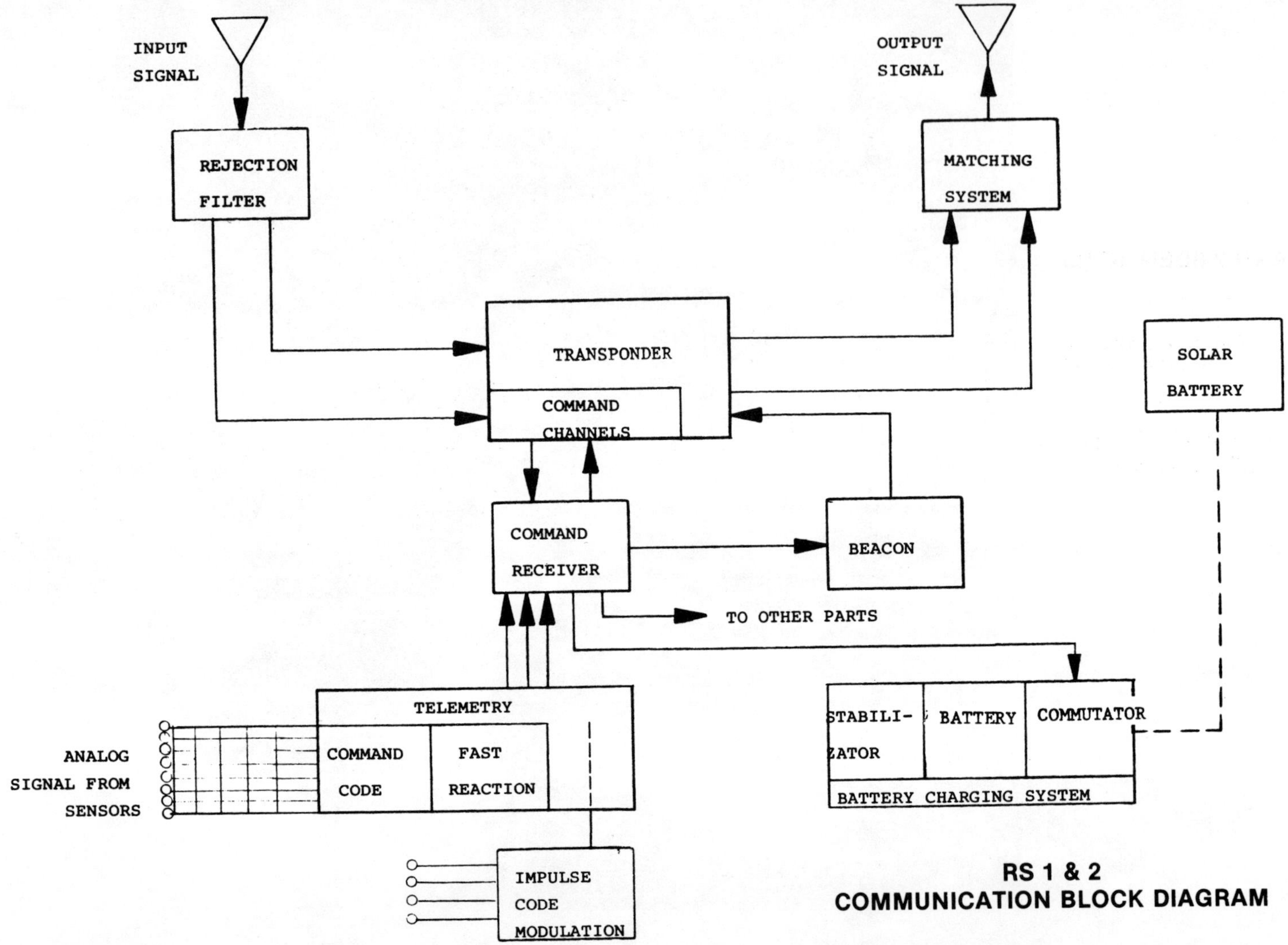

RS 1 & 2
COMMUNICATION BLOCK DIAGRAM

LEGEND:

The command receiver is common with the uplink receiver, but the command signals are taken out separately from the IF strip whereon they are decoded by the logic, deciphered, and go out to the correspondng equipment, indicated by the telemetry.The receiver uses a -90dB rejection filter in the front end, which is in class"A", to produce a 8.4 MHz IF. A 40 k Hz wide quartz filter is employed, giving better than -40 dB rejection outside the 40 k Hz passband .A crystal osc. is used, multiplied byx5 to give a 154.3 MHz signal at the mixer stage. A ring balanced mixer converter takes the signal to two linear cascades, the first amplifier on ten meters working in class"A", with the output stage in class "B". All signals above 40MHz are attenuated by greater than -100dB from the output.

Gustavo D. Moreno
IEEE Compendium

SECTION VIII

REFERENCES

TO

IEEE Compendium of Communication and Broadcast Satellites

I. INTERNATIONAL

A. INTERNATIONAL TELECOMMUNICATIONS SATELLITE ORGANIZATION

1. INTELSAT I (Early Bird)

S. Metzger. "The Commercial Communications Satellite System -- 1963 to 68," Astronautics and Aeronautics, Vol. 6, No. 4, April 1968.

R. M. Bentley, "Early Bird," Astronautics and Aeronautics, Vol. 3, No. 3, March 1965.

R. M. Bentley, "Early Bird Experimental Results," Proceedings of the 16th International Astronautical Congress, 1965, Vol 4: Meteorological and Communication Satellites, 1966.

J. M. Barstow, "Satellite Communications Systems," Microwave Journal, Vol. 9, No. 10, November 1966.

M. J. Votaw, "The Early Bird Project," IEEE Transactions on Communications Technology, Vol. 14, No. 4, August 1966.

A. T. Owens, "The Early Bird Communications System," Space Electronics Symposium, AAS Science and Technology Series, Vol. 6, 1965.

S. B. Bennett, "Early Bird I Communications Parameters," AIAA Paper 66-263, AIAA Communications Satellite Systems Conference, May 1966. Reprinted in Communication Satellite Systems Technology, Progress in Astronautics and Aeronautics, Vol. 19, R. B. Marsten (ed.), Academic Press, 1966.

L. F. Gray, "Experimental Performance of the Early Bird Communication System," AIAA Paper 66-264, AIAA Communications Satellite Systems Conference, May 1966. Reprinted in Communication Satellite Systems Technology, Progress in Astronautics and Aeronautics, Vol. 19, R. B. Marsten (ed.), Academic Press, 1966.

2. INTELSAT II

S. Metzger, "The Commercial Communications Satellite Systems 1963 to 68, "*Astronautics and Aeronautics,* Vol. 6, No. 4, April 1968.

E. J. Martin and W. S. McKee, "Commercial Satellite Communications Experience," *IEEE Spectrum*, Vol. 4, No. 7, July 1967.

H. Shinkawa, "Satellite Communication," *Electronics and Communications in Japan*, Vol. 50, No. 10, October 1967.

Intelsat 2 Communications Satellite," *Telecommunications Journal,* Vol. 34, No. 2, February 1967.

J. Arnaud, "Progress in Intelsat," *IEEE International Conference on Communications: ICC '68,* June 1968.

3. INTELSAT III

C. O. Meredith, "Lessons Learned from the Intelsat III Satellite Program," AIAA Paper 72-534, AIAA 4th Communications Satellite Systems Conference, April 1972. Reprinted in *Communications Satellite Systems,* Progress in Astronautics and Aeronautics, Vol. 32, P. L. Bargellini (ed.), 1974.

M. Feigen, "The Intelsat III Satellite," *IEEE International Conference on Communcations: ICC '68*, June 1968.

W. L. Glomb and L. Feit, "Overall System," *Electrical Communication,* Vol. 45, No. 4, 1970.

A. J. Grey, "Communication, Telemetry, and Command Subsystem," *Electrical Communications,* Vol. 45, No. 4, 1970.

4. INTELSAT IV

Special Issue on INTELSAT IV, *COMSAT Technical Review,* Vol. 2, No. 2, Fall 1972.

S. B. Bennett and I. Dostis, "Design of the Intelsat IV Transponder," AIAA Paper 72-535, AIAA 4th Communications Satellite Systems Conference, April 1972. Reprinted in _Communications Satellite Technology,_ Progress in Astronautics and Aeronautics, Vol. 33, P. L. Bargellini (ed.), MIT Press, 1974.

C. A. Blackwell and M. P. Brown, Jr., "Communication Satellite System Design," _IEEE Communication Systems and Technology Conference_, April 1974.

W. L. Pritchard and P. L. Bargellini, "Trends in Technology for Communications Satellites," _Astronautics and Aeronautics,_ Vol. 10, No. 4, April 1972.

A. A. McKenzie, "Special Report - Communications: What's Up in Satellites," _IEEE Spectrum,_ Vol. 9, No. 5, May 1972.

Intelsat IV Launch Handbook, Hughes Aircraft Co., December 1970 and June 1972.

T. Wilding-White, "Intelsat IV," _Flight International_, 18 February 1971.

M.P. Brown, Jr., "The International Telecommunications Satellite Organization (INTELSAT)", _Communications Satellite Systems: An Overview of the Technology,_ IEEE Press, 1976. Edited by R.G. Gould and Y.F. Lum, pp. 23-28.

5. INTELSAT IVA

Astrain, S., "Early Bird to Intelsat IVA (a decade of growth)," _Telecommunications Journal_, Vol. 42, No. 10, (October 1975).

L. S. Pilcher, "Intelsat IV-A as a Communication Capability," AIAA Paper 74-473, _AIAA 5th Communications Satellite Systems Conference_, April 1974.

J. Dicks and M.P.Brown, Jr., "Intelsat IVA-A Satellite Transmission System Design, "COMSAT Technical Review, Vol. 5, No. 1, Spring 1975.

J. Dicks and M. P. Brown, Jr., "Intelsat IV-A Satellite Transmission Design," AIAA Paper 74-474, AIAA 5th Communications Satellite Systems Conference, April 1974.

G. E. LaVean and E. J. Martin, "Communication Satellites: The Second Decade, "Astronautics and Aeronautics, Vol. 12, No. 4, April 1974.

B. I. Edelson and R. W. Rostron, "Technological Trends in Commercial Satellite Design," Spaceflight, Vol. 15, No. 8, August 1973.

First Intelsat-IVA Launched," Telecommunications Journal, Vol. 42, No. 10, October 1975.

6. INTELSAT V

E. A. Robertson, "Planning for the Intelsat V System, "National Telecommunications Conference: NTC '74, December 1974.

R. J. Rusch, J. T. Johnson, and W. Baer, "Intelsat V Spacecraft Design Summary," AIAA Paper 78-528, AIAA 7th Communications Satellite Systems Conference, April 1978.

G. T. Hayes, N. I. Totah, and J. W. Young, "Design of the Antenna Module Structure for Intelsat V Spacecraft," AIAA Paper 78-592, AIAA 7th Communications Satellite Systems Conference, April 1978.

R. J. Rusch and D. G. Dwyre, "Intelsat V Spacecraft Design," Acta Astronautica, Vol. 5, Nos. 3-4, March-April 1978.

J. L. Dicks and M. P. Brown, Jr. "Intelsat V Satellite Transmission Design, International Conference on Communications: ICC '78, June 1978.

J. C. Fuenzalida, P. Rivalan, and H. J. Weiss, "Summary of the Intelsat V Communications Performance Specifications," Comsat Technical Review, Vol. 7, No. 1, Spring 1977.

C. F. Hoeber, "Intelsat V System Design," Paper 5/1, WESCON '77 Conference Record, September 1977.

E. A. Robertson, "Communications Requirements for the Intelsat V Spacecraft," EASCON '76 Conference Record, September 1976.

K. G. Reseck and D.G. Dwyre, "Intelsat V - Transition to the Shuttle, "Astronautics and Aeronautics, Vol. 15, No. 6, June 1977.

E. Podraczky and N.K.M. Chitre, "Future Intelsat Services and Operations, "AIAA Paper 76-231, AIAA/CASI 6th Communications Satellite Systems Conference (April 1976).

B. MARISAT

D. W. Lipke and D.W. Swearingen, "Communication System Planning for Marisat, "International Conference on Communications: ICC '74, June 1974.

F. Giorgio, I. Knight, and R. Matthew, "A Maritime Mobile Terminal for Commercial Communications Satellite Application," International Conference on Communications: ICC '74, June 1974.

D. Swearingen and D. Lipke, "Marisat Multiple Access Capabilities, "International Conference on Communications: ICC '76, June 1976.

D. W. Swearingen, "Marisat Commercial Communications System Status, "EASCON '75 Convention Record, September 1975.

C. DeVore, "Marisat: Launching a New Era in Marine Communications," Signal, Vol. 30, No. 6, March 1976.

D. W. Lipke, "Marisat Program Status," AIAA Paper 75-281, AIAA 11th Annual Meeting and Technical Display, February 1975.

C. Dorian, T. O. Calvit, and D.W. Lipke, "Marisat: Design and Operational Aspects," Journal of the British Interplanetary Society, Vol. 29, No. 5, May 1976.

W. J. Gribbin and R. S. Cooperman, "Comsat General Satellite Technical Control Network," Comsat Technical Review, Vol. 7, No. 1. Spring 1977.

D. W. Lipke, "Marisat - A Maritime Satellite Communications System, "Comsat Technical Review, Vol. 7, No. 2, Fall 1977.

D. W. Swearingen, "Multiple Shore Station Interworking in Marisat," AIAA Paper 78-550, AIAA 7th Communications Satellite Systems Conference, April 1978.

D. W. Swearingen, "Marisat Program," EASCON '76 Conference Record, September 1976.

E. J. Martin and D. W. Lipke, "Performance of the Marisat Communications System, "IEE Conference Publication No. 160, Conference on Maritime and Aeronautical Satellite Communication and Navigation, March 1978.

R. Svensson and A. Synek, "Swedish Experience from the Marisat System," IEE Conference Publication No. 160, Conference on Maritime and Aeronautical Satellite Communication and Navigation, March 1978.

C. MAROTS/MARECS

D. McLauchlin and R. Stainforth, "MAROTS Communication Satellite, "Journal of the British Interplanetary Society, Vol. 29, No. 5, May 1976.

T. F. Howell, "Marots - The Mission," AIAA Paper 76-258, AIAA/CASI 6th Communications Satellite Systems Conference, April 1976.

J. A. Vandenkerckhove, "The ESRO MAROTS Programme," Journal of the British Interplanetary Society, Vol. 27, No. 10, October 1974.

P. J. Conchie, "The Adoption of OTS to the MAROTS Role," Journal of the British Interplanetary Society, Vol. 27, No. 10, October 1974.

W. M. Lovell, "The Marots Communications Payload," Proceedings of the Conference on Satellite Communications Systems Technology, IEE Conference Publication No. 126, April 1975.

T. F. Howell, "The Marots Maritime Satellite Programme," International Conference on Communications: ICC '77, June 1977.

G. R. Stette, "A Multiple Access and Access Control System for Marots, "European Conference on Electrotechnics: EUROCON '77, Conference Proceedings on Communications, May 1977.

D. INTELSAT V (MCS)

None, to-date.

II. INTERNATIONAL/DOMESTIC

A. LOUTCH

Anon, "INTELSAT Faces Numerous Challenges," Aviation Week and Space Technology, October 17, 1977.

B. MOLNIYA

A. D. Fortunshenko, "The Soviet Communication Satellite Molniya 1, "Telecommunications Journal, Vol. 32, No. 10, October 1965.

L. Ya. Kantor, V. A. Polukhin, and N. V. Talyzin, "New Relay Stations of the Orbita-2 Satellite Communications System," Telecommunications and Radio Engineering, Vol. 27, No. 5, May 1973.

M. P. Brown, Jr., "The USSR Domestic System (Molniya/Orbita)," Communications Satellite Systems: An Overview of the Technology, IEEE Press, 1976. Edited by R. G. Gould and Y. F. Lum, pp. 29-32.

"Active Communications - Satellite Systems" - Report 207-2, Vol IV, Part 2, XII Plenary Assembly of CCIR, New Delhi, 1970 (Green Book), pp 51-54.

"Soviet Space Programs, 1966-70," Staff Report for the Committee on Aeronautical and Space Sciences, U.S. Senate, 92nd Congress, 1st Session, Doc. No. 92-51, December 9, 1971.

"Soviet Space Programs, 1971," Staff Report for the Committee on Aeronautical and Space Sciences, U.S. Senate, 92nd Congress, 2nd Session, April 1972.

List of Space Radiocommunication Stations and Radio Astronomy Stations, "ITU International Frequency Registration Board (IFRB), 5th Edition, 1975 .

I. Petrov, "Intersputnik, International Space Communication System Organization," Telecommunications Journal, Vol 39, November 1972, pp. 679-682.

C. STATSIONAR/EKRAN/GORIZONT/RADUGA

"Synchronous Molniya," Aviation Week and Space Technology, August 5, 1974.

S. V. Borodich, "Development of Satellite Communications and Broadcasting in the Soviet Union," Telecommunication Journal, October 1978.

T. Pirard, "From Molniya to Ekran", Spaceflight, January 1978.

Anon, Aviation Week & Space Technology, June 29, 1977 (photo of Ekran).

K. Johnson, "Soviets Plan Seven Satellite Global Communications Net," Aviation Week & Space Technology, December 15, 1975.

D. VOLNA

Anon, "Soviet Marine/Communications Satellite Network Plans Detailed", Aviation Week & Space Technology, Nov 14, 1977.

Anon, "New Soviet Horizon Satellite Exhibited", Aviation Week & Space Technology, June 25, 1979.

III. REGIONAL

ECS (Eurosat, see also MARECS)

E. W. Bassett, "European Plan Regional Communications Satellite", Aviation Week & Space Technology, May 28, 1979.

R. Morris, "A Family Approach to Communications Satellites", Communications International, Dec. 1978

IV. DOMESTIC

A. COMSTAR

R. J. Rush, "Comsat General Domestic Communications Satellite," National Telecommunications Conference: NTC '74, December 1974.

D. C. Cox, "Design of the Bell Laboratories 19 and 28 GHz Satellite Beacon Propagation Experiment, "International Conference on Communications: ICC '74, June 1974.

A. J. E. vanHover and W. J. Gribben "Design of a Ground Control System to Operate Domestic and Maritime Satellites," AIAA Paper 74-483, AIAA 5th Communications Satellite Systems Conference, April 1974.

R. D. Briskman, "The Comstar Program," Comsat Technical Review, Vol. 7, No. 1, Spring 1977.

G. E. A. Abutaleb, "The Comstar Satellite System," Comsat Technical Review, Vol. 7, No. 1, Spring 1977.

L. Pollack, "Centimeter Wave Beacons for the Comstar Satellites," Comsat Technical Review, Vol. 7, No. 1, Spring 1977.

M. C. Kim, "Comsat General's Domestic Satellite System (Comstar)," Paper 9/4, WESCON Technical Papers, September 1976.

D. T. Nakatani, and G. G. Kuhn, "Comstar 1 Antenna System," IEEE International Symposium on Antennas and Propagation, June 1977.

L. Zahalka, "Frequency Reuse in GTE Earth Stations with Beam Waveguide Feed," Paper 8.4, International Conference on Communications: ICC '77, June 1977.

W. J. Gribbin and D. J. Lee, "Technological Development in Spacecraft Command and Control Systems," Paper 32.2, National Telecommunications Conference: NTC '77, December 1977.

W. J. Gribbin and R. S. Cooperman, "Comsat General Satellite Technical Control Network," Comsat Technical Review, Vol. 7, No. 1, Spring 1977.

B. INSAT

M. G. K. Menon, "Insat in Perspective," AIAA Paper 72-583, AIAA 4th Communications Satellite Systems Conference, April 1972.

P. P. Kale, R. L. Nickelson, and R. W. Sarles, Jr., "A Design for Insat," AIAA Paper 72-576, AIAA 4th Communications Satellite Systems Conference, April 1972.

U. R. Rao, "Educational Television in India," AAS Paper 73-106, AAS 19th Annual Meeting, June 1973. Reprinted in Advances in the Astronautical Sciences, Vol. 30, 1974.

Articles in Flight International:

a. 3 April 1976, p. 852
b. 23 October 1976, p. 1278
c. 6 August 1977, p. 439
d. 23 September 1978, p. 1190

Articles in Aviation Week and Space Technology:

a. 23 August 1971, p. 70.
b. 11 December 1972, p. 20.
c. 10 October 1977, p. 24.
d. 24 October 1977, p. 41
e. 7 August 1978, p. 45.

D. S. Kushwah, "Television Broadcasting in India," Telecommunications Journal, Vol. 44, No. 4, April 1977.

C. PALAPA

"Commercial Applications Satellites," <u>Flight International,</u> 19 December 1974.

"Indonesia" A Satellite Network for a Scattered Nation, "<u>Business Week,</u> 24 August 1974.

J. Sutanggar Tengker, "Indonesian Domestic Satellite System," Paper 11, <u>EASCON '76 Conference Record,</u> September 1976.

P. Hogwood, "Palapa - Indonesia to the Fore," <u>Journal of the British Interplanetary Society,</u> Vol. 30, No. 4, April 1977.

A. P. Djiwatampu, "Palapa - The Indonesian Domestic Satellite Communications Systems, "AIAA Paper 78-613, <u>AIAA 7th Communications Satellite Systems Conference</u>, April 1978.

C. C. Sanderson and B. R. Elbert, "Communication System Design of the Indonesian Domestic Satellite System," Paper 9/2, <u>WESCON Technical Papers</u>, September 1976.

D. SATCOM

J. Napoli and J. Christopher, "RCA SATCOM System;," <u>National Telecommunications Conference: NTC '74,</u> December 1974.

M. V. O'Donovan, "A Lightweight Transponder Design for the U.S. Domestic Communications Satellite, "<u>International Conference on Communications: ICC '73,</u> June 1973.

M. V. O'Donovan, "Design of a Light-Weight Microwave Repeater for a 24-Channel Domestic Satellite System," <u>RCA Review,</u> Vol. 34, No. 3, September 1973.

J. L. Rivard, "The Domestic Satellite Program in Alaska," <u>National Telecommunications Conference: NTC '74,</u> December 1974.

P. Schneider, "New Approaches Make RCA Satcom Most Cost Effective System for Domestic Satellite Communications for United States," EASCON '75 Convention Record, September 1975.

H. R. Hawkins, "RCA Satcom System Begins New Communications Era in United States," Signal, Vol. 28, No. 7, March 1974.

J. E. Keigler, "RCA Satcom: An Example of Weight Optimized Satellite Design for Maximum Communications Capacity," Acta Astronautica, Vol. 5, Nos. 3-4, March-April 1978.

J. Christopher and D. Greenspan, "RCA Satcom Communication System," Paper 7-4, EASCON '77 Conference Record, September 1977.

H. W. Rice, "RCA Americom's Domestic Satellites Pay Off," AIAA Paper 77-351, AIAA 13th Annual Meeting and Technical Display, January 1977.

J. E. Keigler and C. R. Hume, "The RCA Satcom Satellite," Journal of the British Interplanetary Society, Vol. 29, No. 9, September 1976.

A. W. Brook, "RCA Satcom System," RCA Engineer, Vol. 22, No. 1, June/July 1976.

J. E. Keigler, "RCA Satcom - Maximum Communication Capacity per Unit Cost," RCA Engineer, Vol. 22, No. 1, June/July 1976.

J. Christopher, D. Greenspan, and P. H. Plush, "The Launch and In-Orbit Test Elements of the Satcom System," RCA Engineer, Vol. 22, No. 1, June/July 1976.

J. Cuddihy and J. M. Walsh, "RCA Satcom Earth-Station Facilities," RCA Engineer, Vol. 22, No. 1, June/July 1976.

J. Lewin, "Ground-Control System for Satcom Satellites," AIAA Paper 78-539, AIAA 7th Communications Satellite Systems Conference, April 1978.

R. M. Lansey and M. R. Freeling, "RCA's Satellite Distribution System for Small-Dish Earth Terminals, "EASCON '78 Conference Record, September 1978.

E. SBS

M. T. Lyons and P. C. Dougherty, "Spacecraft Design for the SBS System," AIAA Paper 78-545, AIAA 7th Communications Satellite Systems Conference, April 1978.

J. D. Barnla and F. R. Zitzmann, "Digital Communications Satellite System of SBS," Paper 7-2, EASCON '77 Conference Record, September 1977.

D. Kittiver and F. R. Zitzmann, "The SBS System - An Innovative Domestic Satellite System for Private-Line Networks," AIAA Paper 76-307, AIAA/CASI 6th Communications Satellite Systems Conference, April 1976.

H. A. Rosen, "The SBS Communication Satellite - An Integrated Design, "EASCON '78 Conference Record, September 1978.

B. Goode, "Demand Assignment as Part of the SBS TDMA Satellite Communication System, "EASCON '78 Conference Record, Septrember 1978.

F. STATSIONAR-T (EKRAN)

(See STATSIONAR)

G. TELESAT

R.F. Chinnick, "Canadian Satellite Telecommunications Plans for the 1980s," IAF XXVIII Congress, Prague, Czech. (September 25-October 1, 1977).

R. F. Chinnick, "The Canadian Telecommunications Satellite Systems," Journal of the British Interplanetary Society, Vol. 26, No. 4, April 1973.

Canadian Domestic Communications Satellite," "Ideas and Achievements" Section, _Telecommunications Journal,_ Vol. 40, No. 1, January 1973.

D. E. Weese and F. H. Smart, "Measured Communication Performance of the Telesat Satellite System," AIAA Paper 74-455, _AIAA 5th Communications Satellite System Conference_, April 1974.

M. J. Houterman, "Anik' Satellite Communications System," _International Telemetering Conference: ITC '72,_ October 1972.

L. Harrison, "Canadian Domestic Satellite (Telesat), A General Description," _International Conference on Communications: ICC '71,_ June 1971.

A. D. D. Miller, "Operational Experience with Small Unattended Television Receive Earth Stations," AIAA Paper 74-454, _AIAA 5th Communications Satellite Systems Conference_, April 1974.

R. K. Kwan, "The Telesat TDMA System," _International Conference on Communications: ICC -75,_ June 1975.

H. Kowalik, "Telesat Satellite Control System," AIAA Paper 74-451, _AIAA 5th Communications Satellite Systems Conference_, April 1974.

J. W. Crawford, "Operating Experience in the Canadian Domestic Satellite Systems," AIAA Paper 78-541, _AIAA 7th Communications Satellite Systems Conference_, April 1978.

D. A. Gray, "Telesat's Sixty-One Mbps TDMA System Operational Experience," Paper 11.3, _National Telecommunications Conference: NTC '78,_ December 1978.

D. E. Weese, "The Canadian Domestic Satellite Communication System - Present and Future," Paper 7-3 _EASCON '77 Conference Record,_ September 1977.

A. R. Raab and K. Farrell, "A Shaped Beam Multifeed 14/11 GHz Antenna for Anik-B," *1978 AP-S International Symposium*, May 1978.

R. M. Lester, "Telesat Canada Plans for New Satellite Systems," AIAA Paper 78-544, *AIAA 7th Communications Satellite Systems Conference*, April 1978.

R. W. Hoedemaker and D. G. Thorpe, "Anik B, The New Canadian Domestic Satellite," Paper 9/3, WESCON Technical Papers, September 1976.

F. H. Smart, "The Anik C 90 Mb/s Digital Service," *4th International Conference on Digital Satellite Communications*, Montreal, Canada, October 1978.

D. G. Thorpe, "Evolution of the TELESAT Canada System into the Mid-1980s," No. 76-220, *AIAA/CASI 6th Communications Satellite Systems Conference*, Montreal, Canada (April 5-8, 1976).

H. WESTAR

D. B. Nowakowski, "The Western Union Integrated Satellite/Voice/Data Network," Paper 18B, *National Telecommunications Conference: NTC '73*, November 1973.

S. N. Verma, "Westar Communication Characteristics," *National Telecommunications Conference: NTC '74*, December 1974.

E. D. Hilburn, "How Westar Will Affect U.S. Domestic Communications," *Signal*, Vol. 28, No. 7, March 1974.

H. R. Johnson, "Western Union's Domestic Satellite Program," *Signal*, Vol. 28, No. 9, May/June 1974.

S. N. Verma, "U.S. Domestic Communication System Using Westar Satellites," *World Telecommunication Forum Conference Proceedings*, Geneva, ITU, October 1975.

G. L. Sarver, "Satellite Communications for Off-Shore Oil Operations Using Westar," National Telecommunications Conference: NTC '75, December 1975.

J. Ramasastry, "Western Union's Satellite-Switched TDMA Advanced Westar System," AIAA Paper 78-602, AIAA 7th Communications Satellite System Conference, April 1978.

C. L. Washburn, "Westar Operations as Part of the Western Union Integrated Tranmission System," AIAA Paper 78-540, AIAA 7th Communications Satellite Systems Conference, April 1978.

J. E. D. Ball and P. Rubin, "Communication Satellites for Public Television," IEEE Transactions on Broadcasting, Vol. 24, No. 2, June 1978.

J. W. VanCleve, "Operation and Control of an Integrated Satellite/Terrestrial Transmission Network," Paper 25.4, International Conference on Communications: ICC '77, June 1977.

J. Ramsastry, "Advanced Westar SS/TDMA System," 4th International Conference on Digital Satellite Communications, Montreal, Canada, October 1978.

S. N. Verma and D. Fraley, "Sixty-Two Mb/s Transmission via Westar Satellites," Paper 11.2, National Telecommunications Conference: NTC '78, December 1978.

W. N. Verma, J. Ramsastry, and W. R. Monsees, "Digital Speech Interpolation Applications for Domestic Satellite Communications," Paper 14.4, National Telecommunications Conference: NTC '78, December 1978.

I. ADVANCED WESTAR/TDRSS

G. Clark, K. Fellerman, and J. Schwartz, "Current Concepts for a Tracking and Data Relay Satellite System," Paper 28A, National Telecommunications Conference: NTC 1972, December 1972.

P. F. Barritt, "Tracking and Data Relay Satellite System (TDRSS)," Paper 7B, National Telecommunications Conference: NTC 1973, November 1973.

R. D. Godfrey, "The Evolution of the Tracking and Data Relay Satellite Systems," EASCON 1975 Convention Record, September 1975.

L. F. Deerkoski, "Tracking and Data Relay Satellite System (TDRSS) Telecommunication Services," EASCON 1975 Convention Record, September 1975.

L. Deerkoski and P. C. Walker, "Tracking and Data Relay Satellite System (TDRSS)," Paper 5/2, WESCON 1977 Conference Record, September 1977.

W. M. Holmes, "TDRSS System Design," Paper 9-2, National Telecommunications Conference" NTC '77, December 1977.

M. Melnick and C. Dixon, "TDRSS Frequency Management," Paper 19-5, National Telecommunications Conference: NTC '77, December 1977.

C. C. Chen and J. W. Burnett, "TDRS Multiple Access Channel Design," Paper 19-2, National Telecommunications Conference: NTC '77, December 1977.

W. A. Imbriale and G. G. Wong, "An S-Band Phased Array for Multiple Access Communications," Paper 19-3, National Telecommunications Conference: NTC '77, December 1977.

B. C. Tankersley and H. E. Bartlett, "Tracking and Data Relay Satellite Single Access Deployable Antenna," Paper 19-4, National Telecommunications Conference: NTC '77, December 1977.

B. H. Batkson, S. W. Novosaad, and T. W. Sheehan, "Space Shuttle Utilization of TDRSS Services," Paper 9-4, National Telecommunications Conference: NTC '77 December 1977.

W. M. Holmes, "Status Report on TDRSS," International Telemetering Conference Proceedings, November 1978.

S. B. Franklin, "TDRSS-User Satellite Acquisition and Tracking," International Telemetering Conference Proceedings, November 1978.

V. MILITARY

A. DSCS II

A. T. Finney, "A Phase II Satellite for the Defense Satellite Communications System," AIAA Paper 70-493, AIAA 3rd Communications Satellite Systems Conference, April 1970. Reprinted in Communication Satellites for the 70s: Systems, Progress in Astronautics and Aeronautics, Vol. 26, N. E. Feldman and C. M. Kelly (eds.), MIT Press, 1971.

W. D. DeHart, "Defense Satellite Communication System - Phase II," EASCON '70 Convention Record, October 1970.

G. E. LaVean, "The Defense Satellite Communications System," AIAA Paper 74-457, AIAA 5th Communications Satellite Systems Conference, April 1974.

V. W. Wall, "Military Communication Satellites," International Telemetering Conference Proceedings, October 1973.

B. DSCS III

L. S. Haas and A. T. Finney, "The DSCS III Satellite - A Defense Communication System for the 80's," AIAA Paper 78-580, AIAA 7th Communications Satellite Systems Conference, April 1978.

K. R. Swimm and J. A. Loftus, "DSCS III - Flexible Communications," Paper 5/3, WESCON '77 Conference Record, September 1977.

A. W. Weinrich, A. Horvath, and A. Harcar, "DSCS III Communications Satellite Performance," International Telemetring Conference Proceedings, November 1978.

C. FLEETSATCOM

N. L. Wardle, "U.S. Navy Fleet Satellite Communications," AIAA Paper 74-458, AIAA 5th Communications Satellite Systems Conference, April 1974.

C. E. Reid, "A New Era in Worldwide Tactical Communications," International Telemetering Conference Proceedings, October 1973.

A. Shostake, "Navy Telecommunications Past and Present," Navy Research Reviews, Vol. 28, No. 12, December 1975.

M. J. Friedenthal and E. K. Heist, "Fleet Satellite Communications Spacecraft," Paper 9/5, WESCON Technical Papers, September 1976.

J. A. Nooney, "Demand Assignment for Narrowband Tactical UHF Satellite Channels," EASCON '77 Conference Record, September 1977.

J. A. Nooney, "UHF Demand Assignment Multiple Access (DAMA) System for Tactical Satellite Communications," Paper 45.5 International Conference on Communications: ICC '77, June 1977.

F. S. McCartney and E. K. Heist, "FLTSATCOM Program Review: Requirements, Design and Performance," EASCON '78 Conference Record, September 1978.

D. GALS

Anon, "Four New Soviet Satcoms Planned," Aviation Week & Space Technology, Nov. 21, 1977.

Anon, "USSR Satcom Net," Aviation Week & Space Technology, Nov. 14, 1977.

E. IDSCS

W. L. Pritchard, "The Initial Defense Communications Satellite," Microwave Journal, Vol. 9, No. 11, November 1966.

V. W. Wall, "Military Communication Satellites," Astronautics and Aeronautics, Vol. 6, No. 4, April 1968.

J. W. O'Neill "Military Communication and Navigation Satellites," TRW Space Log, Vol. 9, No. 2, Summer/Fall 1969.

H. B. Kucheman, W. L. Pritchard, and V. W. Wall, "The Initial Defense Communication Satellite Program," AIAA Paper 66-267, AIAA Communications Satellite Systems Conference, May 1966.

F. LES

1. LES 1 to 7

H. Sherman, et al., "The Lincoln Experimental Satellite Program (LES-1, -2, -3, -4)," Journal of Spacecraft and Rockets, Vol 4, No. 11, November 1967.

H. Sherman, et al., "The Lincoln Experimental Satellite Program (LES-1, -2, -3, -4)," AIAA Paper 66-271, AIAA Communications Satellite Systems Conference, May 1966.

D. MacLellan, H. MacDonald, and P. Waldron, "Lincoln Experimental Satellites 5 and 6," AIAA Paper 70-494, AIAA 3rd Communications Satellite Systems Conference, April 1970. Reprinted in Communication Satellites for the 70s: Systems, Progress in Astronautics and Aernautics, Vol. 26, N. E. Feldman and C. M. Kelly (eds.), MIT Press, 1971

R. Berg, R. Chick, and D. Snider, "LES-7 Transponder," AIAA Paper 70-511, AIAA 3rd Communications Satellite Systems Conference, April 1970.

A. R. Dion, "Variable-Coverage Communications Antenna for LES-7," AIAA Paper 70-423, AIAA 3rd Communications Satellite Systems Conference, April 1970.

A. R. Dion and L. J. Richard, "A Variable-Coverage Satellite Antenna System," Proceedings of the IEEE, Vol. 59, No. 2, February 1971.

2. LES 8 and 9

W. W. Ward, "Developing, Testing, and Operating Lincoln Experimental Satellites 8 and 9 (LES-8/9)," EASCON '78 Conference Record, September 1978.

LES-8/9: Communications Capabilities, MIT Lincoln Laboratory Project Report SC-21, 15 March 1976.

A. R. Dion, "Satellite Crosslink K-Band Antenna," NEREM 72 Record.

F. W. Sarles, Jr., et al, "The Lincoln Experimental Satellites LES-8 and -9 ," Paper 21.-1, EASCON '77 Conference Record, September 1977.

L. J. Collins, "LES-8/9 Communications System Test Results," AIAA Paper 78-599, AIAA 7th Communications Satellite Systems Conference, April 1978.

F. J. Solman, et al, "The Ka-Band Systems of the Lincoln Experimental Satellites LES-8 and LES-9," AIAA Paper 78-562, AIAA 7th Communications Satellite Sytems Conference, April 1978.

D. M. Snider and D. B. Coomber, "Satellite-to-Satellite Data Transfer and Control," AIAA Paper 78-596, AIAA 7th Communications Satellite Systems Conference, April 1978.

L. J. Collins, Jones, L. R. McElroy, D. R. Siegel, D. A. Ward, W. W. and William, D.K., "LES-8/9 Communications System Test Results," AIAA Paper No. 78-599, AIAA 7th Communications Satellite Systems Conference, April 1978

G. LEASAT

David Braverman and C. J. Wayland, "LEASAT Communication Services, ICC 1979, Boston, Mass. June 1979.

T. F. White, "Fleet Satellite Communications/Leased Satellite Communications Operations." ICC 1979.

C. J. Wayland, "Leased Military Space Communication Systems," 16th Space Congress, Orlando, Florida, April 1979.

R. Rhoads, "LEASAT: Contractor/Government Relationship." ICC 1979.

H. NATO

1. NATO II

See SKYNET I

2. NATO III

"Philco Wins NATO 3 Satellite," Microwave Systems News, Vol. 3, No. 3, May 1973.

Anon, "NATO Comm Satellite Backstops Military Net," Electronic Engineering Times, Dec. 11, 1978.

H. A. Kissinger, "NATO Satellite Communications," Signal, Vol. 30, No. 6, March 1976.

E. T. Bobak and R. G. Clabauh, "NATO Phase III Satellite Design," Paper 15-2, EASCON '77 Conference Record, September 1977.

I. SKYNET

1. SKYNET 1

V. W. Wall, "Satellites for Military Communications," AIAA Paper 74-272, AIAA 10th Annual Meeting, January 1975.

I. C. Wright and P. McLellan, "Defense Satellite Communication System - An Operational System," Paper 31.1, International Conference on Communications: ICC '77, June 1977.

J. Bailey, and V. Chewy, "Ground Mobile Forces Tactical Satellite SHF Ground Terminals," AIAA Paper 78-582, AIAA 7th Communications Satellite System Conference, April 1978.

B. E. Tyree, "An Overview of the Small SHF Satellite Ground Terminal Development Program," RCA Engineer, Vol. 22, No. 1, June/July 1976.

R. D. Rosner, "Communications System Control for the Defense Communications System," Paper 31.1, International Conference on Communications: ICC '78, June 1978.

D. G. Dwyre, "IDSCP/A Satellite: Concept and Performance," AIAA Paper 70-492, AIAA 3rd Communications Satellite Systems Conference, April 1970. Reprinted in Communication Satellites for the 70s: Systems, Progress in Astronautics and Aeronautics, Vol. 26, N. E. Feldman and C. M. Kelly (eds.), MIT Press, 1971.

N. Simmons, "The United Kingdom Programme of Communication Satellites," AIAA Paper 72-548, AIAA 4th Communications Satellite Systems Conference, April 1972. Reprinted in Communications Satellite Systems, Progress in Astronautics and Aeronautics, Vol. 32, P. L. Bargellini (ed.), MIT Press, 1974.

2. SKYNET II

G. W. Carter, "An Experimental Satcom Control System for DSCS," Paper 42:1, National Telecommunications Conference: NTC '77, December 1977.

D. E. Kendall, "Development of the Defense Satellite Communication System - Phase II," EASCON '78 Conference Record, September 1978.

W. M. Lovell, "Design of the Skynet II Communications Satellite," Journal of Science and Technology, Vol. 39, No. 1, 1972.

W. M. Lovell, "The Skynet System, The Satellite Communications Network Built in Britain," Journal of the British Interplanetary Society, Vol. 29, No. 1, January 1976.

J. TACSAT

R. Brandes, "The Tactical Communications Satellites," IEEE Transactions on Aerospace and Electronic Systems, Vol. AES-6, No. 4, July 1970.

VI. EXPERIMENTAL

A. ATS

1. ATS-1 thru 5

Technical Data Report for the Applications Technology Satellite Program, Goddard Space Flight Center, issued 3 March 1967, revised periodically until 20 April 1971, 6 volumes.

R. H. Pickard, "The Applications Technology Satellite," Proceedings of the 16th International Astronautical Congress, 1965, Vol. 4: Meteorological and Communication Satellites, 1966.

P. J. McCeney, "Applications Technology Satellite Program," Acta Astronautica, Vol. 5, Nos. 3-4, March-April 1978.

"Space Systems Summaries," Astronautics and Aeronautics, Vol. 13, No. 2, February 1975.

2. ATS-6

P. J. McCeney, "Applications Technology Satellite Program,", Acta Astronautica, Vol. 5, Nos. 3-4, March-April 1978.

J. P. Corrigan, "The Next Steps in Satellite Communications," Astronautics and Aeronautics, Vol. 9, No.9, September 1971.

A. B. Sabelhaus, "Applications Technology Satellites F and G Communications Subsystem," Proceedings of the IEEE, Vol. 59, No. 2, February 1971.

W. N. Redisch and R. L. Hall, "ATS 6 Spacecraft Design/Performance," EASCON '74 Conference Record, October 1974.

W. A. Johnston, "ATS-6 Experimental Communications Satellite: A Report on Early Orbital Results," National Telecommunications Conference: NTC '74, December 1974.

W. N. Redisch, "ATS-6 Description", International Conference on Communications: ICC '75, June 1975, and EASCON '75 Convention Record, September 1975.

Special Issue on ATS 6, IEEE Transactions on Aerospace and Electronic Systems, Vol. 11, No. 6, November 1975.

A. A. Whalen and W. A. Johnson, Jr., "ATS-6 - A Satellite for Human Needs," AIAA Paper 75-900, AIAA Conference on Communication Satellites for Health/Education Applications, July 1975.

The ATS-6 Satellite," Telecommunications Journal, Vol. 41, No. 10, October 1974.

L. H. Westerlund, "ATS-F Comsat Millimeter Wave Propagation Experiment," COMSAT Technical Review, Vol. 3, No. 2, Fall 1973.

A. L. Berman, "The ATS-F Comsat Propagation Experiment Transponder," COMSAT Technical Review, Vol. 3, No. 2, Fall 1973.

J. L. Levatich and J. L. King, "ATS-F Comsat Millimeter Wave Propagation Experiment," Paper 8A, National Telecommunications Conference: NTC '72, December 1972.

J. L. King and G. Hyde, "The Comsat 13 and 18 GHz Propagation Experiment," International Conference on Communications: ICC '75, June 1975.

L. J. Ippolito, "The GSFC 20 and 30 GHz Millimeter Wave Propagation Experiment," International Conference on Communications: ICC '75, June 1975, and EASCON '75 Convention Record, September 1975.

V. F. Henry and G. Schaefer, "System Design of the ATS-F RFI Measurement Experiment," Paper 38D, National Telecommunications Conference: NTC '72, December 1972.

J. G. Potter and J. M. Janky, "The ATS-F Health-Education Technology Communications System," International Conference on Communications: ICC '73, June 1973.

A. A. Whalen, "Health Education Telecommunications Experiment, "International Conference on Communications: ICC '75, June 1975.

J. R. Burke, "Experimental Systems in Applications Technology SAtellite F and G," AIAA Paper 72-578, AIAA 4th Communications Satellite Systems Conference, April 1972.

W. A. Johnson, "ATS-6 Experimental Communications Satellite - Report on Early Orbital Results," Journal of Spacecraft and Rockets, Vol. 13, No. 2, February 1976.

E. V. Chitnis and J. E. Miller, "Social Implications of Satellite Instructional Television Experiment," International Conference on Communications: ICC '76, June 1976.

F. O. Vonbum, P. D. Argentiero, and P. E. Schmid, "Orbit Determination Accuracies Using Satellite-to-Satellite Tracking," IEEE Transactions on Aerospace and Electronic Systems, Vol. 14, No. 6, November 1978

B. BS (or BSE)

T. Ishida, K. I. Tsukamaoto, M. Hirai, and Y. Ichikawa, "Program of Medium-Scale Broadcasting Satellite for Experimental Purpose," AIAA Paper 76-255, AIAA/CASI 6th Communications Satellites Systems Conference, April 1976.

W. Johnson and H. Reichert, "Development of a Medium Scale Broadcasting Satellite for Experimental Purposes," WESCON Technical Papers, September 1975.

L. T. Seaman, "Japanese Broadcast Satellite," _Acta Astronautica,_ Vol. 5, Nos. 5-6, May-June 1978.

L. T. Seaman, "Japanese Broadcast Satellite," _EASCON '76 Conference Record,_ September 1976.

Y. Ichikawa, "The Results of Initial Checkup of Japanese Broadcasting Satellite kfor Experimental Purpose," _IEEE Transactions on Broadcasting,_ Vol. 24, No. 4, December 1978.

K. Tsukamoto, "Technical Aspects of the Japanese Broadcasting Satellite Experiments," _IEEE Transactions on Broadcasting,_ Vol. 24, No. 4, December 1978.

H. Kaneda and K. Tsukamoto, "Experiments in the Program of Japanese Medium-Scale Broadcasting Satellite for Experimental Purpose," AIAA Paper 78-583, _AIAA 7th Communications Satellite Systems Conference_, April 1978.

C. CAS-C/CTS (Hermes)

C. Franklin and E. Davison, "A High Power Communications Technology Satellite for the 12 and 14 GHz Bands," AIAA Paper 72-580, AIAA 4th Communications Satellite Systems Conference, April 1972. Reprinted in _Communications Satellite Systems,_ Progress in Astronautics and Aeronautics, Vol. 32, P. L. Bargellini (ed.), MIT Press, 1974.

V. O'Donovan, "Design of a 14/12 GHz Transponder for the Communications Technology Satellite," AIAA Paper 72-734, _CASI/AIAA Meeting:_ Spacke-1972 Assessment, July 1972.

P. L. Donoughe, "United States Societal Experiments via the Communications Technology Satellite," _International Conference on Communications: ICC '76,_ June 1976.

L. J. Ippolito, "Characterization of the CTS 12 and 14 GHz Communication Links - Preliminary Measurements and Evaluation," International Conference on Communications: ICC '76, June 1976.

L. D. Braun and M. V. O'Donovan, "Characteristics of a Communications Satellite Transponder," Microwave Journal, Vol. 17, No. 12, December 1974.

J. Day, "CTS Communications Experiments," Paper 35B, National Telecommunications Conference: NTC '72, December 1972.

D. L. Wright and J.W. B. Day, "The Communications Technology Satellite and the Associated Ground Terminals for Experiments," AIAA Paper 75-904, AIAA Conference on Communications Satellites for Health/Education Applications, July 1975.

E. F. Miller, J. L. Fiala, and I. G. Hansen, "Performance Characteristics of the 12 GHz, 200 Watt Tranmitter Experiment Package for CTS," EASCON '75 Convention Record, September 1975.

G. H. Booth, "The Canadian/U.S. High Power Communications Technology Satellite," Satellite Systems for Mobile Communications and Surveillance, IEE Conference Publication No. 95, March 1973.

H. R. Raine, "The Communications Technology Satellite Flight Performance," Acta Astronautica, Vol. 5, Nos. 5-6, May-June 1978.

R. E. Alexovich, "On-Orbit Performance of the 12 GHz, 200 Watt Transmitter Experiment Package for CTS," Paper 1.3, International Conference on Communications: ICC '78, June 1978.

N. G. Davies, J.W. B. Day, and M.V. Patriarche, "The Transition from CTS/Hermes Communications Experiments to Anik-B Pilot Projects," EASCON '78 Conference Record, September 1978.

D. COURIER

G. F. Senn and P. W. Siglin, "Courier Satellite Communication System", IRE Transactions on Military Electronics, Vol. MIL-4, No. 4, October 1960.

P. W. Siglin and G. F. Senn, "The Courier Satellite," Communications Satellites, Proceedings of a Symposium Held in London, L. J. Carter (ed.), 1962.

E. Imboldi and D. Hershberg, "Courier Satellite Communication System," Advances in the Astronautical Sciences, Vol. 8, AAS, 1961.

E. CS

T. Ishida, K. I. Tsukamoto, M. Hirai, and H. Okamoto, "Program of Medium-Capacity Communications Satellite for Experimental Purpose," AIAA Paper 76-244, AIAA/CASI 6th Communications Satellite Systems Conference, April 1976.

H. Hyams, "Design of the CS Communications Subsystem", AIAA Paper 76-293, AIAA/CASI 6th Communications Satellite Syustesm Conference, April 1976. Reprinted in Satellite Communications: Advanced Technologies, Progress in Astronautics and Aeronautics, Vol. 5, D. Jarett (ed.), 1977.

M. Kudo, Y. Nemoto, A.N. Wickert, C.C. Han, and D. Ford, "The Design of the Communications Antenna for CS," AIAA Paper 76-252, AIAA/CASI 6th Communications Satellite Systems Conference, April 1976. Reprinted in Satellite Communications: Advanced Technologies, Progress in Astronautics and Aeronautics, Vol. 55, D. Jarett (ed.), 1977.

H. Kaneda and K. Tsukamoto, "Experiments in the CS Program," AIAA Paper 78-616, AIAA 7th Communications Satellite Systems Conference, April 1978.

C. L. Cuccia, "Transponder and Antenna Design Problems at Millimeter Wavelengths for 20-30 GHz Communication Satellites," Symposium on Advanced Satellite Communications Systems, Genoa, Italy, December 1977.

M. Hirai and H. Uda, "Experiment Programme for the Japanese Communications Satellite," Symposium on Advanced Satellite Communications Systems, using the 20-30 GHz bands, ESA, Genoa, Italy, December 1977.

F. ECHO

Special Issue on Project Echo, Bell System Technical Journal, Vol. 40, No. 4, July 1961.

Satellite Communications (Military-Civil Roles and Relationships), Second Report by the Committee on Government Operations, U.S. House of Representatives, House Report No. 178, 17 March 1968.

H. S. Black, "Latest Results on Project Echo," Advances in the Astronautical Sciences, Vol. 8, AAS 1961.

J. R. Burke, "Passive Satellite Development and Technology," Astronautics and Aerospace Engineering, Vol. 1, No. 8, September 1963.

L. Jaffe, "Project Echo Results," Astronautics, Vol. 6, No. 5, May 1961.

W. C. Nyberg, "Experiments to Determine Communication Capability of the Echo II Satellite," Publications of Goddard Space Flight Center 1964, Vol. II.

G. ECS (Japanese)

T. Ishida, "Program of Experimental Communication Satellite (ECS) of Japan," AIAA Paper 78-614, AIAA 7th Communications Satellite Systems Conference, April 1978.

M. Ohara, "The Satellite Transponder Performance for the Experimental Communications Satellite (ECS)," AIAA Paper 78-563, AIAA 7th Communications Satellite Systems Conference, April 1978.

E. W. Matthews, L. F. Brokish, and G.F. Will, "The Communications Antenna System on the Japanese Experimental Communications Satellite," AIAA Paper 78-584. AIAA 7th Communications Satellite Systems Conference, April 1978.

H. ETS II

N. Fugono and R. Hayashi, "Propagation Experiment in 1.7, 11.5 and 34.5 GHz with Engineering Test Satellite Type II," AIAA Paper 78-623, AIAA 7th Communications Satellite Systems Conference, April 1978.

Article in Aviation Week and Space Technology, 1 March 1976, p. 52.

I. OTS

R. C. Collette and B. Stockwell, "The OTS Project," AIAA Paper 74-495, AIAA 5th Communications Satellite Systems Conference, April 1974.

C. Wearmouth, "The Current Status of the Orbital Test Satellite Programme," World Telecommunication Forum Proceedings, Geneva, ITU, October 1975.

H. Falk, "European Satellites to 'Fly,'" IEEE Spectrum, Vol. 11, No. 8, August 1974.

S. Tirro and A. Bayliss, "The Utilization Programme of the Orbital Test Satellite," AIAA Paper 76-247, AIAA/CASI 6th Communications Satellite Systems Conference, April 1976.

M. Lopriore, H. K. Ball, and E. Mondre, "Design of the 14/11 GHz Repeater for the European Orbital Test Satellite," International Conference on Communications: ICC '73, June 1973.

P. Bartholome, "The European Communications Satellite System - A Review of Current and Planned Activities," AIAA Paper 76-243, AIAA/CASI 6th Communications Satellite Systems Conference, April 1976. Reprinted in Satellite Communications: Future Systems, Progress in Astronautics and Aeronautics, Vol. 54, D. Jarett (ed.), AIAA, 1977.

H. K. Ball and E. Mondre, "The Modular Repeater of the European Communication Satellite (ECS) and the Orbital Test Satellite (OTS)," Journal of the British Interplanetary Society, Vol. 27, No. 5, May 1974.

R. C. Collette, "The European Communication Satellite Programme and the Orbital Test Satellite," Journal of the British Interplanetary Society, Vol. 29, No. 5, May 1976.

A. J. Bayliss and A. Dickinson, "The Orbital Test Satellite of ESA and Its Associated Test Programme," Acta Astronautica, Vol. 5, Nos. 5-6, May-June 1978.

C. Wearmouth and D. E. McLaurin, "The Development of the Orbital Test SAtellite," AIAA Paper 76-246, AIAA/CASI 6th Communications Satellite Systems Conference, April 1976, Reprinted in Satellite Communications: Future Systems, Progress in Astronautics and Aeronautics, Vol. 54, D. Jarett (ed.), AIAA, 1977.

R. D. McQueen, "The Impact of the Communications Mission on the System Design of the Orbital Test Satellite," Journal of the British Interplanetary Society, Vol. 30, No. 4, April 1977.

J. RELAY

L. Jaffe, "The NASA Communications Satellite Program Results and Status," *Proceedings of the 15th International Astronautical Congress*, 1964, Vol. 2: Satellite Systems, 1965.

S. Metzger and R. H. Pickard, "Relay," *Astronautics and Aerospace Engineering*, Vol. 1, No. 8, September 1963.

"Development of the Relay Communications Satellite," *Interavia*, Vol. 17, June 1962.

"Final Report on Relay," *NASA Document SP-76*, 1965.

K. SCORE

S. P. Brown and G. F. Senn, "Project Score," *Proceedings of the IRE*, Vol. 48, No. 4, April 1960.

S. P. Brown, "Project SCORE: Signal Communication by Orbiting Relay Equipment," *IRE Transactions on Military Electronics*, Vol. MIL-4, No. 2-3, April-July 1960.

M. I. Davis and G. N. Krassner, "SCORE - First Communication Satellite," *Journal of the American Rocket Society*, Vol. 4, May 1959.

L. SIRIO

F. Carassa, "The Italian Satellite Sirio," AIAA Paper 70-501, *AIAA 3rd Communications Satellite Systems Conference*, April 1970. Reprinted in *Communication Satellites for the 70s: Systems*, *Progress in Astronautics and Aeronautics*, Vol. 26, N. E. Feldman and C. M. Kelly (eds.), MIT Press, 1971.

P. Fanti and S. Tirro, "The Italian Sirio Experiments: Satellite and Ground Equipment," AIAA Paper No. 70-502, AIAA 3rd Communications Satellite Systems Conference, April 1970. Reprinted in Communication Satellites for the 70s: Systems, Progress in Astronautics and Aeronautics, Vol. 26, N. E. Feldman and C. M. Kelly (eds.), MIT Press, 1971.

Article in Aviation Week and Space Technology, 23 August 1971, p. 92.

"Space Programmes Around the World: 2. Italy," Interavia, Vol. 26, June 1971.

"The World of Aerospace" Section, Interavia, Vol. 29, January 1974.

G. Perrotta, "The Italian Sirio 12-18 GHz Experiment: The Forerunner of 20-30 GHz Preoperational Satellites," AIAA Paper 78-631, AIAA 7th Communications Satellite Systems Conference, April 1978.

F. Carassa, "The Sirio Programme," Acta Astronautica, Vol. 5, Nos. 5-6, May-June 1978.

M. SYMPHONIE

B. Pfieffer and P. Viellard, "The Experimental Telecommunication Satellite Project Symphonie," Journal of the British Interplanetary Society, Vol. 26, No. 2, February 1973.

G. Pfeiffer and P. Viellard, "The Franco-German Telecommunication Satellite Symphonie," AIAA Paper 70-406, AIAA 3rd Communications Satellite Systems Conference, April 1970. Reprinted in Communication Satellites for the 70s: Systems, Progress in Astronautics and Aeronautics, Vol. 26, N. E. Feldman and C. M. Kelly (eds.), MIT Press, 1971.

G. Moesel and J. Muller, "In-Orbit Performance and Experimental Utilization of the Symphonie Satellites," AIAA Paper 76-306, AIAA/CASI 6th Communications Satellite Systems Conference, April 1976. Reprinted in Satellite Communications: Future Systems, Progress in Astronautics and Aeronautics, Vol. 54, D. Jarett (ed.), 1977.

W. Schroeter, "SHF Performances of the Symphonie Satellite," World Telecommunication Forum, Proceedings, Geneva, ITU, October 1975.

F. Hierquist, "The Earth Segment in the Symphonie Project," AIAA Paper 72-549, AIAA 4th Communications Satellite Systems Conference, April 1972. Reprinted in Communications Satellite Systems, Progress in Astronautics and Aeronautics, Vol. 32, P. L. Bargellini (ed.), MIT Press, 1974.

"Symphonie for Emergency Aid," Interavia, Vol. 29, No. 12, December 1975.

B. R. K. Pfeiffer and W. G. Schroter, "The Symphony Satellite System," Acta Astronautica, Vol. 5, Nos. 3-5, March-April 1978.

M. R. E. Bichara, "The Symphonie Project," Communications Satellite Systems: An Overview of the Technology. Edited by R. G. Gould and Y. F. Lum, IEEE Press, 1976, pp. 19-22.

N. SYNCOM

P. E. Norsell, "Syncom," Astronautics and Aerospace Engineering, Vol. 1, No. 8, September 1963.

W. H. Edwards and J. S. Smith, "Experience of the Defense Communications Agency in Operating Pilot Satellite Communications," AIAA Paper 66-268, AIAA Communications Satellite Systems Conference. May 1966. Reprinted in Communication Satellite Systems Technology, Progress in Astronautics, Vol. 19, R. B. Marsten (ed.), Academic Press, 1966.

F. P. Alder, "Syncom," Proceedings of the 14th International Astronautical Congress, 1963, Vol. 2: Satellites and Spacecraft, 1965.

O. TELSTAR

Special Telstar Issue, Bell System Technical Journal, Vol. 42, No. 4, July 1963. Reprinted as Telstar I, NASA SP-32, Vols. 1-3, July 1963 and Vol. 4 (including Telstar II supplement), December 1965.

K. W. Gatland, Telecommunication Satellites, Prentice Hall, New York, 1964.

I. Welber, "TELSTAR", Astronautics and Aerospace Engineering, Vol. 1, No. 8, September 1963.

J. Holahan, "Telstar, Toward Long-Term Communications Satellites." Space/Aeronautics, Vol. 37, No. 5, May 1962.

P. WESTFORD

Special Issue on Project WestFord, Proceedings of the IEEE, Vol. 52, No. 5, May 1964.

VII. AMATEUR

A. OSCAR AND AMSAT

"Radio Amateurs" Section, Telecommunications Journal:

a. Vol. 39, No. 2, February 1972.

b. Vol. 39, No. 3, March 1972.

c. Vol. 39, No. 7, July 1972.

d. Vol. 39, No. 8, August 1972.

e. Vol. 41, No. 1, January 1974.

f. Vol. 42, No. 2, February 1975.

g. Vol. 42, No. 3, March 1975.

h. Vol. 42, No. 10, October 1975.

i. Vol. 43, No. 1, January 1976.

j. Vol. 43, No. 9, September 1976.

k. Vol. 44, No. 5, May 1977.

l. Vol. 45, No. 3, March 1978.

M. Davidoff, "The Future of the Amateur Satellite Service", Ham Radio Magazine, August 1977.

H. Gabrielson, "The OSCAR Satellite", QST Magazine, pp 21-24, February 1962.

G. Jacobs, "OSCAR 3 - An Active Communications Satellite for Radio Amateurs", CQ Magazine, Part 1, pp 54-92, October 1964.

W. Orr, "The OSCAR III VHF Translator Satellite, QST Magazine, pp 54-92, February 1963.

G. Jacobs and J. King, "Australia-OSCAR Progress Report", CQ Magazine, pp 64-65, December 1969.

G. Jacobs, "Australia - OSCAR: Amateur Radio's Next Satellite in Space", CQ Magazine, pg 63, August 1969.

G. Jacobs, "OSCAR - 6: It May Already be in Orbit, CQ Magazine, pp 48-50, November 1972.

G. Jacobs, "Summer Launch Planned for OSCAR-6", CQ Magazine, pp 40-41, May 1972.

G. Jacobs and P. Klein, "Satellites in the Amateur Radio Service," Telecommunications Journal, Vol. 38, No. 5, May 1971.

P. I. Klein and J. A. King, "Results of the Amsat-Oscar 6 Communicatioins Satellite Experiment," Paper 6/3, 1974 IEEE Intercon Record, March 1974.

P. I. Klein and J. A. King, "The AMSAT-OSCAR-B Series of Radio Amateur Satellites," AIAA Paper 72-521, AIAA 4th Communications Satellite Systems Conference, April 1972.

P. R. Hammer, "The Development of Amateur Communications Satellites," Proceedings of the IREE, Vol. 36, No. 5, May 1975.

J. Kasser and J. A. King, "OSCAR 7 and Its Capabilities," QST Magazine, February 1974.

Article in Aviation Week and Space Technology, 28 October 1974, p. 20.

J. Kasser, "OSCAR 7 and Its Capabilities," Radio Communication, November 1973.

G. Sassoon, "In Touch With Oscar," Spaceflight, Vol. 17, No. 10, October 1975.

P. I. Klein and R. Soifer, "Intersatellite Communication Using the Amsat-Oscar 6 and Amsat-Oscar 7 Radio Amateur Satellites," Proceedings of the IEEE, Vol. 63, No. 10, October 1975.

W. I. Dunkerley, "Oscar and Friends: Useful Yet Frugal," IEEE Spectrum, Vol. 15, No. 12, December 1978.

Article in Flight International, 23 October 1976, p. 1278.

J. E. Kasser and J.A. King, 'OSCAR 7 and its capabilities" - Radio Communication, November 1973, pp. 747-799.

G. Jacobs, "OSCAR 7 - !Should be in orbit this month!" - CS Magazine, October 1974, pp 17-21.

P. I. Klein and J.A. King, "The AMSAT - OSCAR-B Series of Radio Amateur Satellites, AIAA 4th Commucations Satellite System Conference, Washington D.C., April 24-26 1972.

S. Place, "AMSAT-OSCAR Phase III on the Horizon" - Part one, QST Magazine, Vol XIII, No.12, December 1979, pp. 61-64.

B. RS

Article in Aviation Week and Space Technology, 6 November 1978, p. 22.

Anon, "Soviet Amateur Satellites", Telecommunication Journal, December 1977.

P. J. Gowen, "R.S. The amateur radio satellites of the Soviety Union" AMSAT Newsletter, Vol X, No. 4, December 1978, pp 4-8.

P. J. Gowen, "More about R.S." - AMSAT Newsletter, No. 1, Vol XI, March 1979, pp. 16.

P. J. Gowen, "The Soviet 'RS' Amateur Radio Satellies," Telecommunication Journal, Volume 46, No. 4, April 1979, p. 239.

VIII. GENERAL

G. Falworth, Satellite News, Lancs, England PR1 OSX.

D. H. Martin, Communication Satellites, 1958 to 1982, SAMSO Report TR-79-078, 10 September 1980.

Telecommunications Journal, "Tables of Artificial Satellites Launched in 1977, 1978, and 1979."

T. M. Wilding - White, Jame's Pocketbook of Space Exploration, MacMillan Publishing Company, Inc., 1976.

MEMBERSHIP

IEEE/AESS
COMMUNICATION AND BROADCAST SATELLITE SYSTEMS COMMITTEE

Martin P. Brown Jr., Chairman
INTELSAT

Norman Abramson
University of Hawaii

Marcel R. E. Bichara
University of Geneva

Richard W. Duesing
INTELSAT

B. G. Evans
University of Essex

Hayden W. Evans
Bell Telephone Laboratories

Richard G. Gould
Telecommunications Sytems

D. Wayne Hanson
National Bureau of Standards

James M. Janky
Stanford University

Robert K. Kwan
COMSAT

David W. Lipke
COMSAT General

James G. Potter
Public Service Satellite Consortium

Orest S. Roscoe
Canadian Department of Communications

Terry Simms
Bell Northern Research

Harvey L. Werstiuk
Canadian Communications Research Center

Michael J. Robusto
INTELSAT

Jai P. Singh
India Space Research Organization

Jahangir A. Tehrani
INTELSAT

J. W. Unger
Bell Telephone Laboratories

IEEE

30 GROUPS AND SOCIETIES

Acoustics, Speech and Signal Processing (01)

Aerospace and Electronic Systems Society (10)

Antennas and Propagation (03)

Broadcast, Cable and Consumer Electronics (02)

Circuits and Systems (04)

Communications (19)

Computer (16)

Control Systems (23)

Education (25)

Electrical Insulation (32)

Electromagnetic Compatability (27)

Electron Devices (15)

Engineering in Medicine and Biology (18)

Engineering Management (14)

Geoscience Electronics (29)

Industrial Electronics and Control Instrumentation (13)

Industry Applications (34)

Information Theory (12)

Instrumentation and Measurement (09)

Magnetics (33)

Manufacturing Technology (35)

Microwave Theory and Technique (17)

Nuclear and Plasma Sciences (05)

Parts, Hybrids and Packaging (21)

Power Engineering (31)

Professional Communication (26)

Reliability (07)

Sonics and Ultrasonics (20)

Systems, Man, and Cybernetics (28)

Vehicular Technology (06)

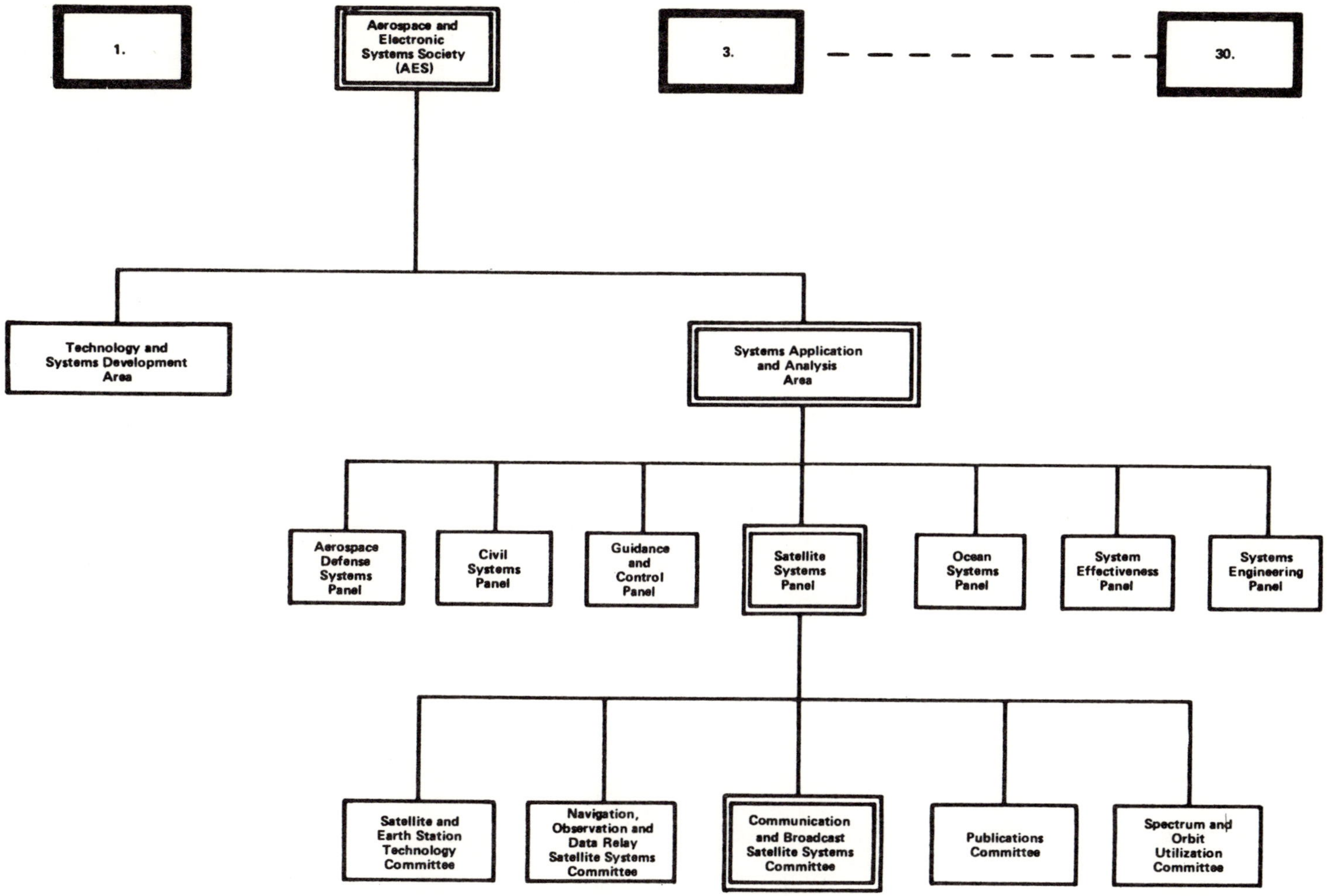

() = IEEE Code for this Group or Society

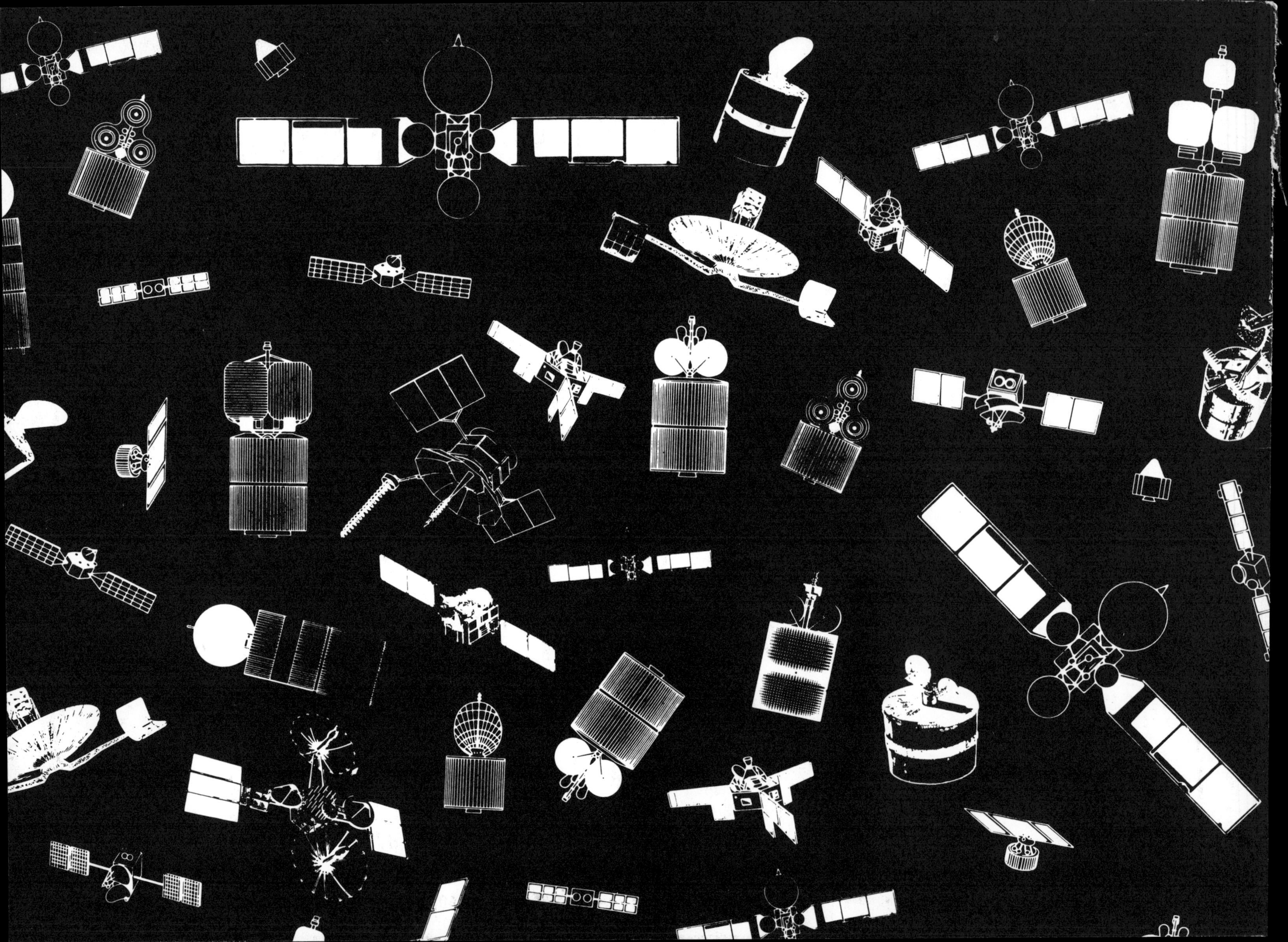